Animal Sciences: Biology, Health and Rearing

Animal Sciences: Biology, Health and Rearing

Edited by Violet Paige

SYRAWOOD
PUBLISHING HOUSE

New York

Published by Syrawood Publishing House,
750 Third Avenue, 9th Floor,
New York, NY 10017, USA
www.syrawoodpublishinghouse.com

Animal Sciences: Biology, Health and Rearing
Edited by Violet Paige

© 2019 Syrawood Publishing House

International Standard Book Number: 978-1-68286-836-2 (Hardback)

Cataloging-in-Publication Data

Animal sciences : biology, health and rearing / edited by Violet Paige.
    p. cm.
Includes bibliographical references and index.
ISBN 978-1-68286-836-2
1. Zoology. 2. Animal health. 3. Animal culture. I. Paige, Violet.
QL45.2 .A56 2019
590--dc23

TABLE OF CONTENTS

PREFACE

Animal science studies the biology, health, production and management of animals, which are of economic interest to humans. It integrates the studies of genetics, animal physiology, reproduction and microbiology. There are different areas of focus in animal science, such as food and fiber production, breeding, animal agribusiness, nutrition, animal behavior and welfare, etc. The branch of medicine which is concerned with the management, treatment and prevention of animal diseases and injuries is known as veterinary medicine. Research in animal science and veterinary medicine transcends all species boundaries and also works at human-animal interfaces such as zoonotic diseases, food safety, ecosystem health, etc. This book is compiled in such a manner, that it will provide in-depth knowledge about the theories and principles in the field of animal sciences. The topics covered in this extensive book deal with the core aspects of the biology, health and rearing of animals. For all readers who are interested in animal sciences, the case studies included in this book will serve as an excellent guide to develop a comprehensive understanding.

The researches compiled throughout the book are authentic and of high quality, combining several disciplines and from very diverse regions from around the world. Drawing on the contributions of many researchers from diverse countries, the book's objective is to provide the readers with the latest achievements in the area of research. This book will surely be a source of knowledge to all interested and researching the field.

In the end, I would like to express my deep sense of gratitude to all the authors for meeting the set deadlines in completing and submitting their research chapters. I would also like to thank the publisher for the support offered to us throughout the course of the book. Finally, I extend my sincere thanks to my family for being a constant source of inspiration and encouragement.

Editor

Effect of Different Levels of Live Yeast in a High Concentrate Diet on Performance, Blood Constituents and Immune System Status of Zandi Lambs

M. Raghebian[1*], A. Babaei Yazdi[2], N. Dabiri[2], A. Hajimohammadi[3], P. Hatami[4], A. Raghebian[5], J. Shomeyzi[2] and M.J. Bahrani[2]

[1] Department of Animal Science, Faculty of Agriculture and Natural Resources, Science and Research Branch, Islamic Azad University, Tehran, Iran
[2] Department of Animal Science, Karaj Branch, Islamic Azad University, Karaj, Iran
[3] Department of Animal Science, Faculty of Agricultural Science, University of Guilan, Rasht, Iran
[4] Department of Animal Science, Faculty of Agriculture and Natural Resources, University of Tehran, Karaj, Iran
[5] Department of Animal Science, Saveh Branch, Islamic Azad University, Saveh, Iran

*Correspondence E-mail: majid.raghebian@yahoo.com

ABSTRACT

A feeding trial with twenty-seven male weaned Zandi lambs (initial body weight 27.1±0.38 kg) was conducted to evaluate the effect of different levels of *Saccharomyces cerevisiae* SC47 in diet containing high concentrate (85%) on the growth performance, blood parameters and immune system status. Lambs were allocated to one of three treatment diets in a completely randomized design with 3 replicates and 3 observations per replicate including: 1) basal diet without yeast, control diet; (CD) 2) basal diet with 3 g yeast per lambper day, low yeast; (LY) and 3) basal diet supplemented with 4.5 g yeast per lambper day, high yeast; (HY). Regarding dry matter intake (DMI), there was no significant difference among treatments (P>0.05). Average daily gain (ADG) was greater in HY group, but differences among treatments were not significant (P>0.05). Feed conversion ratio (FCR) was not significantly affected by dietary treatments (P>0.05). Differences between concentrations of total protein, globulin and albumin/globulin ratio (A/G) were significant (P<0.05). The highest amount of total protein and globulin was observed in LY group (P>0.05). Lambs in CD group had the highest amount of A/G ratio. No significant differences were found for the hematology results (P>0.05). No significant differences were detected in differential white blood cells, except neutrophil band that was greater in HY group at the twelfth week (P<0.05).It is concluded that the use of dietary live yeast in high concentration can improve performance (P>0.05), plasma biochemical metabolites(P<0.05) and hematological parameters (P>0.05) in Zandi lambs.

KEY WORDS blood constituents, fattening Zandi lambs, high concentrate, immune system, live yeast, performance.

INTRODUCTION

Different feed additives were added to animal's diet with the aim of increase breeding, health and quality of products in animal breeding for a long time. For several years antibiotics were the most important feed additives that was used, but in the last decade, the use of antibiotics in animal husbandry is in question, due to the development of antibiotic resistance. Research shows an association between the use of subtherapeutic dose of antibiotics and antibiotic resistance in organisms (Amabile-Cuevas *et al.* 1995). In an effort to replace antibiotics in animal feeds, many additives have been proposed. Recently, alternatives for substituting these traditional growth promoters have been evaluated, probiotics, prebiotics and ionophores are examples of these promoters (Heinrichs *et al.* 2003). The probiotics are classically defined as live microbial dietary supplements that when administered through the digestive tract, cause a positive impact on the host's health by improving gut micro flora (FAO, 2002; Salminen *et al.* 1996; Fuller, 1989). Studies on the beneficial impact of probiotics on animal performance have indicated that probiotic supplementation could have positive effects (Miles *et al.* 1981). One of the most common probiotics in ruminants, are *Saccharomyces cerevisiaeyeast* (SC). The use of SC as a probiotic, began during the 1940's and 1950's (Beeson and Perry, 1952). Live yeast consumes free oxygen in the rumen with respiration, so provides an anaerobic environment that proper for rumen metabolic function (Newbold *et al.* 1995). Enzymes, vitamins, saccharides [β-glucans (βGs) and mannan oligosaccharides (MOSs)] found in yeast cell walls have immunomodulative properties. MOSs are capable of neutralizing pathogenic bacteria, and they support βGs in the process of stimulating defense mechanisms (Małaczewska and Milewski, 2010). Other metabolites produced from yeast fermentation may have benefit on performance and health of animals. The main target of using yeast for growing lamb is to increase the breakdown of dietary fiber and protein that lead to increase microbial protein as a main source of amino acids in the small intestinal, consequently, improve growth. Moreover, SC yeast has biologically valuable proteins, vitamin B-complex, important trace minerals and several unique plus factors. Many other beneficial effects identified such as improve performance (Glade and Sist, 1988; Martin *et al.* 1989) and feed efficiency (Onifade and Babatunde, 1996), affects on ruminal pH by reducing activity of lactic acid producer bacteria and thus reduce ruminal acidosis and metabolic disorders (Williams, 1989; Guedes *et al.* 2008; Thrune *et al.* 2009), enhance the immune response (Keyser *et al.* 2007), ability to enhancement of phosphorus availability (Glade and Biesik, 1986; Brake, 1991; Moore *et al.* 1994) and utilization by animals (Erdman, 1989; Pagan, 1990), reduction in cases of disease infection (Line *et al.* 1997). However, the results of using the SC in ruminants are contradictory because of this fact that in many cases no influence or opposing and many unclear results have been shown (Masek *et al.* 2007). The lack of positive results: (Mikulec *et al.* 2010), can be related to biotic factors such as amount and yeast viability and to abiotic factors such as diet sources and animal management (Sales, 2011). One of the most important reasons for such inconsistent results is diet composition. Young animals with high growth potential, need the diets with high protein and energy content according to (NRC, 1985), which can hardly be achieved in an exclusive forage diet. So, to obtain high performance, they should be fed with a high proportion of concentrate diets. Yeasts are most efficient when animals are fed diets overloaded in energy and thus easily fermented by rumen microorganisms (Williams *et al.* 1991) or diets poor in nutrient supply (Jouany *et al.* 1998). Consumption of large amounts of readily fermentable carbohydrates can change the rumen fermentation pattern. This function can increase production of short chain fatty acidsand lactate, and decrease pH, which affect the amounts of cellulolytic bacteria and reduces fiber digestibility and the production of microbial mass (Mackie *et al.* 2002). Therefore, it is necessary to control fermentation and use additives to maintain rumen health and improve animal production. So, the objective of current study was to investigate the effect of different levels of live yeast SC47 in diet containing high concentrate (85%) on the growth performance, blood constituents and immune system status of Zandi lambs.

MATERIALS AND METHODS

Animals, diets and experimental design

Feeding trial with twenty-seven male weaned Zandi lambs (initial body weight 27.1±0.38 kg and 90±5 days-old), that were grouped based on body weight, were conducted to evaluate the effect of different levels of *Saccharomyces cerevisiae* SC47 (biosaf probiotic) in supplemental diet on the growth performance, blood constituents and immune system status. Lambs were grouped based on body weight, ear tagged and vaccinated against internal and external parasites, and were allocated to one of three dietary treatments in a completely randomized design with 3 replicates and 3 observations per replicate including: 1) basal diet without yeast, control diet; (CD) 2) basal diet with 3 g yeast per lambper day, low yeast; (LY) and 3) basal diet supplemented with 4.5 g yeast per lambper day, high yeast; (HY). Basal diet was consisted of commercial concentrate and hays. Ingredients and chemical composition of basal diet according to the dietary nutrient requirements for lambs (NRC, 1985) are provided in Table 1. The lambs were adapted to feed about 2 weeks. During these 2 weeks, feed intake was restricted to 3.5% of body weight (BW), based on the average BW within a pen, to allow animals to adapt to the change in diet and to prevent the occurrence of digestive disorders. After the adaption period to the end of experiments (twelve weeks), they were fed three times a day (7:00 a.m., 13:00 p.m. and 19:00 p.m.) with a total mixed ration (TMR) diet. Each pen had an automatic water cup so they had free access to water all times.

Sampling, measurement and analyses

Food intake was measured daily (before the morning feeding) and DMI was calculated.

Table 1 Ingredients and chemical composition of basal diet

Ingredients	Amount (%)	Chemical composition	
Alfalfa hay[1]	13	Dry matter (%)	88.6
Wheat straw[1]	2	Crude protein (%)	16.6
Barley	60	Total digestible nutrients (%)	77.9
Soybean meal	6	Metabolizable energy (Mcal/kg)	2.81
Wheat	12	Calcium (%)	0.54
Wheat bran	5	Phosphorus (%)	0.32
Limestone	0.5	Potassium (%)	0.89
Mineral and vitamin premix[2]	1	Ca/P ratio	1.68
Sodium bicarbonate	0.5	-	-

[1] Hays approximately chopped into particles of 3 cm.
[2] The mineral-vitamin premix contained per kg: Cu: 500 mg; Zn: 6500 mg; I: 100 mg; Co: 10 mg; Se: 10 mg; Mn: 1000 mg; Fe: 4000 mg; Antioxidant: 12500 mg; vitamin A: 2000000 IU; vitamin D: 220000 IU and vitamin E: 2500 IU.

This value was expressed as grams DMI per day (g/day). Lambs live weight was measured at the beginning of the experiment and every two weeks interval (before the first meal) and ADG was calculated. This value was expressed as grams ADG per day (g/day). Blood samples were collected at the days of 0, 42 and 84, at 10:00 h, through jugular vein (10 mL into sterile tubes containing ethylenediaminetetraacetic acid (EDTA) solution) and analyzed for concentration of plasma biochemical indicators (total protein, albumin, globulin, A/G ratio, blood urea nitrogen (BUN) and cholesterol), hematological parameters [white blood cells (WBC), red blood cells (RBC), hemoglobin (Hb), platelets (PLT) and differential white blood cells].

Statistical analysis

Data were analyzed by the general linear model (GLM) procedure of the Statistical Analysis System software (SAS, 1997), in a completely randomized design with three treatments and three replicates. The shapiro-wilk and kolmogorov-smirnov tests were used to confirm normal distribution of data. Initial body weight and the first series of each parameters of concentration of plasma biochemical indicators, hematological parameters (blood cellular elements) and differential white blood cells, were used as a covariant. Means were obtained by LSMEANS procedures and PDIFF was used to compare means. Effects between the control and experimental groups were considered significant when (P<0.05) and finally results were presented as least square means with standard error of the means (SEM).

RESULTS AND DISCUSSION

Performance

Least square means and SEM for effect of different levels of yeast supplement on performance of Zandi lambs are presented in Table 2. Considering DMI, there was no significant differences among the dietary treatments (P>0.05). So we can conclude that, live yeast (*Saccharomyces cerevisiae* SC47) has no significant effect on feed intake and appetite, and this may explain the lack of effect for the av-

erage daily gain and feed conversion ratio, because the dry matter intake and available nutrients concentration in food, determine the amount of nutrients used to meet demands for maintenance and production. In agreement with our results, other authors also did not find any improvement in DMI after yeast addition to lamb diets (Mikulec *et al.* 2010; Hernandez *et al.* 2009).

But in some study an increase in dry matter intake was observed when yeast was fed to bulls (Galina *et al.* 2006) and lambs (Desnoyers *et al.* 2009; Rezaeian, 2004). Average daily gain was greater in HY group compared to the others throughout the experiment, but the differencewas notstatistically significant (P>0.05).

Similar to the results of our study, Mandour *et al.* (2009) and Antunovic *et al.* (2006) reported higher, but not significant, ADG for lambs fed diet enriched with probiotics. In contrast, Milewski *et al.* (2009) and Ding *et al.* (2008) reported higher ADG in fatting lambs fed diet enriched with probiotics. Feed conversion ratio was not significantly affected by treatments (P>0.05), but at the end of trial, lambs in group HY consistence with ADG had the best total FCR. In Agreement with our results, some of authors also did not find any improvement in FCR after yeast addition to lamb diets (Khalid *et al.* 2011; Mikulec *et al.* 2010). In contrast to our results, Ding *et al.* (2008) and Masek *et al.* (2007) reported better FCR in fatting lambs fed diet enriched with probiotics.

Plasma biochemical indicators

Results of different levels of yeast supplement on concentration of plasma biochemical indicators of Zandi lambs are presented in Table 3. Except for total protein in the twelfth week, for the other periods the highest and lowest amounts of total protein and globulin was observed in LY and CD groups, respectively (P<0.05). In general, sub-acute acidosis can cause intestinal inflammation in animals that this situation will have a negative effect on the digestion and absorption, which one of the consequences is impaired digestion and absorption of protein and thus reduction in plasma total protein.

Table 2 Effect of different levels of yeast supplement on performance of Zandi lambs

Item	Treatments			SEM
	Control	Low yeast	High yeast	
Initial weight (kg)	27.35	27.65	26.25	0.379
DMI (g/day)				
First 4 weeks	1185.3	1161.1	1173.9	5.76
Second 4 weeks	1379.0	1361.2	1374.6	4.32
Third 4 weeks	1580.5	1562.0	1573.4	4.49
Total DMI	1381.6	1361.4	1374.0	4.49
ADG (g/day)				
First 4 weeks	183.3	220.8	243.3	15.75
Second 4 weeks	269.5	252.3	278.3	9.84
Third 4 weeks	273.4	261.4	279.7	12.74
Total ADG	242.1	244.8	267.1	10.20
FCR				
First 4 weeks	6.61	6.35	6.14	0.620
Second 4 weeks	5.18	5.80	5.04	0.239
Third 4 weeks	6.46	6.26	5.89	0.361
Total FCR	5.89	5.86	5.33	0.250

DMI: dry matter intake; ADG: average daily gain AND FCR: feed conversion ratio.
SEM: standard error of the means.

Table 3 Effect of different levels of yeast supplement on concentration of plasma biochemical indicators of Zandi lambs

Item	Treatment			SEM
	Control	Low yeast	High yeast	
Sixth week				
Total protein (g/dL)	6.95[b]	7.57[a]	7.15[b]	0.105
Albumin (g/dL)	3.95	3.98	3.93	0.067
Globulin (g/dL)	3.06[b]	3.63[a]	3.10[b]	0.085
A/G ratio	1.31[a]	1.09[b]	1.27[ab]	0.040
BUN (mg/dL)	17.77	16.76	16.62	0.457
Twelfth week				
Total protein (g/dL)	6.25	6.52	6.30	0.132
Albumin (g/dL)	3.73	3.56	3.23	0.098
Globulin (g/dL)	2.48[b]	2.94[a]	2.92[a]	0.079
A/G ratio	1.50[a]	1.21[b]	1.11[b]	0.050
Cholesterol (mg/dL)	60.50	68.83	59.16	2.577
BUN (mg/dL)	18.56	18.30	18.23	0.680

A/G ratio: albumin/globulin and BUN: blood urea nitrogen.
SEM: standard error of the means.
The means within the same row with at least one common letter, do not have significant difference (P>0.05).

The results showed that the yeast had no significant effect on plasma albumin (P>0.05) concentrations. Lambs in CD group had the highest A/G ratio and BUN, while the lowest amounts of these plasma biochemical indicators were seen in groups supplemented with yeast. Low but not significant concentrations of BUN in response to probiotic supplements can be due to increase ability of the rumen microflora in trapping ammonia (Abo El-Nor and Kholif, 1998). BUN is an indicator of the protein status in ruminants (Sykes, 1978) and its concentration is related to the level of ammonia absorption from the rumen and the deamination of amino acids not deposited in the tissues (Deaville and Galbraith, 1992).

Another possibility for the lower BUN concentration is that additives promote the utilization and deposition of nitrogen in tissues. No significant differences were detected in cholesterol levels between the different groups (P>0.05). Probiotics that can reduce the intestinal pH, could contribute to the regulation of serum cholesterol concentrations by its relationship with bile acids. In the acidic conditions of the intestine, because of the inability to re-absorption of bile acids at the end of intestine, excretion of bile acids is enhanced and therefore serum cholesterol, as a precursor of bile acids, used to rebuild the bile (De Smet *et al.* 1994). Since in the current study *Saccharomyces cerevisiae* SC47 did not reduce the intestinal pH, we did not found any significant differences in serum cholesterol levels between the different groups. In agreement with our results, other authors reported similar results for some concentration of plasma biochemical indicators after yeast addition to lamb diets (Khalid *et al.* 2011; Mukhtar *et al.* 2010; Bruno *et al.* 2009; Masek *et al.* 2008).

Hematological parameters and differential WBC

Results of different levels of yeast supplement on concentration of hematological parameters and differential WBC of Zandi lambs are presented in Tables 4 and 5, respectively. Overall no significant effects were recorded on the hematology results (P>0.05). However, the highest amount of WBC was seen in yeast-supplemented groups (P>0.05). There was an increase in values for hemoglobin in LY and HY groups at the twelfth week (P>0.05).

Probiotics can reproduce and develop in the gut wall as a living cell, or can capture antigens released by dead microorganisms, and by different ways, stimulate the immune system and its components (Fuller, 1977). Live yeast did not modify value of RBC and PLT in the experiments and we observed only a slight increase in RBC in group HY at the end of trial (P>0.05). No significant differences were detected in differential white blood cells, except neutrophil band that was greater in HY group at the twelfth week (P<0.05). Probiotics can stimulate and strengthen the immune system by increase of macrophage activity, which this act appears by increasing the ability of phagocytosis of microorganisms. One of this phagocytosis of organisms is neutrophil, that increase in the number of neutrophils band, in response to supplementation of probiotic, can be stimulated macrophage activity (Tizard, 2008; Fuller, 1992).

Table 4 Effect of different levels of yeast supplement on hematological parameter (blood cellular elements) of Zandi lambs

Item	Treatment			SEM
	Control	Low yeast	High yeast	
Sixth week				
WBC ($\times 1/\mu L$)	8474	8556	11393	708.3
RBC ($\times 10^6/\mu L$)	13.65	13.51	13.43	0.285
Hb (g/dL)	11.01	11.04	11.02	0.201
PLT ($\times 10^5/\mu L$)	4.57	3.95	4.41	0.404
Twelfth week				
WBC ($\times 1/\mu L$)	9259	9277	12612	1009.6
RBC ($\times 10^6/\mu L$)	13.72	13.80	14.32	0.288
Hb (g/dL)	10.88	11.39	11.40	0.175
PLT ($\times 10^5/\mu L$)	4.61	4.31	4.62	0.449

WBC: white blood cells; RBC: red blood cells; Hb: hemoglobin and PLT: platelets.
SEM: standard error of the means.
The means within the same row with at least one common letter, do not have significant difference (P>0.05).

Table 5 Effect of different levels of yeast supplement on differential white blood cells of Zandi lambs

Item	Treatment			SEM
	Control	Low yeast	High yeast	
Sixth week				
Neutrophil segmented (%)	33.36	35.12	33.67	1.317
Neutrophil band (%)	0	0	0	0
Lymphocyte (%)	64.54	63.73	64.05	1.303
Abnormal lymphocytes (%)	0	0	0	0
Basophil (%)	0	0	0	0
Monocyte (%)	0.50	0.33	0.33	0.143
Eosinophil (%)	1.66	1	1.66	0.217
PCV (%)	33.50	33.33	32.88	0.704
Twelfth week				
Neutrophil segmented (%)	29.28	32.34	34.53	1.456
Neutrophil band (%)	0[b]	0[b]	0.50[a]	0.090
Lymphocyte (%)	68.99	66.56	63.61	1.477
Abnormal lymphocytes (%)	0	0	0	0
Basophil (%)	0	0	0	0
Monocyte (%)	0.33	0.16	1	0.202
Eosinophil (%)	1.33	0.83	0.50	0.227
PCV (%)	30.42	32.66	32.48	0.573

SEM: standard error of the means.
The means within the same row with at least one common letter, do not have significant difference (P>0.05).

The immune stimulating effect of SC was ascribed to the activity of βGs and MOs presented in yeast cell walls (Milewski *et al.* 2007). This mechanism involves the stimulation of immune competent cells, mainly by βGs (Xiao *et al.* 2004; Siwicki *et al.* 2004). βGs activate intercellular defense mechanisms where macrophages, T-cells and NK cells play the key role (Demir *et al.* 2007).

The specific ability of MOs to bind selected pathogenic microbes has a profound effect on the organism's health status. MOs blocks microbial lectins and prevent pathogens from colonizing the host's gastrointestinal system (Sharon, 2008). MOs are not degraded by the digestive enzymes of the small intestine, therefore, the attached pathogens are more easily excreted (Spring *et al.* 2000). In this study, where the experiment was conducted, was cleared from contamination and infection and the possibility of create any stress for the lambs was the lowest rate, that it could be one of the factors of lack of response expected for health and immune system. In agreement with our results, Mohamadi and Dabiri (2012) reported that WBC and differential white blood cells (neutrophil, monocyte and lymphocyte) were unaffected by probiotic, prebiotic and synbiotic in Holstein female calves (P>0.05).

Mandour *et al.* (2009) reported that WBC, RBC, neutrophil, lymphocyte, hemoglobin and albumin were unaffected by bio-nutra probiotics in Awassi, Najdi and Najdi crossbred male weanded lambs (P>0.05) and Shim (2005) reported too that, hematological traits (WBC count, neutrophil, monocyte, lymphocyte and hemoglobin) were unaffected by prebiotic, multi-strain probiotic and synbiotic in weaned pigs (P>0.05). In contrast, the results of experiments conducted on suckling lambs (Milewski *et al.* 2009) and cattle (Dobicki *et al.* 2005; Dobicki *et al.* 2007) indicate that yeast preparations have a favorable effect on the animals immune system and values of WBC, Hb and lymphocyte were affected by yeast (P<0.05). Onifade *et al.* (1999) and Onifade (1997) reported a positive correlation between dietary levels of SC with the hematological parameters like WBC, RBC and hematocrit or packed cell volume (PCV) in rabbit and broiler chickens. They suggested that theses correlations may be an additional mechanism growth promotion by supplemental yeast. Recent studies indicated that yeast components may interact with immune systems and triggering immune responses (Muchmore *et al.* 1990; Davis *et al.* 2004).

CONCLUSION

In the present study, the use of live yeast (*Saccharomyces cerevisiae* SC47) in diet containing high concentrate could improve performance (whit not significantly different, (P>0.05)), concentration of plasma biochemical indicators (whit significantly different, (P<0.05)) and hematological parameters (whit not significantly different, (P>0.05)) of Zandi lambs. Greater ADG and better FCR were seen in the group of lambs that fed with 4.5 gr yeast per lambs (HY group), that could probably be due to improved cellulolytic bacteria in the rumen of lambs fed probiotics supplemented diets and reduce the risk of acidosis compared to other groups specially control group, but differences between groups have not been statistically significant (P>0.05). Blood parameters which related to immune system, were unaffected by treatments. This could be due to good hygienic provided to the lambs. The differences in the type and amount of food consumed, type of probiotic or how to use it can be the reason of difference between results of this study and results of other researchers.

ACKNOWLEDGEMENT

This study was supported by Pars Jivar Soufi Company. Also we are grateful to Mrs Leila Rostami for providing us equipment.

REFERENCES

Abo El-Nor S.A.H. and Kholif M.A. (1998). Effect of supplementation of live yeast culture in the diet on the productive performance of lactating buffaloes. *Milchwissenschaft.* **53,** 663-666.

Amabile-Cuevas C., Cardenas-Garcia M. and Ludgar M. (1995). Antibiotic resistance. *J. Anim. Sci.* **83,** 320-332.

Antunovic Z., Speranda M., Amidzic D., Seric V., Steiner Z., Doma-Cinovic N. and Boli F. (2006). Probiotic application in lambsnutrition. *Krmiva.* **4,** 175-180.

Beeson W.M. and Perry T.W. (1952). Balancing the nutritional deficiencies of roughages for beef steers. *J. Anim. Sci.* **11,** 501-509.

Brake J. (1991). Lack of effect of all live yeast culture on broiler, breeders and progeny performance. *J. Poultr. Sci.* **70,** 1037-1039.

Bruno R.G.S., Rutigliano H.M., Cerri R.L., Robinson P.H. and Santos J.E.P. (2009). Effect of feeding *Saccharomyces cerevisiae* on performance of dairy cows during summer heat stress. *Anim. Feed Sci. Technol.* **150,** 175-186.

Davis M.E., Brown D.C., Maxwell C.V., Johnson Z.B., Kegley E.B. and Dvorak R.A. (2004). Effect of phosphorylated mannans and pharmacological additions of zinc oxide on growth and immunocompetence of weaning pigs. *J. Anim. Sci.* **82,** 581-587.

De Smet I., Van Hoorde L., De Saeyer Van de Woeslyne M. and Verstraele W. (1994). *In vitro* study of bile salt hydrolase (BSH) activity of BSH isogonics *Lactobacillus plantarum* 80 strains and estimation of cholesterol lowering through enhanced BSH activity. *Microbial Ecol. Health Dis.* **7,** 315-329.

Deaville E.R. and Galbraith H. (1992). Effect of dietary protein level and yeast culture on growth, blood prolactin and mohair fibre characteristics of British Angora goats. *Anim. Feed Sci. Technol.* **38,** 123-133.

Demir G., Klein H.O., Mandel-Molinas N. and Tuzuner N. (2007). Beta glucan induces proliferation and activation of monocytes in peripheral blood of patients with advanced breast cancer. *Int. J. Immunol. Pharmacol.* **7,** 113-116.

Desnoyers M., Giger-Reverdin S., Bertin G., Duvaux-Ponter C. and Sauvant D. (2009). Meta-analysis of the influence of *Saccharomyces cerevisiae* supplementation on ruminal parameters and milk production of ruminants. *J. Dairy Sci.* **92,** 1620-1632.

Ding J., Zhou Z.M., Ren L.P. and Meng Q.X. (2008). Effect of monensin and live yeast supplementation on growth performance, Nutrient digestibility, carcass characteristics and ruminal fermentation parameters in lambs fed steam-flaked corn-based diets. *Asian-Australas J. Anim. Sci.* **21,** 547-554.

Dobicki A., Preś J., Łuczak W. and Szyrner A. (2005). Influence of dried brewery's yeast on body weight gains, physiological and biochemical indicators of blood and development of the rumen micro-organisms in calves. *Med. Wet.* **61,** 946-949.

Dobicki A., Preś J., Zachwieja A., Mordak R. and Jakus W. (2007). Influence of yeast preparations on chosen biochemical blood parameters and the composition of cow milk. *Med. Wet.* **63,** 955-959.

Erdman J.W. (1989). Phytic acid interactions with divalent cations in foods and in gastro intestinal tract. Pp. 161-170 in Mineral Absorption in Monogastric Gastro-Intestinal Tract. F.R. Dintizisand and J.A. Laszlo, Eds. Plenum Press, New York.

FAO. (2002). Probiotics in Food: Health and Nutritional Properties and Guidelines for Evaluation. Food and Agriculture Organization of the United Nations World Health Organization. Rome, Italy.

Fuller R. (1989). Probiotics in man and animals. *J. Appl. Bacterial.* **66,** 365-374.

Fuller R. (1977). The Importance of lactobacilli in maintaining normal microbial balance in the crop. *Br. Poult. Sci.* **18,** 84-94.

Fuller R. (1992). History and development of probiotics. Chapman and Hall, London, UK.

Galina M.A., Delgado M. and Ortíz M. (2006). Effect of a lactic probiotic on kids growth. Pp. 54 57th Ann. Meet. European Assoc. Anim. Prod. Wageningen, Netherlands.

Glade M.J. and Biesik L.M. (1986). Enhanced nitrogen retention in yearling horses supplemented with yeast culture. *J. Anim. Sci.* **62,** 1635-1642.

Glade M.J. and Sist M.O. (1988). Dietary yeast culture supplementation enhances urea recycling in equine large intestine. *Nutr. Reprod. Int.* **37,** 11-17.

Guedes C.M., Goncalves D., Rodrigues M.A.M. and Diasda-Silva A. (2008). Effects of a *Saccharomyces cerevisiae* yeast on ruminal fermentation and fibre degradation of maize silages in cows. *Anim. Feed Sci. Technol.* **145,** 27-40.

Heinrichs A.J., Jones M. and Heinrichs B.S. (2003). Effects of mannan oligosaccharide or antibiotic in neonatal diets on health and growth of dairy calves. *J. Dairy Sci.* **86,** 4064-4069.

Hernandez R., Gonzalez S.S., Pinos-Rodriguez J.M., Ortega M.E., Hernandez A., Bueno G. and Cobos M. (2009). Effect of a yeast culture on nitrogen balance and digestion in lambs fed

early and mature Orchard grass. *J. Appl. Anim. Res.* **35,** 53-56.

Jouany J.P., Mathieu F., Senaud J., Bohatier J., Bertin G. and Mercier M. (1998). The effects of *Saccharomyces cerevisiae* and *Aspergilus oryzae* on the digestion of the cell wall fraction of a mixed diet in defaunated and refaunated sheep rumen. *Reprod. Nutr. Dev.* **38,** 401-416.

Keyser S.A., McMeniman J.P., Smith D.R., MacDonald J.C. and Galyean M.L. (2007). Effects of *Saccharomyces cerevisiae* subspecies boulardii CNCM I-1079 on feed intake by healthy beef cattle treated with flor fenicol and on health and performance of newly received beef heifers. *J. Anim. Sci.* **85,** 1264-1273.

Khalid M.F., Sarwar M., Mahr Un N. and Zia-Ur R. (2011). Response of growing lambs fed on different vegetable protein sources with or without probiotics. *Int. J. Agric. Biol.* **13,** 332-338.

Line J.E., Bailey J.S., Cox N.A. and Stern N.J. (1997). Yeast treatment to reduce Salmonella and Campylobacter population associated with broiler chickens subjected to transport stress. *Poult. Sci.* **76,** 1227-1231.

Mackie R.I., Mcsweeney C.S. and Klievea V. (2002). Microbiologyofthe rúmen. Pp. 95-118 in Sheep Nutrition. M. Freer and H. Dove, Eds. CSIRO Publishing, Canberra, Australia.

Małaczewska J. and Milewski S. (2010). Immunomodulating effect of Inter Yeast S on the non-specific and specific cellular and humoral immunity in lambs. *Pol. J. Vet. Sci.* **13,** 163-170.

Mandour M.A., Al-Shami S.A. and Altabari G. (2009). The effect of feeding probiotics on the productive performance of Saudi Arabia sheep breeds during. *Mansoura Vet. Med. J.* **11(1),** 87-103.

Martin S.A., Nisbet B.J. and Dean R.G. (1989). Influence of a commercial yeast supplement on the *in vitro* ruminal fermentation. *Nutr. Reprod. Int.* **40,** 395-403.

Masek T., Mikulec Z., Valpotić H., Kušće L., Mikulec N. and Antunac N. (2008). The influence of live yeast cells (*Saccharomyces cerevisiae*) on the performance of grazing dairy sheep in late lactation. *Veterinarski Arhi.* **78,** 95-104.

Masek T., Mikulec H., Valpotic Snježan P., Stipetic B. and Perkic D. (2007). Influence of yeast culture (*Saccharomyces cerevisiae*) on performance of fattening lambs fed ground or whole grain diet. *Krmiva.* **4,** 179-187.

Mikulec Ž., Mašek T., Habrun B. and Valpotić H. (2010). Influence of live yeast cells (*Saccharomyce scerevisiae*) supplementation to the diet of fattening lambs on growth performance and rumen bacterial number. *Vet. Arhiv.* **80(6),** 695-703.

Miles R.D., Arafa I.S., Harms R.H., Carson C.W., Reid B.L. and Crawford J.S. (1981). Effects of a living non freeze dried lactobacillus acidophilus culture on performance, egg quality and gut microflora in commercial layers. *Poult. Sci.* **60,** 993-1004.

Milewski S., Wójcik R., Małaczewska J., Trapkowska S. and Siwicki A.K. (2007). Effect of β-1.3/1.6-D-glucan on meat performance and non-specific humoral defense mechanisms in lambs. *Med. Wet.* **63,** 360-363.

Milewski S., Brzostowski H., Tański Z., Zaleska B., Ząbek K. and Kosińska K. (2009). Effect of yeast preparations *Saccharomyces cerevisiae* on meat performance traits and hematological indices in sucking lambs. Pp. 101 in Proc. Ann. Meet. European Assoc. Anim. Prod. Wageningen, Netherlands.

Mohamadi P. and Dabiri N. (2012). Effects of probiotic and prebiotic on average daily gain, fecal shedding of *Escherichia coli* and immune system status in newborn female calves. *Asian-Australas J. Anim. Sci.* **25(9),** 1255-1261.

Moore B.E., Newman K.E., Spring P. and Chandler F.E. (1994). The effect of yeast culture (Yea Sace 1026) in microbial population's digestion in the cecum and colon of the equine. *J. Anim. Sci.* **72,** 1-10.

Muchmore A.V., Sathyamoorthy N., Decker J. and Sherblom A.P. (1990). Evidence that specific high mannose oligosaccharides can directly inhibit antigen-driven T-cell responses. *J. Leukoc. Biol.* **48,** 457-464.

Mukhtar N., Sarwar M., Nisa M.U. and Sheikh M.A. (2010). Growth response of growing lambs fed on concentrate with or without ionophores and probiotics. *Int. J. Agric. Biol.* **12,** 734-738.

Newbold C.J., Wallace R.J. and Chen X.B. (1995). Differents trains of *Saccharomyces cerevisiae* differ in their effects on ruminal bacterial numbers *in vitro* and in sheep. *J. Anim. Sci.* **73,** 1811-1818.

NRC. (1985). Nutrient Requirements of Poultry, 6th Rev. Ed. National Academy Press, Washington, DC., USA.

Onifade A.A. and Babatune G.M. (1996). Supplemental value of dried yeast in a high fiber diet for broiler chicks. *Anim. Feed Sci. Technol.* **62,** 91-96.

Onifade A.A., Obiyan R.I., Onipede E., Adejumo O.A., Abu O.A. and Babatune G.M. (1999). Assessment of the effects of supplementing rabbit diets with a culture of *Saccharomyces cerevisiae* using growth performance, blood composition and clinical enzyme activities. *Anim. Feed Sci. Technol.* **77,** 25-32.

Onifade A.A. (1997). Growth performance, carcass characteristics, organ measurements and hematology of broiler chickens fed a high fiber diet supplemented with antibiotics or dietary yeast. *Die Nahrung.* **41,** 370-374.

Pagan J.D. (1990). Effect of yeast culture supplementation on nutrient digestibility in mature horses. *J. Anim. Sci.* **68,** 371-380.

Rezaeian M. (2004). Effect of yeast culture supplementation on the performance of finishing Shal lambs. *Proc. Br. Soc. Anim. Sci.* **128,** 211-121.

Sales J. (2011). Effectsof *Saccharomyces cerevisiae* supplementation on ruminal parameters, nutrient digestibility and growth in sheep: a meta-analysis. *Small Rumin. Res.* **100,** 19-29.

Salminen S., Isolauri E. and Salminen E. (1996). Clinical uses of probiotics for stabilizing the gut mucosal barrier: successful strains and future challenges. *Antonievan Leeuwenhoek.* **70,** 347-358.

SAS Institute. (1997). SAS®/STAT Software, Release 6.11. SAS Institute, Inc., Cary, NC. USA.

Sharon N. (2008). Lectins: past, present and future. *Biochem. Soc. Trans.* **36,** 1457-1460.

Shim S.B. (2005). Effects of prebiotics, probiotics and synbiotics in the diet of young pigs. Ph D. Thesis. Wageningen University and Research Center, Wageningen, Netherlands.

Siwicki A.K., Kazuń K., Głąbski E., Terech-Majewska E., Baranowski P. and Trapkowska S. (2004). The effect of beta-1.3/1.6-glucan in diets on the effectiveness of anti-*Yersinia ruckeri* vaccine-an experimental study in rainbow trout (*Oncorhynchus mykiss*). *Pol. J. Food Nutr. Sci.* **54,** 59-61.

Spring P., Wenk C., Dawson K.A. and Newman K.E. (2000). The effects of dietary manna oligosaccharides on cecal parameters and the concentration of enteric bacteria in the ceca of *Salmonella* challenged broiler chicks. *Poult. Sci.* **79,** 205-211.

Sykes A.R. (1978). An assessment of the value of plasma urea nitrogen and albumin concentrations as monitors of the protein status of sheep. Pp. 143-154 in The Use of Blood Metabolites in Animal Production. O.C.C. Publications, British, UK.

Thrune M., Bach A., Ruiz-Moreno M., Stern M.D. and Linn J.G. (2009). Effects of *Saccharomyces cerevisiae* on ruminal pH and microbial fermentation in dairy cows: yeast supplementation on rumen fermentation. *Livest. Sci.* **124,** 261-265.

Williams P.W. (1989). Understanding the biochemical mode of action of yeast culture. Pp. 79-99 in Biotechnology in the Feed industry. T.P. Lyons, Ed. Alltech Technical Publications, Nicholasville, Kentucky.

Williams P.E., Tait C.A., Innes G.M. and Newbold C.J. (1991). Effects of the inclusion of yeast culture in the diet of dairy cows on milk yield and forage degradation and fermentation patterns in the rumen of steers. *J. Anim. Sci.* **69,** 3016-3026.

Xiao Z., Trincado C.A. and Murtaugh M.P. (2004). Beta-glucan enhancement of T-cell IFN gamma response in swine. *Vet. Immunol. Immunopathol.* **102,** 315-320.

In vitro Assessment of the Effect of Plant Extracts on Digestibility, Estimated Energy Value, Microbial Mass and Rumen Fermentation Kinetics

V. Naseri[1*], F. Kafilzadeh[1] and H. Jahani-azizabadi[2]

[1] Department of Animal Science, Faculty of Agriculture, Razi University, Kermanshah, Iran
[2] Department of Animal Science, College of Agriculture, University of Kurdistan, Sanandaj, Iran

*Correspondence E-mail: w.naseri@razi.ac.ir

ABSTRACT

Three ethanol extracts, chamomile (CHA), clove (CLO) and tarragon (TAR), were tested at five doses (0, 250, 500, 750 and 1000 µL/L) to determine their effects on *in vitro* organic matter digestibility (IVOMD), metabolizable energy (ME), net energy of lactation (NEL), short-chain fatty acids (SCFA), microbial mass (MM) and rumen fermentation kinetics of a 40:60 forage: concentrate diet using *in vitro* gas production. These three extracts had significant effects on gas production kinetics. CHA (at 500 µL/L dose) and CLO (at 1000 µL/L dose) decreased ($p<0.05$) potential gas production. The initial gas production rate constants (*c*) was increased ($p<0.05$). However, CHA, CLO and TAR ethanol decreased ($p<0.05$) later gas production rate constants (*d*). Lag time (h) was decreased ($p<0.05$) due to addition of CHA and TAR ethanol (at 750 µL/L dose), and CLO (at 500 and 1000 µL/L doses). TAR and CLO ethanol did not affect fermentation rate (h^{-1}), but CHA at 1000 µL/L increased it. The TAR and CLO ethanol did not affect IVOMD, ME, NEL, SCFA and microbial mass. However, organic matter digestibility, ME, NEL, SCFA and microbial mass were increased by addition of CHA ethanol at 750 and 1000 µL/L doses. Results suggest that CHA, CLO and TAR ethanol extracts at appropriate doses may have potential to improve the rumen fermentation kinetics and nutritive value of ruminant diets due to secondary metabolites contents.

KEY WORDS *in vitro* gas production, plant ethanol extract, rumen fermentation kinetics.

INTRODUCTION

In the recent years, the use of plant extracts in dairy cattle rations have been considered worldwide by ruminant nutritionists especially after the prohibition of growth promoting antibiotics by the (EC number 1831/2003; European Union, 2003), because plant extracts were believed to be natural, safe and efficient without negative side effects. Secondary metabolites present in the natural plant extracts can modify rumen fermentation kinetics and improve milk production in dairy cattle (Alexander *et al.* 2007; Benchaar *et al.* 2008; Hart *et al.* 2008; Naseri *et al.* 2012; Naseri *et al.* 2015). It

has also been observed that secondary metabolites suppressed protozoal populations, increased bacterial and fungal populations, propionate production, microbial yield and efficiency of microbial protein synthesis (EMPS), increased dietary dry matter (DM), organic matter (OM) and neutral detergent fibre (NDF) degradation and reduced dietary crude protein (CP) degradation and methanogenesis. A number of fast and cost-effective *in vitro* gas measurement methods have been used by several groups to evaluate the nutritional value of feedstuffs and kinetics of rumen fermentation (Getachew *et al.* 1998; Getachew *et al.* 2004; Makkar, 2005; Mirzaei-Aghsaghali *et al.* 2011a; Naseri *et al*. 2015). These methods can provide useful data on fermentation kinetics of feedstuffs, prediction of feed intake (Khazaal *et al*. 1995; Mirzaei-Aghsaghali *et al*. 2011b), digestibility, and microbial nitrogen supply, amount of short-chain fatty acids, carbon dioxides and metabolizable energy of feeds for ruminants (Menke and Steingass, 1988; Babayemi, 2007; Mirzaei-Aghsaghali *et al*. 2008b; Mirzaei-Aghsaghali *et al*. 2008a; Maheri-Sis *et al*. 2008; Maheri-Sis *et al*. 2007). The ease of measuring fermentation end-products makes these methods more preferable (Makkar, 2005).

This work aimed to evaluate the *in vitro* gas production kinetics and estimate the *in vitro* organic matter digestibility (IVOMD), metabolizable energy (ME), short-chain fatty acids (SCFA), net energy of lactation (NEl) and microbial protein production of high-concentrate diet for dairy cattle after supplementing the feed material with ethanol extract of chamomilla (*Matricaria chamomilla*), clove (*Syzygium aromaticum*) and tarragon (*Artemisia dracunculus*).

MATERIALS AND METHODS

Selection of plants

Three medicinal plants: chamomile, clove and tarragon were selected on the basis of their traditional usage for the various digestive ailments, and in the light of recent literature (Patra, 2011).

Preparation of plant extract

Chamomile and tarragon leaves used in this study were collected at vegetative stage from Abidar Mountains and clove buds were purchased from local markets in Sanandaj (longitude 46.99 °E, latitude 35.32 °N and Köppen-Geiger climate), Iran. Approximately 100 g of fresh chamomile and tarragon leaves were cut into small pieces, placed into a blender (Saya Quick, QMC-20) and added 80 mL 70% ethanol then they were well blended three times for 5 minutes per time. The blended material was squeezed through four layers of muslin cloth into the labeled beaker and fibrous materials discarded. The combined filtrate was filtered using Whatman No.1 filter paper, and then transferred to a round-bottom Buchi flask. Also, the clove buds crushed into small pieces, oven-dried at 39 °C and ground to pass a 1mm screen. Fifty of ground sample was weighed into a 250 mL conical flask and added 200 mL 70% ethanol. The extraction was completed by placing the flasks in a shaker at 22 °C and 200 rpm for 24 h. Contents of the flask were squeezed through four layers of muslin cloth into the labeled beaker and fibrous materials discarded. The combined liquid phase was filtered using Whatman No.1 filter paper and then transferred to a round-bottom Buchi flask. Finally, ethanol was evaporated by using a vacuum evapo-

rator (Heidolph Laborota 4011 digital) at 40-50 °C until the ethanol-streak stopped on the side of the bottle. The remaining concentrate was resuspended in 10 mL water, transferred into 10 mL sterile anaerobic crimped serum vials, and stored at -20 °C.

Inoculum and substrate

The inoculum was prepared according to the method of Tilley and Terry (1963). Briefly, rumen fluid was obtained from three rumen cannulated rams before the morning feeding. The rumen fluid was mixed on volume basis then it was bubbled with CO_2 for approximately 2 min and strained through four layers of cheese cloth. The incubation inoculum was prepared by diluting the fluid inoculum with the buffer (Tilley and Terry, 1963) in a 1:4 (V/V) ratio and stirring in a water bath at 39 °C with purging CO_2 until its use. The ration of the rams consisted of 40% alfalfa, 35% barley grain, 15% corn grain, 9% soybean meal, 0.5% salt and 0.5% vitamin-mineral premix. The substrate used in the *in vitro* ruminal fermentation was at 40:60 forage:concentrate ratio, formulated for dairy cattle (Table 1), oven dried (at 39 °C for 72 h) and finely ground to pass through a 1 mm screen.

In vitro gas production

The method used for gas production measurements was as described by Theodorou *et al*. (1994). Approximately 250 mg dry matter (DM) of substrate was weighed into 100 mL sterile tubes, kept at 39 °C. Plant extracts were added at different volumes (0, 250, 500, 750 and 1000 μL/L). Each sample was incubated in three replicates. Thirty milliliters of incubation inoculum (in the proportion of 20% rumen fluid+80% buffer) prepared (as described in the inoculum and substrate) and by flushing CO_2 before was anaerobically dispensed in each tube at 39 °C. The samples were swirled to mix the contents and placed in ashaker incubator (Thermoshaker Gerhardt) at 39 °C (Blümmel and Ørskov, 1993). The pressure of gas produced in each tube was recorded using a pressure transducer (Testo 512; Testo Inc., Germany) at 0, 2, 4, 8, 16, 24, 48 and 72nd h of incubation. To estimate the kinetics of gas production, data on cumulative gas volume produced were fitted using the generalized Mitscherlich model, proposed by France *et al*. (1993):

$$G = A\left(1 - e^{-c(t-L)-d(\sqrt{t}-\sqrt{L})}\right)$$

Where:
G (mL): denotes cumulative gas production at time t.
A (mL): asymptotic gas production.
c (h^{-1}): initial gas production rate constant.
d ($h^{-1/2}$): later gas production rate constant rate constants.
L (h): lag time.

Calculation

The half-life ($t_{1/2}$, h) of the degradable fraction of substrate was calculated as the time taken for gas accumulation to reach 50% of its asymptotic value. The fractional degradation rate at $t_{1/2}$ ($\mu_{1/2}$, h^{-1}) was calculated as:

$$\mu_{1/2} = c + \frac{d}{2\sqrt{t_{1/2}}}$$

The metabolizable energy (MJ/kg DM) content of the substrate and *in vitro* organic matter digestibility were calculated using the equations below (Menke *et al.* 1979) as:

ME (MJ/Kg DM)= 2.20 + 0.136 GP + 0.0057 CP + 0.00029 EE^2

IVOMD (%)= 14.88 + 0.889 GP + 0.45 CP + 0.0651 XA

Where:
GP: 24 h net gas production (mL/250 mg^{-1}).
CP: crude protein (%).
EE: ether extract (%).
XA: ash content (%).

Short-chain fatty acid (SCFA) content was calculated using the equation of Makkar (2005); Maheri-Sis *et al.* (2007) and Maheri-Sis *et al.* (2008):

SCFA (mmol)= 0.0222 × GP – 0.00425 (Makkar, 2005).

Where:
GP: 24 h net gas production (mL/250 mg^{-1}).

Net energy for lactation (NEL) was calculated using the equation of Abas *et al.* (2005) as follows:

NEL (MJ/kg DM)= 0.115 GP + 0.0054 CP + 0.014 EE - 0.0054 CA - 0.36

Microbial mass (mg) was estimated using equation of Blummel *et al.* (1997):

Microbial mass (mg)= mg substrate truly degraded (OMD) - (GP×stoichiometrical factor)

The stoichiometrical factor was 2.20.

Chemical analysis

The substrate was analysed for DM (24 h at 103 °C), ash and organic matter (OM) (4 h at 550 °C), CP content was adapted for an automatic distiller Kjeldahl apparatus (Kjeltec Auto 1030 Analyser; Tecator, Höganäs, Sweden)

and using $CuSO_4$/Se as catalyst instead of $CuSO_4$/TiO_2, ether extract using petroleum ether for distillation instead of diethyl ether (AOAC, 1990). The neutral detergent fibre (NDF) contents were determined as described (Van Soest *et al.* 1991).

Statistical analysis

Data were subjected to analysis of variance (ANOVA) using the general linear model (GLM). Significant differences between individual means were identified using Duncan's test (all pairwise multiple comparison procedures). All statements of significance were based on a probability of ($p < 0.05$) (SAS, 1996).

RESULTS AND DISCUSSION

Chemical composition

The chemical composition of diet which used as fermentation substrate is shown in Table 1.

Table 1 Chemical composition (g/kg DM) of substrate used for *in vitro* gas production

Parameters	Substrate
Dry matter	945
Organic matter	900
Ash	100
Crude protein	160
Ether extract	40
Neutral detergent fibre (NDF)	300

Effect of plant ethanol extracts on *in vitro* rumen fermentation kinetics

Effect of ethanol extracts of chamomille, clove and tarragon on *in vitro* fermentation kinetics is presented in Tables 2, 3 and 4, respectively. Potential gas production (A) decreased (by 7%) significantly ($p < 0.05$) by the addition of chamomile and clove extracts at 500 and 1000 (μl/L) doses, respectively.

In addition, 500 and 750 μL/L doses of tarragon extract were also found to be effective in decreasing potential gas production (A) by 8% (P=0.07).

The main active compounds of chamomile, clove and tarragon extract were terpenoids α-bisabolol and chamazulene, eugenol (phenylpropanoid) and methyleugenol, respectively (Janmejai *et al.* 2010; Jamalian *et al.* 2012; Renata and Grażyna, 2014).

These active compounds are known as of plant secondary metabolites, which include terpenoids, alkaloids and phenolics present in the essential oil fraction of many plants (Sallam *et al.* 2011).

Essential oils have antimicrobial activities against both gram-negative and gram-positive bacteria, a property that has been attributed to the presence of terpenoid and phenolic compounds (Conner, 1993; Dorman and Deans, 2000; Calsamiglia *et al.* 2007).

Table 2 Parameters estimated by fitting generalized mitscherlich model to gas production values, recorded for a high-concentrate diet for dairy cattle treated with different levels (0, 250, 500, 750 and 1000 µL/L) of ethanol Chamomile (*Matricaria chamomilla*) extract

Different levels of ethanol Chamo-milla extract (µL/L)	Kinetics parameters					
	A (mL/250 mg DM)	c (h⁻¹)	d (h⁻¹ᐟ²)	L (h)	Half-life (h)	Fermentation rate (h⁻¹)
0	68.34[ab]	0.017[c]	0.330[a]	0.740[b]	7.24[d]	0.080[bc]
250	65.18[bc]	0.043[b]	0.193[b]	0.667[b]	7.84[ab]	0.077[c]
500	64.46[c]	0.023[c]	0.333[a]	1.010[a]	7.38[cd]	0.080[bc]
750	66.36[abc]	0.077[a]	0.060[c]	0.417[c]	8.06[a]	0.087[ab]
1000	69.35[a]	0.063[a]	0.143[b]	0.757[b]	7.65[bc]	0.093[a]
SEM	0.63	0.01	0.03	0.06	0.09	0.01
P-value	0.0324	< 0.0001	< 0.0001	0.0031	0.0004	0.0070

A: asymptotic gas production; c (h⁻¹): initial gas production rate constant; d (h⁻¹ᐟ²): later gas production rate constant rate constants and L (h): lag time.
The means within the same column with at least one common letter, do not have significant difference (P>0.05).
SEM: standard error of the means.

Table 3 Parameters estimated by fitting generalized mitscherlich model to gas production values, recorded for a high-concentrate diet for dairy cattle treated with different levels (0, 250, 500, 750 and 1000 µL/L) of ethanol Clove (*Syzygium aromaticum*) extract

Different levels of ethanol Clove extract (µL/L)	Kinetics parameters					
	A (mL/250 mg DM)	c (h⁻¹)	d (h⁻¹ᐟ²)	L (h)	Half-life (h)	Fermentation rate (h⁻¹)
0	68.34[a]	0.017[b]	0.330[a]	0.740[a]	7.24[c]	0.080
250	65.48[ab]	0.073[a]	0.073[c]	0.503[ab]	7.91[a]	0.083
500	66.58[ab]	0.057[a]	0.120[bc]	0.380[b]	7.91[a]	0.080
750	66.42[ab]	0.053[a]	0.177[b]	0.570[ab]	7.61[b]	0.083
1000	63.80[c]	0.073[a]	0.073[c]	0.283[b]	7.79[ab]	0.087
SEM	0.62	0.01	0.03	0.05	0.07	0.01
P-value	0.2275	0.0004	0.0002	0.0311	0.0005	0.3818

A: asymptotic gas production; c (h⁻¹): initial gas production rate constant; d (h⁻¹ᐟ²): later gas production rate constant rate constants and L (h): lag time.
The means within the same column with at least one common letter, do not have significant difference (P>0.05).
SEM: standard error of the means.

Table 4 Parameters estimated by fitting generalized mitscherlich model to gas production values, recorded for a high-concentrate diet for dairy cattle treated with different levels (0, 250, 500, 750 and 1000 µL/L) of ethanol Tarragon (*Artemisia dracunculus*) extract

Different levels of ethanol Tarra-gon extract (µL/L)	Kinetics parameters					
	A (mL/250 mg DM)	c (h⁻¹)	d (h⁻¹ᐟ²)	L (h)	Half-life (h)	Fermentation rate (h⁻¹)
0	68.34[ab]	0.017[c]	0.330[ab]	0.740[b]	7.24[b]	0.080
250	69.19[a]	0.013[c]	0.380[a]	0.920[a]	7.12[b]	0.087
500	63.29[b]	0.030[b]	0.297[b]	0.713[b]	7.11[b]	0.087
750	63.49[b]	0.053[a]	0.190[c]	0.527[c]	7.30[ab]	0.087
1000	67.09[ab]	0.027[b]	0.273[b]	0.673[b]	7.49[a]	0.080
SEM	0.89	0.01	0.02	0.04	0.05	0.01
P-value	0.0743	< 0.0001	0.0002	0.0015	0.0201	0.1705

A: asymptotic gas production; c (h⁻¹): initial gas production rate constant; d (h⁻¹ᐟ²): later gas production rate constant rate constants and L (h): lag time.
The means within the same column with at least one common letter, do not have significant difference (P>0.05).
SEM: standard error of the means.

Debashis-Roy *et al.* (2015) have reported that eugenol hadamore effective antimicrobial potential in comparison with other non phenolic plant secondary metabolites because of the presence of a hydroxyl group in its phenolic structure and resulted in the loss of integrity of bacterial cell membrane and ultimately in reduction in glucose-uptake of bacteria. It has also been demonstrated that α-bisabolol and Chamazulene had the strongest activity against both gram-positive and gram-negative bacteria (Janmejai *et al.* 2010). However, decrease in potential gas production may be due to their secondary metabolites. In the present study, it was evidenced that other kinetic parameters of fermentation also affected.

Overall, initial gas production rate constant (*c*) increased (P<0.05) due to addition of plant ethanol extracts to medium.

But, ethanol extracts decreased (P<0.05) later gas production rate constant (*d*). Chamomile extract at 750 µL/L, clove extract at 1000 µL/L and tarragon extract at 750 µL/L had the lowest lag time, resulting in a faster rate of fermentation.

Effect of plant ethanol extracts on *in vitro* OM digestibility, estimated energy value and microbial mass
In vitro OM digestibility, estimated energy value and microbial mass results were presented (Tables 5, 6 and 7).

Table 5 Predictions of *in vitro* organic matter digestibility (IVOMD), metabolizable energy (ME), short-chain fatty acids (SCFA), net energy lactation (NEL) and microbial mass estimation (MM) for a high-concentrate diet for dairy cattle treated with different levels (0, 250, 500, 750 and 1000 µl/L) of ethanol Chamomile (*Matricaria chamomilla*) extract

Different levels of ethanol Chamomile extract (µL/L)	IVOMD (%)	ME (MJ/Kg DM)	SCFA (mmol)	NEL (MJ/Kg DM)	Microbial mass (mg)
0	73.14[bc]	10.87[bc]	1.26[bc]	6.25[bc]	58.10[b]
250	71.04[bc]	10.55[bc]	1.20[bc]	5.98[bc]	58.05[b]
500	70.85[c]	10.52[c]	1.19[c]	5.95[c]	58.04[b]
750	74.05[ab]	11.01[ab]	1.28[ab]	6.36[ab]	58.13[ab]
1000	76.75[a]	11.42[a]	1.34[a]	6.72[a]	58.19[a]
SEM	0.68	0.10	0.02	0.09	0.02
P-value	0.0072	0.0073	0.0069	0.0070	0.0094

The means within the same column with at least one common letter, do not have significant difference ($P>0.05$).
SEM: standard error of the means.

Table 6 Predictions of *in vitro* organic matter digestibility (IVOMD), metabolizable energy (ME), short-chain fatty acids (SCFA), net energy lactation (NEL) and microbial mass estimation (MM) for a high-concentrate diet for dairy cattle treated with different levels (0, 250, 500, 750 and 1000 µl/L) of ethanol Clove (*Syzygium aromaticum*) extract

Different levels of ethanol Clove extract (µL/L)	IVOMD (%)	ME (MJ/Kg DM)	SCFA (mmol)	NEL (MJ/Kg DM)	Microbial mass (mg)
0	73.14	10.87	1.26	6.25	58.10
250	73.67	10.95	1.27	6.32	58.12
500	73.33	10.90	1.26	6.27	58.11
750	73.27	10.89	1.26	6.27	58.11
1000	72.20	10.73	1.23	6.13	58.08
SEM	0.34	0.05	0.01	0.04	0.01
P-value	0.7805	0.7842	0.7134	0.7904	0.7561

The means within the same column with at least one common letter, do not have significant difference ($P>0.05$).
SEM: standard error of the means.

Table 7 Predictions of *in vitro* organic matter digestibility (IVOMD), metabolizable energy (ME), short-chain fatty acids (SCFA), net energy lactation (NEL), and microbial massestimation (MM) for a high-concentrate diet for dairy cattle treated with different levels (0, 250, 500, 750 and 1000 µl/L) of ethanol Tarragon (*Artemisia dracunculus*) extract

Different levels of ethanol Tarragon extract (µL/L)	IVOMD (%)	ME (MJ/Kg DM)	SCFA (mmol)	NEL (MJ/Kg DM)	Microbial mass (mg)
0	73.14	10.87	1.26	6.25	58.10
250	71.39	10.60	1.21	6.02	58.06
500	70.52	10.47	1.19	5.91	58.04
750	72.09	10.71	1.23	6.11	58.08
1000	72.76	10.81	1.25	6.20	58.09
SEM	0.43	0.06	0.01	0.05	0.01
P-value	0.3194	0.3178	0.3083	0.3189	0.3793

The means within the same column with at least one common letter, do not have significant difference ($P>0.05$).
SEM: standard error of the means.

Chamomile extract at 1000 µL/L increased (approximately 5 to 8%) significantly ($P<0.05$) IVOMD, metabolizable energy, SCFA, net energy lactation and microbial mass. Ethanol extracts of clove and tarragon did not affect *in vitro* OM digestibility of substrate, estimated energy value and microbial mass.

The results of GP measurement revealed that chamomile ethanol extract at 1000 µL/L resulted in an increase in GP compared with the control, which was consistent with an increase in IVOMD, metabolizable energy, SCFA and NEL. However, an increase in OM digestibility because of the addition of chamomile ethanol extract at high dose could also be attributed to stimulated bacterial activity (Naseri *et al.* 2012), which results in an increase in poten-tial gas production. Generally, medicinal plants or their extracts usually yield complex mixtures of biochemical so that identification of the phytochemical fractions that might be involved in the effects observed was not possible (Scehovic, 1999).

However, three explanations can be made as follows: (1) the inhibitory or stimulatory action of plant secondary metabolites (PSM) on some rumen microorganisms; (2) the effect of the degradation products of PSM and (3) direct action of other secondary metabolites. Therefore, in the current study, our observations possibly might have resulted from the inhibitory or stimulatory action of PSM, especially from the presence of essential oils (EOs) on some rumen microorganisms.

CONCLUSION

In vitro effect of ethanol extracts of chamomilla (*Matricaria chamomilla*), clove (*Syzygium aromaticum*) and tarragon (*Artemisia dracunculus*) at differing concentrations on organic matter digestibility, estimated energy value, microbial mass, and rumen fermentation kinetics of a high-concentrate diet for dairy cattle, suggested that chamomile, clove and tarragon extracts have potential to alter rumen fermentation kinetics. However, these findings should be considered preliminary and further investigation should be undertaken which also use *in vivo* methods in order to better assess the value of these plant extracts as feed additives to improve the yield of dairy products.

ACKNOWLEDGEMENT

The authors thank the University of Razi (Kermanshah, Iran) for the financial support.

REFERENCES

Abas I., Ozpinar H., Can-Kutay H. and Kahraman R. (2005). Determination of the metabolizable energy (ME) and net energy lactation (NEL) contents of some feeds in the Marmara region by *in vitro* gas technique. *Turkish J. Vet. Anim. Sci.* **29,** 751-757.

Alexander G., Singh B., Sahoo A. and Bhat T.K. (2007). *In vitro* screening of plant extracts to enhance the efficiency of utilization of energy and nitrogen in ruminant diets. *Anim. Feed Sci. Technol.* **145,** 229-242.

AOAC. (1990). Official Methods of Analysis. Vol. I. 15th Ed. Association of Official Analytical Chemists, Arlington, VA, USA.

Babayemi O.J. (2007). *In vitro* fermentation characteristics and acceptability by West African dwarf goats of some dry season forages. *African J. Biotechnol.* **6(10),** 1260-1265.

Benchaar C., Calsamiglia S.,Chaves A.V., Fraser G.R., Colombatto D., McAllister T.A. and Beauchemin K.A. (2008). A review of plant-derived essential oils in ruminant nutrition and production. *Anim. Feed Sci. Technol.* **145,** 209-228.

Blümmel M. and Ørskov E. R. (1993). Comparison of *in vitro* gas production and nylon bag degradability of roughagesin predicting feed intake in cattle. *Anim. Feed Sci. Technol.* **40,** 109–119.

Blümmel M., Makkar H.P.S. and Becker K. (1997). The relationship between *in vitro* gas production, *in vitro* microbial mass yield and 15N incorporation and its implications for the prediction of voluntary feed intake of roughages. *British J. Nutr.* **77,** 911-921.

Calsamiglia S., Busquet M., Cardozo P.W., Castillejos L. and Ferret A. (2007). Invited review: essential oils as modifiers of rumen microbial fermentation. *J. Dairy Sci.* **90,** 2580-2595.

Conner D.E. (1993). Naturally occurring compounds. Pp. 441-468 in Antimicrobials in Foods. P.M. Davidson and A.L. Branen,

Eds. Marcel Dekker, New York, USA.

Debashis R.,Tomar S.K. and Vinod K. (2015). Rumen modulatory effect of thyme, clove and peppermint oils *in vitro* using buffalo rumen liquor. *Vet. World.* **8(2),** 203-207.

Dorman H.J.D. and Deans S.G. (2000). Antimicrobial agents from plants: antibacterial activity of plant volatile oils. *J. Appl. Microbiol.* **88,** 308-316.

European Union. (2003). Regulation (EC) No 1831/2003 of the European Parliament and the Council of 22 September 2003 on additives for use in animal nutrition.

France J., Dijkstra J., Dhanoa M.S., Theodorou M.K., Lister S.J., Davies D.R. and Isac D.A. (1993). A model to interpret gas accumulation profiles associated with *in vitro* degradation of ruminant feeds. *J. Theor. Biol.* **163,** 99-111.

Getachew G., Blummel M., Makkar H.P.S. and Becker K. (1998). *In vitro* gas measuring techniques for assessment of nutritional quality of feeds: a review. *Anim. Feed Sci. Technol.* **72,** 261-281.

Getachew G., Depeters E.J. and Robinson P.H. (2004). *In vitro* gas production provides effective method for assessing ruminant feeds. *California Agric.* **58(1),** 54-58.

Hart K.J., Yanez-Ruiz D.R., Duval S.M., McEwan N.R. and Newbold C.J. (2008). Plant extracts to manipulate rumen fermentation. *Anim. Feed Sci. Technol.* **147,** 8-35.

Jamalian A., Shams-Ghahfarokhi M., Jaimand K., Pashootan N., Amani A. and Razzaghi-Abyaneh M. (2012). Chemical composition and antifungal activity of *Matricaria recutita* flower essential oil against medically important dermatophytes and soil-borne pathogens. *J. Mycol. Med.* **22,** 308-315.

Janmejai K., Eswar S. and Sanjay G. (2010). Chamomile: a herbal medicine of the past with bright future. *Mol. Med. Rep.* **3(6),** 895-901.

Khazaal K., Dentinho M.T., Ribeiro R. and Ørskov E.R. (1995). Prediction of apparent digestibility and voluntary intake of hays fed to sheep: comparison between using fibre components, *in vitro* digestibility or characteristics of gas production or nylon bag degradation. *J. Anim. Sci.* **61,** 527-538.

Maheri-Sis N., Chamani M., Sadeghi A.A., Mirza-Aghazadeh A. and Aghajanzadeh-Golshani A. (2008). Nutritional evaluation of kabuli and desi type chickpeas (*Cicer arietinum*) for ruminants using *in vitro* gas production technique. *African J. Biotechnol.* **7(16),** 2946-2951.

Maheri-Sis N., Chamani M., Sadeghi A.A., Mirza-Aghazadeh A. and Safaei A.A. (2007). Nutritional evaluation of chickpea wastes for ruminants using *in vitro* gas production technique. *J. Anim. Vet. Adv.* **6(12),** 1453-1457.

Makkar H.P.S. (2005). *In vitro* gas methods for evaluation of feeds containing phytochemicals. *Anim. Feed Sci. Technol.* **123,** 291-302.

Menke K.H. and Steingass H. (1988). Estimation of the energetic feed value obtained from chemical analysis and *in vitro* gas production using rumen fluid. *Anim. Res. Dev.* **28,** 47-55.

Menke K.H., Raab L., Salewski A., Steingass H., Fritz D. and Schneider W. (1979). The estimation of the digestibility and metabolisable energy content of ruminant feedstuffs from the gas production when they are incubated with rumen liquor. *J. Agric. Sci.* **93,** 217-222.

Mirzaei-Aghsaghali A. and Maheri-Sis N. (2008a). Nutritive value of some agro-industrial by-products for ruminants: a review. *World J. Zool.* **3(2)**, 40-46.

Mirzaei-Aghsaghali A., Maheri-Sis N., Mansouri H., Ebrahim R.M., Aghajanzadeh-Golshani A. and Cheraghi H. (2011b). Evaluating nutrItional value of sugar beet pulp for ruminant animals using *in vitro* gas production technique. *Int. J. Res.* **3(2)**, 147-152.

Mirzaei-Aghsaghali A., Maheri-Sis N., Mirza-Aghazadeh A., Safaei A.R. and Aghajanzadeh-Golshani A. (2008b). Nutritive value of alfalfa varieties for ruminants with emphasis of different measuring methods: a review. *Res. J. Biol. Sci.* **3(10)**, 1227-1241.

Mirzaei-Aghsaghali A., Maheri-Sis N., Mansouri H., Razeghi M.E., Shayegh J. and Aghajanzadeh-Golshani A. (2011a). Evaluating nutritional value of apple pomace for ruminants using *in vitro* gas production technique. *Ann. Biol. Res.* **2**, 100-106.

Naseri V., Hozhabri F. and Kafilzadeh F. (2012). Assessment of *in vitro* digestibility and fermentation parameters of alfalfa hay based diet following direct incorporation of fenugreek seed (*Trigonella foenum*) and asparagus root (*Asparagus officinalis*). *J. Anim. Physiol. Anim. Nutr.* **97(4)**, 773-784.

Naseri V., Kafilzadeh F. and Hozhabri F. (2015). Fenugreek seed (*Trigonella foenum-graecum*) and Asparagus Root (*Asparagus officinalis*) effects on digestion and kinetics of gas production of alfalfa hay using *in vitro* technique. *Iranian J. Appl. Anim. Sci.* **5(3)**, 185-188.

Patra A.K. (2011). Effects of essential oils on rumen fermentation, microbialecology and ruminant production. *Asian J. Anim. Vet. Adv.* **6**, 416-428.

Renata N. and Grażyna Z. (2014). Herb yield and bioactive compounds of Tarragon (*Artemisia dracunculus*) as influenced by plant density. *Acta. Sci. Pol. Hortorum Cultus.* **13(2)**, 207-221.

Sallam S.M.A., Abdelgaleil S.A.M., Bueno I.C.S., Nassera M.E.A., Araujo R.C. and Abdalla A.L. (2011). Effect of essential oils on ruminal fermentation, microbial population and methane emission *in vitro*. *Options Méditerran.* **57**, 149-156.

SAS Institute. (1996). SAS®/STAT Software, Release 6.11. SAS Institute, Inc., Cary, NC. USA.

Scehovic J. (1999). Evaluation *in vitro* de l'activité de la population microbienne du rumen en présence d'extraits végétaux. *Rev. Suisse. Agric.* **31**, 89-93.

Theodorou M.K., Williams B.A., Dhanoa M.S., McAllan A.B. and France J. (1994). A simple gas production method using a pressure transducer to determine the fermentation kinetics of ruminant feeds. *Anim. Feed Sci. Technol.* **48**, 185-197.

Tilley J.M.A. and Terry R.A. (1963). A two-stage technique for the *in vitro* digestion of forage crops. *J. British Grassland. Soc.* **18**, 104-111.

Van Soest P.J., Robertson J.B. and Lewis B.A. (1991). Carbohydrate methodology, metabolism and nutritional implications in dairy cattle. *J. Dairy Sci.* **74**, 3583-3597.

3

Ruminal Kinetics of Nutrients Degradation, Hydration, and Functional Specific Gravity of Three Types of Beet Pulp

A. Teimouri Yansari[1*]

[1] Department of Animal Science, Faculty of Agricultural Science, Sari University of Agricultural Science and Natural Resources, Sari, Iran

*Correspondence E-mail: astymori@yahoo.com

ABSTRACT

Two experiments were conducted to evaluate the relationships between nutrients degradability, kinetics of hydration, functional specific gravity (FSG) of the three types of beet pulp (BP) including fine (FBP), normal (NBP) and pelleted (PBP) BP. In experiment 1, about 3 g of samples was weighed in sealed nylon bags (6 cm×7.5 cm, 40±5 µm pore size), incubated in rumen of two cannulated Holstein steers at 0, 3, 6, 12, 18, 24, 36 and 48 h. The dry matter (DM) degradation was different among the treatments for soluble, slowly and potentially degradable fractions, rate of degradation, and effective degradability. In experiment 2, after ruminal incubation of two bags at 0, 0.5, 1, 1.5, 3, 6, 12, 18, 24, 36 and 48 h, the bags were removed without and with washing, the kinetic of hydration, functional specific gravity (FSG) measured with pycnometer. Hydration rate and water holding capacity (WHC) were different. Grinding and pelleting decreased hydration rate and WHC of BP, but increased initial and final FSG over incubation time. Soluble, slowly degradable, and indigestible fraction of DM explained 82.4, 94.8, 2.7, 54.2, 87.3 and 79.7%; 34.1, 50.0, 2.2, 31.4, 62.2 and 63.4%; and 89.1, 12.2, 68.0, 84.7 and 92.9% of the total variation of the fractional rate of degradation, effective degradability, hydration rate, WHC, initial and final FSG, respectively. In addition, the correlations between digestion and hydration parameters were high. As BP has lower than critical size, can easily pass from the reticulorumen orifice, therefore, its FSG is more important to control ruminal retention time and degradation.

KEY WORDS beet pulp, physical property, physically effective fiber, ruminant feed.

INTRODUCTION

Particle size and functional specific gravity (FSG) are feed characteristics that influenced ruminal mean retention time and ruminal nutrients digestibility. Beet pulp (BP) is economical sources of nutrients in ruminant rations because of initial higher specific gravity (SG) than dried forages and smaller size than the critical size of particles that can retain in the rumen. In addition, BP has high digestible neutral detergent fiber (NDF) and is often used to reduce the content of non-fiber carbohydrate (NFC) in dairy cattle diets. Much of the NFC in BP is pectin. Pectins are the main components of the primary cell walls of dicotyledons where they play a major role in the physico-chemical properties (Ben-Ghedalia et al. 1989). They are complex polysaccharides; their backbone, characterized by a high proportion of galacturonic acid and presence of rhamnose, presents two distinct structures: homogalacturonan or rhamnogalacturonan which has a propensity for acetate versus propionate production in the rumen (Catherine et al. 1999). In addition, the NDF in BP is highly fermentable in the rumen, and it can be used to supply fermentable fiber in the diet. The two processes in the rumen that increase the functional specific gravity (FSG) of particles are liquid uptake and particle size reduction. Therefore, both particle size and specific gravity should be used to define escapable and nonescapable rumen fiber fractions (Allen and Mertens, 1988). The inescapable fraction consists of particles that have an FSG less than the rumen fluid (Allen and Mertens, 1988) and size greater than the critical size (Poppi *et al.* 1980). The escapable fraction consists of particles that are denser than the rumen fluid and are below the threshold size for retention. In addition, digestibility is directly proportional to the digestible fraction of fiber and rate of fiber digestion, but inversely related to the rate of release of particles from nonescapable to an escapable fiber pool and rate of escape. The rate of release from fiber fractions is a function of the rate of change in FSG and the rate of particle size breakdown (Allen and Mertens, 1988). Particle size reduction increases the release rate from the nonescapable fraction, which results in reduced digestibility. Ruminal particulate matters are mostly below the threshold size for escape; therefore, particle size reduction may not be the rate-limiting step in clearance from the reticulorumen (Kaske and Engelhardt, 1990; Kaske and Engelhardt, 1992). High FSG of feeds seems to reduce the amount of fiber in ruminal mat and affects escapable particle retention (Allen and Mertens, 1988). For stimulation of chewing activity and ruminal mat formation and maintenance of mat consistency, particles must be retained in the rumen (Teimouri Yansari *et al.* 2004). However, Kaske and Engelhardt (1990), Kaske and Engelhardt (1992) and Teimouri Yansari *et al.* (2004) found that FSG is a better indicator of retention than particle size. On the otherwise, feed DM can be divided into soluble and insoluble fractions. Soluble DM is easily degraded and, in any case, its rumen passage rate is that of the liquid phase, which is easily measurable. Insoluble DM can be degraded at different rates, and its passage rate can vary widely, depending on physico-chemical properties of feed particles (Ehle and Stern, 1986). The objectives of the current study were to evaluate kinetics of hydration and FSG of different types of BP over time of incubation with ruminal inoculum and to relate these data to kinetics of nutrients degradability *in situ.*

MATERIALS AND METHODS

Sample preparation and composition

Two types including normal beet pulp (NBP) and pelleted beet pulp (PBP) were used in this experiment were prepared from feed manufacture of Khorasan province, Iran. The FBP was prepared by milling NBP, using 1 mm screen pore size of miller. Feeding dried BP chips into the pelleting machine produced pellets. Pellets were obtained from dried by grinding and hardening into a cylindrical shape, about 5 cm long and about 0.5 cm in diameter and are uniform in appearance and texture. Samples were dried at 55 °C,

ground through a Wiley mill (1 mm screen), analyzed for dry and organic matter, Kjeldahl N, ether extract (EE), Ca, P, Cl, Na and K (AOAC, 2002), NDF, acid detergent fiber (ADF; Van Soest *et al.* 1991; using heat stable amylase and sodium sulfite) and ash at 605 °C at 3 h. NFC was calculated by 100 - (% crude protein (CP)+% NDF + % Ash + % EE) (Table 1).

Digestion kinetics

The ruminal nutrients degradation was determined *in situ*, using two cannulated Holstein steers (approximately 1 year old, body weight= 335.2 ± 10.3 kg). Steers were given access to water at all times and were housed in an open front shed. The steers were fed corn silage and concentrates in a ratio of about 65:35 (DM basis) according to their requirements. The diet was fed in two equal meals at 08:00 and 20:00 h. About 3 g of DM equivalent were weighed in each sealed nylon bag (6 cm×7.5 cm, polyamide, 26% porosity, 40±5 μm pore size) that was closed using a heat sealer. Three bags were incubated in the rumen for each of the following periods 0, 3, 6, 12, 18, 24, 36 and 48 h. All incubations started after the morning feeding. Bags were attached to a plastic tube (5 mm diameter) that was fixed to the outside of the fistula with a string. The bags and the tubes had free movement inside the rumen and reticulum. On removal, bags were washed using cold water until the effluent ran clear. The bags were dried in an oven at 60 °C for 48 h, and weighed. Following the weighing, bags were opened and residues from the three bags for each period were homogenized and placed in tightly capped plastic bottles. Samples were ground through a 1 mm sieve, analyzed for Kjeldahl N and NDF (Van Soest *et al.* 1991). All analyses were made on combined residues of the three bags. The analyses were run in duplicate and rerun when differences were greater than 3% and sufficient residue was available. The potentially degradable fraction was calculated as 100 minus the 0-h fraction. Kinetics of DM degradation *in situ* was estimated by the nonlinear regression procedure of SAS (1998). For each TMR and period, the following model was fitted to the percentage of degradation of DM (Ørskov and MacDonald, 1979):

$$P = a + b(1 - \exp^{(-Kdt)})$$

Where:
P: degradability (%).
a: soluble fraction (%).
b: slowly digestible fraction (%).
Kd: fractional rate of degradation (%/h).
t: time of incubation (h).

The effective degradability (ED) was calculated, assuming a passage rate (Kp) of 0.04, 0.05 and 0.06/h. The equation ED= [a + b × Kd / (Kd+Kp)] was used to calculate ED.

In this equation, the Kp represents the flow rate of particles out of the rumen.

Kinetic of hydration and functional specific gravity

The ruminal kinetic of hydration, FSG and changes of FSG of samples were measured *in situ*, using two cannulated Holstein steers that fed as described before. The two bags were incubated in the rumen for each of the following periods 0, 0.5, 1, 1.5, 3, 6, 12, 18, 24, 36 and 48 h. After incubation in rumen and removal of bags, the kinetic of hydration, FSG, and changes of FSG of samples measured with 100 mL pycnometer (Wattiaux, 1990), without washing. All the measurements were made in a separate oven that was maintained at 39 ± 0.15 °C. All of solution, distiller water, sample, small magnet and pycnometer put 5 h before the starting of FSG in the oven at 39 °C. The McDougall's buffer (artificial saliva solution, 1984) was used as hydration solution (McDougall, 1948). The earliest reading of the total weight of pycnometers was taken after 6 min of initial soaking was the shortest interval necessary to eliminate all gas bubbles that considered as FSG. The data obtained during hydration were used to determine the hydration rate and water uptake using NLIN procedures of SAS, (2002); Wattiaux, (1990). A biexponential models could be described by the function below:

$$Y_t = Ae^{-k_a t} + Be^{-k_b t}$$

Where:

Y: water uptake over time (g/g of insoluble DM).
A and B: represent pool sizes of hydration.
k_a and k_b: represent respective fractional hydration rate (min^{-1}).

Some data were fit best to single (i.e., the B component was removed) rather than double exponential models. Total water holding capacity (WHC; g/g of insoluble DM) was calculated as the sum of total solution uptake (sum of A+B) and initial moisture content of the samples. A mean for hydration rate that was weighted for pool sizes from biexponential models was calculated: $[(A \times k_a) + (B \times k_b)] / (A+B)$ (Bhatti and Firkins, 1995). Data were analyzed as a complete randomized design by ANOVA using Proc GLM of SAS (2002). Means were separated using Duncan's multiple range test at an alpha level of 0.05.

Statistical analysis

The data were analyzed by a complete randomize block design as three types of BP and steers were considered as treatment and block, respectively. Data were analyzed by using the GLM procedure of SAS (2002). Pearson correlation analyses of the physical and chemical properties of feed were done with CORR procedure of SAS (2002). In addition, regression analyses of the physical and chemical properties of feed were done with REG procedure of SAS (2002). Means were separated using Duncan's multiple range test with an alpha level of 0.05.

RESULTS AND DISCUSSION

Ruminal degradation

The DM degradation parameters and ED showed important and significant differences among three types of BP (Table 1). The soluble fraction of DM for FBP was higher than in NBP and PBP. However, there was no significant difference of soluble fraction of DM between NBP and PBP. In addition, the slowly degradable fraction in FBP was higher and NBP than PBP (59.00 and 57.00 *vs.* 55.66% of DM, respectively). The potential extent of DM degradation was high and ranged between 78.33 and 90.00% and was significantly different among the samples (P<0.0001). In contrast, indigestible DM fraction was relatively low and had significantly different among the samples (P<0.0001). The DM fractional rate of degradation was relatively large (11.2, 10.3 and 10.1 %/h, respectively) and significantly different (P=0.0007). Regardless the value of Kp, the values of the ED of DM was also significantly different (P<0.0001). The FBP and PBP samples had the highest and lowest values of ED, respectively. Similar to DM, the NDF degradation showed significant differences between three types of BP as for the degradation parameters and for the ED (Table 1). FBP and PBP had the highest and lowest soluble fraction of NDF, respectively. The slowly digestible fraction of FBP was significantly higher than NBP and PBP. However, there was no significant difference between NBP and PBP. The potential extent of NDF degradation was high and was significantly different among the samples (P<0.0001; 86.00, 79.00 and 75.00 for FBP, NBP and PBP, respectively). The indigestible NDF fraction had significant differences among three types of BP (P<0.0001). The NDF fractional rate of degradation was relatively large (9.80, 8.80 and 8.30 %/h, for FBP, NBP and PBP, respectively) and significantly different (P=0.0002). The ED of NDF was also significantly different (P<0.0001). The FBP and PBP samples had the highest and lowest values of ED for NDF, respectively. Degradation of CP for the degradation parameters and for the ED, similar to DM and NDF, had significant differences among three types of BP. FBP and PBP had the highest and lowest soluble protein, respectively. There were no significant difference on the slowly digestible protein between FBP and NBP, but the slowly digestible protein for PBP was significantly lower than FBP and NBP. The potential extent of CP degradation was high and significantly different among the samples (P<0.0001;

87.00, 83.00 and 77.00 for FBP, NBP and PBP, respectively). The indigestible CP fraction was different among three types of BP (P<0.0001). The CP fractional rate of degradation was relatively large (10.80, 10.00 and 9.70 %/h, for FBP, NBP and PBP, respectively) and significantly different (P<0.0001). Regardless the value of Kp, the values of the ED of CP was also significantly different (P<0.0001). The FBP and PBP samples had the highest and lowest values of ED for CP, respectively.

The DM and NDF (DePeters et al. 1997) or DM and CP (Pereira and Gonzalez, 2004) degradation of the BP samples showed similar trends. Torrent et al. (1994) reported that apparent digestibilities of BP were 78.0, 81.5, 81.3 and 78.9% for DM, OM, NDF and ADF, respectively. Using in situ method, Torrent et al. (1994) found that rates of degradation, potentially digestible fractions, and lag time of NDF and ADF fractions of BP were 11.6%/h, 94.1% of NDF, and 0.8 h; and 14.3%/h, 92.6% of ADF, and 1.9 h, respectively. DePeters et al. (1997) found while the NDF content of BP ranged from 33.22 to 42.18%, the rates of NDF digestion ranged from 7.3 to 9.0%/h (with mean K_d=8.3%/h) for BP. In addition, Bhatti and Firkins (1995) reported a much larger range in NDF digestion for BP, 5.5 to 11.6%/h. Richardson et al. (2003) found that proportion of soluble, slowly degradable, and the rate of degradation of the potentially degradable OM and N of unmolassed BP were 6.50%, 87.20% and 7.70%/h; and 0.00%, 92.30%, and 3.20%/h, respectively. These results clearly showed that BP did not contain a rapidly degradable N fraction and had a slower rate of degradation of the potentially degradable fraction than winter barley. BP is commonly added with molasses or vinasses, which had an important effect on DM and protein solubility. However, the most experiments confirmed that rates of degradation for DM, NDF and CP were high and had similar trend. Gonzalez et al. (2001) found that in a BP sample with 49.1, 42.7, and 11.0 (% of DM) NDF, ADF, and CP content the soluble CP, insoluble degradable fraction and potential degradable fraction were 17.7, 65.4 and 83.1% of CP content, respectively. Pereira and Gonzalez (2004) tested 10 samples of dried beet pulp observed a range of variation of the soluble, slowly degradable fraction and fractional degradation rate of DM and CP from 1.53 to 41.3%, 7 to 50%, 3.71 to 6.21%/h; 11.5 to 50.0%, 51.0 to 80.7% and 5.55 to 10.5%/h, respectively. Pereira and Gonzalez (2004) reported that the ED values of DM ranged from 41.0 to 70.8% and those of CP from 34.3 to 73.1% and both values were correlated (r=0.874). Gonzaleze et al. (2001) tested the effects of degradation of soluble or insoluble proteins measured by rumen incubations at 0 h (washout value) and 16 h, respectively. Washing effects on essential amino acid proportions of DBP were limited. However, the DBP sample employed by Pereira and Gon-

zalez (2004) and Gonzaleze et al. (2001) included an important addition of molasses and therefore, its DM, NDF and protein solubility was high. In the present experiment, grounding and pelleting significantly increased and decreased soluble fraction of DM, NDF, and CP (Table 1), therefore, the significant difference on potential degradable fraction and fractional rate of degradation may be result of differences in soluble fraction of DM, NDF, and CP. In addition, the lag time of DM, NDF, and CP of three types of BP were not detected. The most experiments also, have confirmed the current results. However, using in vitro method, Bhatti and Firkins (1995) found that extent, fractional rate and lag time of DM for FBP samples were 67.4%, 8.41%/h and 3.9 h, respectively. In addition, Pereira and Gonzalez (2004) showed the ED values of CP had a high variability (range from 34.3 to 73.1%), therefore, using a mean constant value may lead to major errors to estimate nutritive value of BP. In this experiment, based on chemical composition, the degradation parameters and the ED values were in a range that reported in literature, however, degradation rate of CP was much higher than the others. Pereira and Gonzalez (2004) reported that there were marked differences between samples for all the degradation kinetic parameters and for the ED values. The values of soluble and slowly degradable fraction for CP were closely complementary, since the potential extent of CP degradation was relatively similar (from 86.7 to 95.5%). Consequently, both fractions were closely correlated. Therefore, the ED of CP was closely correlated with both fractions. The variation of the ED values for both DM and CP were mainly caused by the variation of their soluble fractions. Therefore, these values are mainly conditioned by the addition of soluble raw materials and by the cell wall content in DBP, which in turn also depends on the original content of soluble materials in fresh beets and on the efficiency of the extraction process.

These facts explain the close correlations observed between the ED values and those of ash and NDF for the DM degradation or ash, CP, soluble CP and NDIN for CP degradation. The appearance of multiple correlations between ED values and soluble CP or many chemical parameters is logical, because most of these factors are intercorrelated (Pereira and Gonzalez, 2004). The results of this experiment and Pereira and Gonzalez (2004) confirmed that the high and positive correlation between the degradation rate of DM and NDF is result of low degree of lignification of BP that does not seem to be an important barrier to ruminal degradation.

Kinetic of hydration, and functional specific gravity
Kinetics of ruminal hydration of three types of BP is presented in Table 2.

Table 1 The Kinetics of ruminal digestion of three types of beet pulps

Item	Three types of beet pulp			SEM	P-value
	Fine	Normal	Pelleted		
Degradability of dry matter					
Soluble fraction (%)	31.00[a]	25.33[b]	22.67[b]	0.005	*
Slowly digestible fraction (%)	59.00[a]	57.00[ab]	55.66[b]	0.004	NS
Potential extent of dry matter (DM) degradation (a+b)	90.00[a]	82.33[b]	78.33[c]	0.002	***
Indigestible fraction (%)	10.00[c]	17.67[b]	21.67[a]	0.003	***
Fractional rate of degradation (%/h)	11.2[a]	10.3[b]	10.1[b]	0.006	*
Effective degradability (%/h)[1]					
Kp= 0.04	72.90[a]	65.81[b]	65.30[b]	0.002	***
Kp= 0.05	71.75[a]	63.66[b]	59.90[c]	0.002	***
Kp= 0.06	69.37[a]	61.31[b]	57.59[c]	0.002	***
Degradability of neutral detergent fiber					
Soluble fraction (%)	28.00[a]	25.00[b]	22.00[c]	0.003	**
Slowly digestible fraction (%)	58.00[a]	54.00[b]	53.00[b]	0.002	***
Potential extent of NDF degradation (a+b)	86.00[a]	79.00[b]	75.00[c]	0.002	***
Indigestible fraction (%)	14.00[c]	21.00[b]	25.00[a]	0.003	***
Fractional rate of degradation (%/h)	9.8[a]	8.8[b]	8.4[c]	0.001	**
Effective degradability (%/h)					
Kp= 0.04	69.19[a]	62.12[b]	57.90[c]	0.003	***
Kp= 0.05	66.40[a]	59.43[b]	55.22[c]	0.003	***
Kp= 0.06	63.97[a]	57.10[b]	52.91[c]	0.003	***
Degradability of crude protein					
Soluble fraction (%)	21.00[a]	18.00[b]	14.00[c]	0.002	**
Slowly digestible fraction (%)	66.00[a]	65.00[a]	63.00[b]	0.002	*
Potential extent of CP degradation (a+b)	87.00[a]	83.00[b]	77.00[c]	0.002	***
Indigestible fraction (%)	13.00[c]	17.00[b]	23.00[a]	0.002	***
Fractional rate of degradation (%/h)	10.8[a]	10.0[b]	9.7[c]	0.001	***
Effective degradability (%/h)					
Kp= 0.04	69.20[a]	64.39[b]	58.68[c]	0.002	***
Kp= 0.05	66.16[a]	61.29[b]	55.64[c]	0.002	***
Kp= 0.06	63.47[a]	58.58[b]	53.01[c]	0.002	***

[1] The equation ED= [a + b × K_d / (K_d+Kp)] was used to calculate effective degradability (ED). In this equation, Kp represents the flow rate of particles out of the rumen that was considered equal to 0.04, 0.05 and 0.06 (Ørskov and MacDonald, 1979).
The means within the same row with at least one common letter, do not have significant difference (P>0.05).
* (P≤0.05); ** (P≤0.01) and *** (P≤0.001).
NS: not significant.

Table 2 Kinetics of ruminal hydration of three types of beet pulps

Item	Three types of beet pulp			SEM	P-value
	Fine	Normal	Pelleted		
Initial water content (g/g insoluble dry matter (DM))	0.062	0.064	0.064	0.051	NS
Hydration rate (g/g insoluble DM/h)	0.065[b]	0.075[a]	0.064[b]	0.002	***
Water holding capacity (g/g insoluble DM)	3.25[c]	4.21[a]	3.87[b]	0.021	***

The means within the same row with at least one common letter, do not have significant difference (P>0.05).
*** (P≤0.001).
NS: not significant.
SEM: standard error of the means.

Initial water content (g/g insoluble DM) was not significantly different among three types of BP pulps. Hydration rate (g/g insoluble DM/h) and WHC (g/g insoluble DM) were significantly different among the three types of BP. Grinding and pelleting significantly decreased hydration rate and WHC of BP. Therefore, NBP had the highest hydration rate and WHC. In addition, there was no significant difference on hydration rate between NBP and PBP, but WHC was significantly different between NBP and PBP (Table 2).

Changes in FSG over incubation time are shown in Table 3. Grinding and pelleting of BP increased FSG for FBP and PBP over incubation time (Table 3). Initial FSG and final FSG increased when were grinded and pelleted. Over the ruminal incubation, FSG of all typed of BP were higher than ruminal liquid that measured by Wattiaux (1990) that was 1.001 2 to 1.008 9. In addition, after the ruminal incubation to 2 h for FBP and NBP and to 4 h for PBP, the FSG of different types of BP were higher than critical size (Poppi et al. 1980) for passage from reticulorumen orifice.

In ruminant, the voluntary dry matter intake and ruminal filling is related to the bulk density of forages (Wattiaux, 1990), but ruminal retention time and passage rate are related to FSG of feeds (Wattiaux, 1990; Kaske and Engelhardt, 1990; Kaske and Engelhardt, 1992; Teimouri Yansari et al. 2004).

Ramanzin et al. (1994) found that the FSG of DM of original unwashed feedstuffs estimated at 0 h soaking was not significantly related to the various chemical components. However, the FSG of DM of unwashed feedstuffs estimated after 15 h of soaking, as well as that of the insoluble fractions of washed feedstuffs (both after 0 and 15 h of soaking) and that of soluble fractions, showed significantly positive correlations with various fibrous fractions and negative correlations with CP and fat. Correlations were highly significant and positive between WHC and fibrous fractions, except lignin. On average, the FSG of DM of three type of BP was higher than those reported for forage particles (Wattiaux, 1990; Ramanzin et al. 1994; Teimouri Yansari et al. 2004) and were in agreement with the few data available for concentrates (Wattiaux, 1990; Ramanzin et al. 1994). Solubility effect on FSG is important because soluble DM is in many feedstuffs a substantial fraction of total DM and has an FSG different from that of the insoluble fraction (Ramanzin et al. 1994). This result is in agreement with the results of Wattiaux (1990) in forages and confirms that measurements of FSG should be made on the insoluble DM of feedstuffs.

The WHC of three BP samples were 3.25, 4.21 and 3.87 (g/g insoluble DM) in FBP, NBP and PBP, respectively. As mentioned in material and methods, using an exponential model (cure fitting method; Wattiaux (1990); Teimouri Yansari et al. (2004)) the values of WHC were estimated. Using the centrifugation method and polyester bags, Ramanzin et al. (1994) found an average higher WHC (6.44 and 597 g/g insoluble DM) than the filtration methods (Giger-Reverdin, 2000; 5.37 g/g insoluble DM). However, many experiments (Wattiaux, 1990; Teimouri Yansari et al. 2004) explained that the estimated WHC with an exponential model was lower than the WHC that obtained with filtration or the centrifugation method. Giger-Reverdin (2000) found that WHC was highly correlated with NDF (R^2=0.456) and ADF (R^2=0.418), but not with lignin (R^2=0.013). It increased with the cell wall content of feedstuffs and decreased with the bulk density. However, BP had a WHC value higher than other feed because of high pectin content.

By the way, for stimulating chewing activity, particles must be retaining in rumen. The extent of FSG especially for concentrate and by products is a better indicator of ruminal mean retention time than the particle size (Kaske and Engelhardt, 1990; Teimouri Yansari et al. 2004).

When particle reached to threshold of size and FSG, they must pass the rumen. BP has a size lower than critical point that can easily pass from the riticulorumen orifice. It seems that FSG is more important to control ruminal retention time, consequently ruminal degradation of BP.

Relationship of degradation and hydration parameters
As a general conclusion, the correlation between digestion and hydration parameters is significantly high. Soluble fraction of DM, NDF and CP had high positive and negative correlation with slowly digestible and indigestible fractions, respectively. In addition, soluble fraction of DM, NDF and CP had high negative correlation with hydration rate, WHC, initial and final FSG. Slowly digestible fraction of DM, NDF and CP had similar trend to soluble fraction of DM, NDF, and CP (Table 4). In contrary, indigestible fractions of DM, NDF and CP had high positive correlation with hydration rate, WHC, initial and final FSG. Fractional rate of degradation of DM, NDF and CP had high positive correlation with soluble and slowly digestible and negative correlation with indigestible fractions, with hydration rate, WHC, initial and final FSG. The ED of DM had negative correlation with indigestible fractions of DM, NDF and CP, hydration rate, WHC, initial and final FSG, but positive correlation with soluble and slowly digestible fractions of DM, NDF and CP and fractional rate of degradation of DM, NDF and CP (Table 4).

The ED degradability of DM had close correlation with ED of NDF (r=0.966) and CP (r=0.989). Therefore, the values in Table 4 showed only the values of degradability for DM. The ED degradability of DM had negative correlation with soluble, slowly digestible fraction, fractional rate of degradation, ED of DM, hydration rate, WHC, initial and final FSG, indigestible fraction of NDF and CP, but high direct correlation with soluble and slowly digestible fraction of NDF and CP and fractional rate of degradation of NDF and CP. Pereira and Gonzalez (2004) found that the ED of DM was significantly correlated with the soluble, slowly degradable and fractional rate of degradation for all chemical fractions, except ADL and ADIN; however the closest correlations were recorded with NDF (negative) and ash (positive). Both fractions also showed the closest correlations with the soluble, slowly degradable, and fractional rate of degradation parameters. In addition, the ED of CP showed close and direct correlations with the soluble CP, CP and ash content that the closest correlations were recorded with the CP content. On the contrary, the ED of CP showed inverse correlations with the proportion of NDIN and the NDF and ADF contents (Pereira and Gonzalez, 2004). In addition, Haj-Ayed et al. (2000) found that ED of CP in vetch-oat hays was positively correlated at a lower level with the CP content and at high level with NDF.

Table 3 Changes in the functional specific gravity over incubation time

Beet pulp types	Functional specific gravity of beet pulp over the incubation time (h)							
	0.1	2	4	6	12	24	36	48
Fine beet pulp	1.243[a]	1.209[a]	1.148[b]	1.018[d]	1.003[d]	1.094[c]	1.117[c]	1.123[c]
Normal beet pulp	1.285[a]	1.274[a]	1.106[f]	1.056[c]	1.001[d]	1.056[c]	1.118[b]	1.136[b]
Pelleted beet pulp	1.312[a]	1.302[a]	1.235[b]	1.112[c]	1.054[e]	1.098[c]	1.135[c]	1.132[c]

The means within the same row with at least one common letter, do not have significant difference (P>0.05).

Table 4 Correlation coefficients (%, above diagonal) between digestion and hydration parameters and their P-values (below diagonal)

Parameter[2]	1	2	3	4	5	6	7	8	9	10	11	12	13	14	15	16	17
1	-	0.54	-0.95	0.91	0.97	-0.16	-0.74	-0.93	-0.89	0.95	0.88	0.96	0.85	0.83	-0.89	0.92	0.90
2	NS	-	-0.77	0.65	0.72	-0.15	-0.56	-0.79	-0.80	0.64	0.69	-0.69	0.81	0.52	-0.76	0.78	0.78
3	***	*	-	-0.92	-0.99	0.19	0.76	0.99	0.96	-0.94	-0.95	0.98	-0.95	-0.93	-0.81	0.94	-0.98
4	**	NS	**	-	0.94	-0.35	-0.82	-0.92	-0.91	0.87	0.96	-0.95	0.90	0.81	0.77	-0.84	0.94
5	***	*	***	**	-	-0.19	-0.77	-0.99	-0.96	0.95	0.96	-0.99	0.95	0.92	0.83	-0.93	0.98
6	NS	NS	NS	NS	NS	-	0.76	0.73	0.84	-0.60	-0.83	0.73	-0.78	-0.55	-0.41	0.53	-0.84
7	**	NS	*	**	*	**	-	0.73	0.94	-0.94	-0.95	0.97	-0.96	-0.96	-0.81	0.96	-0.97
8	**	*	***	**	***	*	**	-	0.94	-0.94	-0.95	0.98	-0.96	-0.96	-0.81	0.96	-0.97
9	**	*	***	**	***	*	**	**	-	-0.91	-0.90	0.94	-0.91	-0.87	-0.70	0.86	-0.97
10	**	NS	***	**	***	NS	**	**	**	-	0.85	-0.97	0.84	0.92	0.85	-0.95	0.88
11	***	*	***	***	***	*	***	**	**	**	-	-0.96	0.96	0.84	0.82	-0.87	0.98
12	***	*	***	***	***	*	***	***	**	***	***	-	-0.93	-0.92	-0.87	0.94	-0.96
13	**	*	***	**	***	NS	***	***	**	**	***	**	-	0.87	0.77	-0.89	0.98
14	**	*	**	**	**	NS	***	***	**	**	**	**	**	-	0.78	-0.89	0.97
15	**	NS	**	**	**	NS	**	**	**	**	**	**	**	**	-	-0.89	0.77
16	**	*	**	**	**	NS	***	***	**	**	**	**	**	***	**	-	-0.89
17	**	*	***	**	***	**	***	***	***	**	***	***	***	**	**	**	-

1: soluble fraction of DM (%); 2: slowly digestible fraction of DM (%); 3: indigestible fraction of DM (%); 4: fractional rate of degradation of DM (%/h); 5: effective degradability of DM (%/h); 6: hydration rate (g/g insoluble DM/h); 7: water holding capacity (g/g insoluble DM); 8: initial functional specific gravity; 9: final functional specific gravity; 10: soluble fraction of NDF (%); 11: slowly digestible fraction of NDF (%); 12: indigestible fraction of NDF (%); 13: fractional rate of degradation of NDF (%/h); 14: soluble fraction of CP (%); 15: slowly digestible fraction of CP (%); 16: indigestible fraction of CP (%) and 17: fractional rate of degradation of CP (%/h).
* (P≤0.05); ** (P≤0.01) and *** (P≤0.001).
NS: not significant.

For DM degradability, NDF and ADL contents explained 79.4% of the total variation, while for CP degradability; the cellulose content explained 83.9%. Rodriguez (1996) showed that the soluble fraction was linked to the cell content (complementary to the NDF fraction) while the slowly degradable and indegradable fractions were mainly linked to fiber components and lignin, respectively. Consequently, the variability of the soluble and slowly degradable fractions was linked to the variability of the fiber fractions while a higher variability of the undegradable fraction was related to the higher variability of lignin and ADIN.

Using the stepwise procedure, physical characteristics (hydration kinetics and degradation parameters) and chemical fractions as independent variables, we obtained regression equations to estimate relationships between degradation and hydration kinetic with soluble fraction (Figure 1), slowly degradable fraction (Figure 2) and indigestible fraction (Figure 3) of DM for BP samples. For DM degradability, soluble fraction of DM explained 82.4, 94.8, 2.7, 54.2, 87.3 and 79.7% of the total variation of factional rate of degradation, ED, hydration rate, WHC, initial and final FSG, respectively (Figure 1).

In addition, the slowly degradable fraction of DM explained 34.1, 50.0, 2.2, 31.4, 62.2 and 63.4% of the total variation of fractional rate of degradation, ED, hydration rate, WHC, initial and final FSG, respectively (Figure 2).

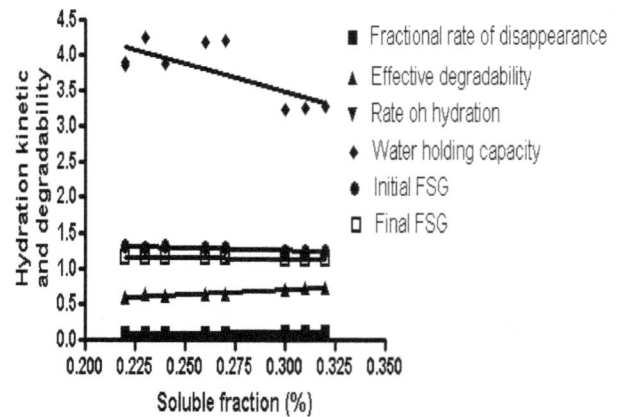

Figure 1 Relationships between degradation and hydration kinetics with the soluble fraction of DM in beet pulp samples. Relationships with the soluble fraction of dry matter are shown with the following equations:
1) The fractional rate of degradation (%/h)= 0.073 + 0.121 × soluble fraction of DM (%) (R^2=0.824; P=0.0007)
2) Effective degradability (%/h)= 0.304 + 1.317 × soluble fraction of DM (%) (R^2=0.948; P=0.0007)
3) Hydration rate (g/g insoluble DM/h)= 0.071 – 0.024 × soluble fraction of DM (%) (R^2=0.027; P=0.6753)
4) Water holding capacity (g/g insoluble DM)= 5.917 – 8.115 × soluble fraction of DM (%) (R^2=0.542; P=0.0238)
5) Initial FSG= 1.472 – 0.730 × soluble fraction for DM (%) (R^2=0.873; P=0.0002)
6) Final FSG= 1.173 – 0.127 × soluble fraction for DM (%) (R^2=0.797; P=0.0012)

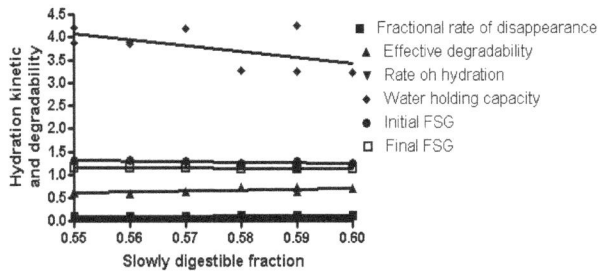

Figure 2 Relationship between degradation and hydration kinetics parameters with slowly digestible fraction of beet pulp samples. Relationships with slowly digestible fraction of dry matter are shown with the following equations:

1) Fractional rate of degradation (%/h)= -0.0003 + 0.184 × slowly digestible fraction of DM (%) (R^2=0.431; P=0.0546)

2) Effective degradability (%/h)= -0.512 + 2.003 × slowly digestible fraction of DM (%) (R^2=0.500; P=0.0333)

3) Hydration rate (g/g insoluble DM/h)= 0.094 – 0.046 × slowly digestible fraction of DM (%) (R^2=0.022; P=0.7063)

4) Water holding capacity (g/g insoluble DM)= 11.214 – 12.991 × slowly digestible fraction of DM (%) (R^2=0.314; P=0.1168)

5) Initial FSG= 2.021 – 1.296 × slowly digestible fraction for DM (%) (R^2=0.622; P=0.0115)

6) Final FSG= 1.277 – 0.239 × slowly digestible fraction for DM (%) (R^2=0.634; P=0.0102)

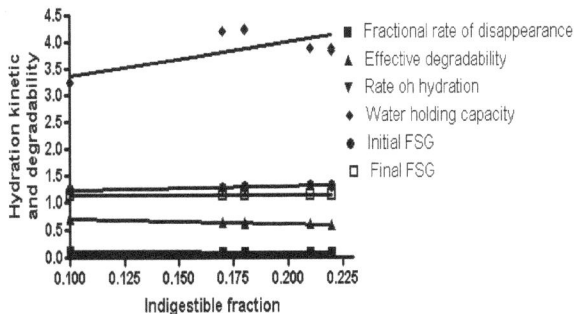

Figure 3 Relationship between degradation and hydration kinetic with indigestible fraction of beet pulp samples. Relationships with indigestible fraction of dry matter are shown with following equations:

1) Fractional rate of degradation (%/h)= 0.120 – 0.094 × indigestible fraction of DM (%) (R^2=0. 856; P=0.0004)

2) Effective degradability (%/h)= 0.819 – 1.021 × indigestible fraction of DM (%) (R^2=0.991; P<0.0001)

3) Hydration rate (g/g insoluble DM/h)= 0.064 + 0.020 × indigestible fraction of DM (%) (R^2=0.031; P=0.6497)

4) Water holding capacity (g/g insoluble DM)= 2.735 + 6.352 × indigestible fraction of DM (%) (R^2=0.517; P=0.0175)

5) Initial FSG= 1.184 – 0.588 × indigestible fraction for DM (%) (%) (R^2=0.986; P<0.0001)

6) Final FSG= 1.233 + 0.104 × indigestible fraction for DM (%) (R^2=0.929; P<0.0001)

For DM degradability, indigestible fraction of DM explained 85.6, 99.1, 3.1, 51.7, 98.6 and 92.9 % of the total variation of fractional rate of degradation, ED, hydration

rate, WHC, initial and final FSG, respectively (Figure 3). In addition, fractional rates of degradation, explained 89.1, 12.2, 68.0, 84.7 and 92.9 % of the total variation of ED, hydration rate, WHC, initial and final FSG, respectively (Figure 4).

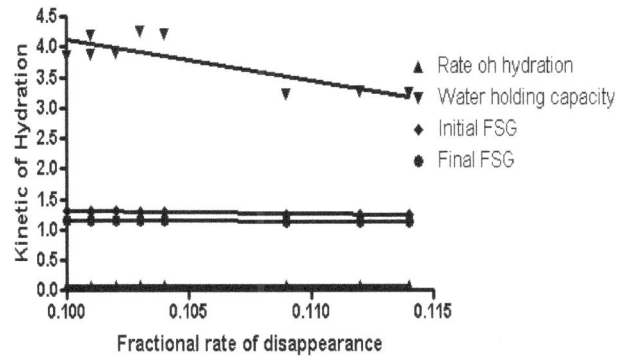

Figure 4 Relationship between fractional rates of degradation and hydration kinetics of DM for beet pulp samples. Relationships with fractional rates of degradation of dry matter are shown with following equations:

1) Effective degradability (%/h)= -0.355 + 9.571 × fractional rates of degradation of DM (%/h) (R^2=0.891; P=0.0001)

2) Hydration rate (g/g insoluble DM/ h)= 0.108 – 0.386 × fractional rates of degradation of DM (%/h) (R^2=0.122; P=0.3572)

3) Water holding capacity (g/g insoluble DM)= 10.948 – 68.192 × fractional rates of degradation of DM (%/h) (R^2=0.680; P=0.0062)

4) Initial FSG= 1.847 – 5.390 × fractional rates of degradation for DM (%/h) (R^2=0.847; P=0.0004)

5) Final FSG= 1.243 – 0.975 × fractional rates of degradation for DM (%/h) (R^2=0.929; P=0.0006)

Using the stepwise procedure, relationships between NDF degradability coefficient and hydration parameters were significantly explained with the following equation

A) Soluble fraction of NDF (a)= 1.327 – 0.841 × initial FSG (R^2=0.878; P=0.0002; equation 1).

B) Slowly digestible fraction of NDF= 1.481 – 0.727 × final FSG (r^2=0.894; P=0.0001; equation 2).

C) Indigestible fraction of NDF= -1.808 + 1.568 × initial FSG (R^2=0.953; P<0.0001; equation 3).

D) Fractional rate of degradation for NDF= 0.349 – 0.203 × initial FSG (R^2 0.926; P<0.0001; equation 4).

E) ED for NDF= 2.690 – 1.609 × initial FSG (R^2=0.966; P<0.0001; equation 5).

F) WHC= 12.155 – 15.227 × slowly digestible fraction of NDF (R^2=0.688; P=0.0057; equation 6).

G) Initial FSG= 1.659 – 0.601 × ED of NDF (R^2=0.966; P<0.0001; equation 7).

H) Final FSG= 1.207 – 0.105 × indigestible fraction for NDF (R^2=0.889; P=0.0001; equation 8).

The best prediction soluble ($R^2=0.878$), slowly degradable ($R^2=0.894$), indegradable fraction ($R^2=0.953$ and 0.994) and fractional rate of degradation ($R^2=0.926$) of NDF were derived from the initial FSG (equation 1, 2, 3, 4 and 5). In addition, the best prediction WHC ($R^2=0.688$), initial FSG ($R^2=0.966$) and final FSG ($R^2=0.889$) were derived from the slowly digestible fraction of NDF (equation 6), ED of NDF (equation 7), and indigestible fraction for NDF (equation 8).

Using the stepwise procedure, relationships between CP degradability coefficient and hydration parameters were significantly explained with the following equation:

A) Soluble Fraction of CP= 1.640 + 0.022 × WHC − 1.208 × initial FSG ($R^2=0.966$; P<0.0001; equation 9).

B) Slowly digestible Fraction of CP= 1.155 − 0.397 × initial FSG ($R^2=0.653$; P=0.0084; equation 10).

C) Indigestible fraction of CP= -1.522 - 1.937 × hydration rate + 1.429 × initial FSG ($R^2=0.985$; P<0.0001; equation 11).

D) Fractional rate of degradation for CP= 0.273 − 0.003 × WHC − 0.224 × initial FSG ($R^2=0.990$; P<0.0001; equation 12).

E) ED for CP= 2.468 + 1.498 × hydration rate − 1.507 × initial FSG ($R^2=0.989$; P<0.0001; equation 13).

F) WHC= 22.432 − 11.296 × indigestible fraction of CP + 163.511 × fractional rate of degradation ($R^2=0.981$; P<0.0001; equation 14).

G) Initial FSG= 1.809 − 2.970 × fractional rate of degradation - 0.353 × ED of CP ($R^2=0.988$; P<0.0001; equation 15).

H) Final FSG= 1.251 − 1.089 × fractional rate of degradation for CP ($R^2=0.938$; P<0.0001; equation 16).

The best prediction soluble and fractional rate of degradation for CP was derived from the WHC and initial FSG ($R^2=0.966$; equation 8 and $R^2=0.990$; equation 13, respectively). Similar to NDF fraction, slowly digestible fraction of CP was derived from initial FSG ($R^2=0.653$; equation 10), but indigestible fraction and ED of CP was derived from Hydration rate and initial FSG ($R^2=0.985$; equation 11 and $R^2=0.989$; equation 13). WHC was a function of indigestible fraction and fractional rate of degradation of CP ($R^2=0.981$; equation 14). Initial FSG of CP was derived from fractional rate of degradation and ED of CP ($R^2=0.988$; equation 15), but final FSG well explained with fractional rate of degradation for CP ($R^2=0.938$; equation 16). The results of current experiment were confirmed with previous experiments (Rodriguez, 1996; Gonzalez *et al.* 2001; Pereira and Gonzalez, 2004). Pereira and Gonzalez (2004) reported that a good prediction of the ED of DM ($R^2=0.933$) was obtained from the contents of NDF and

ADF as the first and second predictive variables. The best prediction ($R^2=0.954$) of the ED of CP was derived from the concentrations of CP and NDF and the CP content allowed to explain 87.8% of the variation.

CONCLUSION

The grinding and pelleting of BP had significant effects on degradation parameters of DM, NDF and CP. The rate and extent of degradation for DM, NDF and CP were high and had similar trend. Grinding and pelleting significantly increased and decreased soluble fraction of DM, NDF and CP, respectively. It seems that the significant difference on potential degradable fraction and fractional rate of degradation may be result of the differences in soluble fraction of DM, NDF and CP. The high and positive correlation between the degradation rate of DM and NDF is result of low degree of lignification of BP that does not seem to be an important barrier to ruminal degradation. Grinding and pelleting significantly decreased hydration rate and WHC, increased initial and final FSG over incubation time. The correlation between digestion and hydration parameters is significantly high. In addition, the correlations between digestion and hydration parameters were significantly high. As beet pulp has size lower than critical size, can easily pass from the reticulorumen orifice, therefore, its functional specific gravity is more important to control ruminal retention time and degradation.

REFERENCES

Allen M.S. and Mertens D.R. (1988). Evaluating constraints on fiber digestion by rumen microbes. *J. Nutr.* **118**, 261-270.

AOAC. (2002). Official Methods of Analysis. Vol. I. 17th Ed. Association of Official Analytical Chemists, Arlington, VA, USA.

Ben-Ghedalia D.E., Yosef J., Miron D. and Est Y. (1989). The effects of starch- and pectin- rich diets on the quantitative aspects of digestion in sheep. *Anim. Feed Sci. Technol.* **24**, 289-298.

Bhatti S.A. and Firkins J.L. (1995). Kinetics of hydration and functional specific gravity of fibrous feed by-products. *J. Dairy Sci.* **73**, 1449-1460.

Catherine M.G., Crepeau C.R.M. and Thibault J. (1999). Glucuronic acid directly linked to galacturonic acid in the rhamnogalacturonan backbone of beet pectins. *Europe J. Biochem.* **266**, 566-574.

DePeters E.J., Fadel J.G. and Arosemena A. (1997). Digestion kinetics of neutral detergent fiber and chemical composition within some selected by-product feedstuffs. *Anim. Feed Sci. Technol.* **67**, 127-140.

Ehle F.R. and Stern M.D. (1986). Influence of particle size and density on particulate passage through alimentary tract of Holstein heifers. *J. Dairy Sci.* **69**, 564-571.

Giger-Reverdin S. (2000). Characterization of feedstuffs for ruminants using some physical parameters. *Anim. Feed Sci. Technol.* **86,** 53-69.

Gonzalez J., Centeno C. and Lamrani F. (2001). *In situ* rumen degradation of amino acids from different feeds corrected for microbial contamination. *Anim. Res.* **50,** 253-264.

Haj-Ayed M., Gonzalez J., Caballero R. and Alvir M.R. (2000). Nutritive value of on-farm vetch-oat hays. I. Voluntary intake and nutrient digestibility. *Ann. Zootech.* **49,** 381-389.

Kaske M. and Englhardet W.V. (1990). The effect of size and density on mean retention time of particles in the gastrointestinal tract of sheep. *Br. J. Nutr.* **63,** 683-704.

Kaske M., Hatiboglu S. and Englhardet W.V. (1992). The influence of density and size of particles on rumination and passage from the reticulo-rumen of sheep. *Br. J. Nutr.* **67,** 235-244.

McDougall E.I. (1948). Studies on ruminant saliva1: the composition and out put of sheep's saliva. *Biochem. J.* **4,** 39-45.

Ørskov E.R. and McDonald I. (1979). The estimation of protein degradability in the rumen from incubation measurements weighted according to rate of passage. *J. Agric. Sci.* **92,** 499-503.

Pereira J.C. and Gonzalez J. (2004). Rumen degradability of dehydrated beet pulp and dehydrated citrus pulp. *Anim. Res.* **53,** 99-110.

Poppi D.P., Norton B.W., Minson D.J. and Hendrickson R.E. (1980). The validity of the critical size theory for particles leaving the rumen. *J. Agric. Sci.* **94,** 275-280.

Ramanzin M., Bailoni L. and Bittante G. (1994). Solubility, water-holding capacity and specific gravity of different concentrates. *J. Dairy Sci.* **11,** 174-781.

Richardson J.M., Wilkinson R.G. and Sinclair L.A. (2003). Synchrony of nutrient supply to the rumen and dietary energy source and their effects on the growth and metabolism of lambs. *J. Anim. Sci.* **81,** 1332-1347.

Rodríguez C.A. (1996). Estudio de la colonización microbiana de los alimentos en el rumen. Implicaciones sobre la estimación de la degradabilidad ruminal de las materias nitrogenadas mediante técnicas *in situ.* Ph D. Thesis. Universidad Politécnica de Madrid, Spain.

SAS Institute. (1998). SAS®/STAT Software, Release 8. SAS Institute, Inc., Cary, NC. USA.

Teimouri Yansari A., Valizadeh R., Naserian A., Christensen D.A., Yu P. and Eftekhari Shahroodi F. (2004). Effects of alfalfa particle size and specific gravity chewing activity, digestibility and performance of Holstein dairy cows. *J. Dairy Sci.* **87,** 3912-3924.

Torrent J., Johnson D.E. and Kujawa M.A. (1994). Co-product fiber digestibility: kinetic and *in vivo* assessment. *J. Anim. Sci.* **72,** 790-795.

Van Soest P.J., Robertson J.B. and Lewis B.A. (1991). Methods for dietary fiber, neutral detergent fiber, and non-starch polysaccharides in relation to animal nutrition. *J. Dairy Sci.* **74,** 3583-3597.

Wattiaux M.A. (1990). A mechanism influencing passage of forage particles through the reticulo-rumen: change in specific gravity during hydration and digestion. Ph D. Thesis. University of Wisconsin, USA.

4

Phenotypic and Genetic Analysis of Lori-Bakhtiari Lamb's Longevity Up to Yearling Age for Autosomal and Sex-Linked Chromosomes

A. Salemi[1], M. Vatankhah[2*] and B. Asadi[1]

[1] Department of Animal Science, Behbahan Branch, Islamic Azad University, Behbahan, Iran
[2] Department of Animal Science Research, Chaharmahal and Bakhtiari Agricultural and Natural Resources Research and Education Center, AREEO, Shahrekord, Iran

*Correspondence E-mail: vatankhah_mah@yahoo.com

ABSTRACT

The data set used in this study contained 8793 records of lamb's longevity (days) from 320 sires and 2349 dams collected during 1989 to 2014, from the Lori-Bakhtiari flock at Shooli station in Shahrekord, Iran. Genetic parameters (partitioned into autosomal, sex-linked and maternal) and breeding values of cumulative lamb's longevity from birth up to yearling age (at 1, 2, 3, 6, 9 and 12 months) were estimated using restricted maximum likelihood procedure. The results showed that the overall mean of the cumulative longevity of lamb up to yearling (12 months of age) was 295.87 days. The effect of fixed factors; year and month of birth, sex of lamb, age of dam, lamb birth weight as quadratic covariate and dam body weight as linear covariate were significant (P<0.05) on lamb's longevity. The heritability estimates of lamb's longevity were low and ranged from 0.01 to 0.08 for autosomal, 0.01 for sex-linked and 0.02 to 0.03 for maternal additive genetic effects. The estimates of autosomal, sex-linked and maternal genetic correlations of lamb's longevity in different ages were high. The pearson and spearman correlation coefficients between autosomal breeding values and sex-linked breeding values in lamb's longevity at different ages were 0.15 to 0.46 and 0.11 to 0.43 respectively. Thus, lamb's longevity up to yearling can be improved by farm management practices and improving environmental factors at first. Genetic analysis using linear models which able to estimate breeding values in direct (autosomal and sex-linked chromosomes separately) and maternal effects, could be more effective to improve longevity in lambs.

KEY WORDS autosomal and sex-linked, longevity, Lori-Bakhtiari lambs.

INTRODUCTION

Improved ewe productivity would be achieved by increasing the number of lambs successfully reared per ewe in a given year. The high mortality rates obtained greatly reduce the efficiency and profitability of a lamb production enterprise. According to Fogarty *et al.* (1985), the mortality of lambs from birth to weaning is a major factor affecting the number of lambs weaned per lambing. Poor survival of lambs up to weaning time is a major source of reproductive inefficiency in most sheep flocks around the world (Hinch

and Brien, 2014). To obtain an ethically acceptable increase in the net reproductive rate, a strategy to reduce lamb mortality is required. Young *et al.* (2014) concluded that research on improving survival of lambs is likely to have a higher payoff than research on improving the number of lambs conceived. Lamb survival is a complex trait influenced by many different factors associated with management, climate, behavior of the ewe and lamb, and genetic effects (Smith, 1977; Christley *et al.* 2003; Everett-Hincks *et al.* 2008). Further, the rate of survival is higher in female lambs compared to male lambs in the same flock.

Nash *et al.* (1996) reported that after adjusting for birth weight, male lambs to be at greater risk of postnatal mortality. The higher mortality rate in ram lambs, compared to females is also reported in the other findings (Mukasa-Mugerwa *et al.* 2000; Mandal *et al.* 2007; Vatankhah and Talebi, 2009).

Lamb survival can be analyzed as a binary trait (Fogarty, 1995), but it suffers from a severe information loss because it ignores the continuity of the survival process and the precise time of death.This simplification implies that dead animals in 5 days or 5 months after birth are treated alike and contribute the same amount of information. However, longevity (days) can be analyzed as a continuous trait and accounted the precise time of death in breeding programs to improve survival rate in lambs.

The Lori-Bakhtiari sheep is one of the most common native breeds in the southwestern part of Iran (the Zagros Mountains), with a population of more than 1.7 million head, having the largest fat-tail size among all of the breeds in Iran. The animals of this breed are usually kept in villages under semi intensive systems. The Ministry of Jihad-Agriculture in Iran has found it important to increase the efficiency of sheep production, because the output of sheep in this system is low and the relative economic weights of lamb survival rate at pre- and post- weaning in Lori-Bakhtiari sheep were higher than growth traits (Vatankhah and Akhoundi, 2015). Improving lamb's longevity in male lambs is economically beneficial as the replacement rate in male lambs is lower than for female lambs and the majority of male lambs are fattened for sale. The lower longevity in male lambs may be due to sex-linked determinants, which have not yet been researched and identified (Mandal *et al.* 2007). Thus, the objective of this study was to determine the non-genetic factors affecting on lamb's longevity and estimation of genetic parameters for lamb's longevity up to yearling age in Lori-Bakhtiari sheep for autosomal, sex-linked and maternal additive genetic effects.

MATERIALS AND METHODS

Data and flock management
The data set used in this study was collected from 8793 lambs descended from 320 rams and 2349 ewes born between 1989 and 2014 in the Shooli station, Shahrekord, Iran (Table 1). The flock is managed under a semi-migratory or village system. The animals are kept in the range on cereal pastures from mid-spring to late-autumn and kept indoors from December to May at the station and fed a ration composed of alfalfa, barley and wheat stubble. The breeding period extends from late August to late October (ewes were assigned randomly to the rams) and consequently, lambing starts in late January.

From 15 days of age, lambs have access to creep feed *ad libitum* and are weaned at an average age of 90 ± 5 days. After weaning, male and female lambs were separated. Surplus male lambs chosen for fattening were separated from the rest of the animals. Female lambs were kept in the pasture of cultivated alfalfa, while the rest of the males were kept indoors and fed a maintenance and growth ration (45% alfalfa hay, 39% barley, 7% beet pulp, 8% cottonseed meal, 1% salt and mineral supplements which containing 13.5% crude protein and 2.5 Mcal/kg metabolizable energy) to 12 months of age. The animals were monitored daily and dates of death in lambs were recorded. Traits studied were longevity of lamb in days from birth up to 1, 2, 3, 6, 9 and 12 months of age with the day of birth equal to 1 day of age. For calculate each cumulative time period, the birth date minus the removed date of lamb from the flock. Then for example the longevity of all lambs up to 2 months set equal to 60 days except for lambs that left the flock before 60 days.

Statistical analysis
The GLM procedure of SAS (2000) was applied to identify the important fixed effects to be considered in the final linear model. The final statistical model included year of birth, age of dam, birth type, month of birth and sex of lamb as fixed effects and the birth weight of lamb (linear and quadratic) and the mother's body weight of each lamb (as linear) were fitted as covariates in the following model.

$$Y_{ijklmn} = \mu + A_i + B_j + T_k + S_l + M_m + b_1(W_{ijklmn} - W_{000000}) + b_2(W_{ijklmn} - W_{000000})^2 + b_3(EW_{ijklmn} - EW_{000000}) + e_{ijklmn}$$

Where:

y_{ijklmn}: observed longevity in days of nth lamb.

μ: overall mean.

A_i: effect of ith age of dam (i=2,...,≥7).

B_j: effect of jth year of lambing (j=1989 to 2014).

T_k: effect of kth birth type (k=single, twin or triplet).

S_l: effect of lth sex of lamb (l=male and 2=female).

M_m: effect of mth month of lamb birth (m=Jan, Feb or Mar).

W_{ijklmn}: birth weight of nth lamb.

W_{000000}: overall mean of lamb birth weight.

EW_{ijklmn}: mother's body weight of nth lamb.

EW_{000000}: overall mean of mother's body weight.

b_1, b_2: linear and quadratic regression coefficients of lamb birth weight, respectively.

b_3: linear regression coefficient of mother's body weight.

e_{ijklmn}: residual effects.

Variance components and genetic parameters were estimated from a linear animal model in a multi-variate analysis using the restricted maximum likelihood (REML) method by WOMBAT program (Meyer, 2013). The animal model included all fixed effects described above, random effects of animal (separated to autosomal and sex-linked) and random maternal effects. The following multi-variate animal model was fitted to analyze the data set:

$$y_i = X_i b_i + Z_{1i} a_i + Z_{1i} s_i + Z_{2i} m_i + e_i$$

Where:

i: longevity of lamb up to i^{th} age (i=1, 2, 3, 6, 9 and 12 months).

y, b, a, s, m and e: vectors of observations, fixed effects, direct additive genetic effects in autosomal chromosomes, direct additive genetic effects in sex-linked, maternal additive genetic effects and residual random effects, respectively.

X, Z_1, Z_2 and Z_3: incidence matrices relating the observations to the respective effects.

The average information (AI) REML algorithm was used to maximize the likelihood (convergence criterion was 10^{-8}) and additional restarts were performed until no further improvement in log likelihood occurred. With one record for each individual, the BLUP breeding values of longevity at different ages of lambs for autosomal effects ($\hat{a}$, sex-linked effects ($\hat{s}$) and maternal effects ($\hat{m}$) are obtained using mixed model equations as follow:

$$
\begin{bmatrix}
X'X & X'Z_1 & X'Z_1 & X'Z_2 \\
Z_1'X & Z_1'Z_1 + A^{-1}\dfrac{\sigma_e^2}{\sigma_a^2} & Z_1'Z_1 & Z_1'Z_2 \\
Z_1'X & Z_1'Z_1 & Z_1'Z_1 + S^{-1}\dfrac{\sigma_e^2}{\sigma_F^2} & Z_1'Z_2 \\
Z_2'X & Z_2'Z_1 & Z_2'Z_1 & Z_2'Z_2 + A^{-1}\dfrac{\sigma_e^2}{\sigma_m^2}
\end{bmatrix}
\begin{bmatrix} b \\ a \\ s \\ m \end{bmatrix}
=
\begin{bmatrix} X'y \\ Z_1'y \\ Z_1'y \\ Z_2'y \end{bmatrix}
$$

Where:

$A\sigma_a^2$: A covariance matrix of $\hat{a}$.

A: matrix of the co-ancestries between relative for autosomal loci (Henderson, 1976). And

σ_a^2 : variance of additive genetic values for autosomal loci.

$S\sigma_F^2$: covariance matrix of $\hat{s}$.

S: a matrix whose elements are functions of co-ancestries between relative for X-chromosomal loci.

σ_F^2 : additive genetic variance for X-chromosomal loci for noninbred females (Fernando and Grossman, 1990).

$A\sigma_m^2$: covariance matrix of $\hat{m}$.

A: matrix of the co-ancestries between relative.

σ_m^2 : variance of maternal additive genetic values.

The inverse of A obtained by an algorithm described by Henderson (1976). The construction of S and its inverse obtained by an algorithm described by Fernando and Grossman (1990). The total direct BLUP breeding values of animals calculated by summation of autosomal and sex-linked breeding values. The correlation coefficients (Pearson and Spearman) between total, autosomal and sex-linked breeding values of animals were obtained using correlation procedure of SAS (2000).

RESULTS AND DISCUSSION

The values in Table 2 indicate that the cumulative longevity decrease by 0.7, 1.88, 3.39, 18.66, 41.06 and 69.13 days up to 1, 2, 3, 6, 9 and 12 months of age respectively (the differences between observed overall mean of longevity from expected longevity if all of lambs were live and mortality were equal to zero). For example, if all of the lambs were live and could reach to yearling age, the mean of longevity had been 365 days, but due to death some of them at different ages the mean of longevity estimated as 295.87 days. These values indicated that the lamb's longevity decrease by 3.39, 15.27, 22.40 and 28.07 days in the first, second, third and fourth quarter of age, respectively.

The least squares mean for the longevity of lamb for different levels of non-genetic factors are shown in Table 2. There was significant (P<0.01) variation in lamb's longevity between birth years, although, no clear trend was observed during 1989 and 2014. The least squares mean of longevity was lower in lambs born from the younger and older ewes compared to 4-5 years old, but these differences were not significant (P>0.05) up to 1, 2 and 3 months of age and significant for 6, 9 and 12 months of ages (P<0.05). The birth type did not have significant effect on lamb's longevity when corrected for lamb birth weight (P>0.05). Although, the longevity of lambs was lower in triplet lambs, but due to the high standard error of the means the differences were not significant with singleton and twin lambs. The month of birth had a significant effect (P<0.01) on lamb's longevity.

The least square means of lamb's longevity decreased with increasing month of birth. However, the differences between these values were not significant for the first and middle third of the lambing period up to 3 months of age, but the differences increased and were significant after 3 months of age. Female lambs were found to have a higher longevity than males at all the considered ages. The differences between lamb's longevity in females and males were significant (P<0.05). This study showed a quadratic relationship between lamb's birth weight and longevity at all ages (Table 2).

Table 1 Pedigree structure for longevity analysis data set in Lori-Bakhtiari lambs

Item	Number	Item	Number
Original animals	9310	Sires with progeny	320
Animals with record	8793	Sires with record and progeny	246
Animals without offspring	6501	Dams with progeny	2349
Animals with offspring	2669	Dams with record and progeny	2046
Animals with offspring and record	2292	Founders	455
Animals with unknown sire	567	Inbred animals	4153
Animals with unknown dam	466	Average inbreeding coefficient in inbred animals	0.025

Table 2 The least squares means (±SE) of fixed effect for longevity (days) in Lori-Bakhtiari lambs

Factor	Number	1 mo	2 mo	3 mo	6 mo	9 mo	12 mo
Overall mean	8793	29.30	58.12	86.61	161.34	228.94	295.87
Birth year		**	**	**	**	**	**
Age of dam (yr)		NS	NS	NS	*	**	**
2	2355	28.91±0.18	57.05±0.43	84.86±0.69	150.96±2.41^{ab}	213.33±4.16^{b}	277.47±6.24^{a}
3	1873	28.96±0.18	57.34±0.41	85.42±0.67	153.70±2.36^{a}	218.85±4.07^{a}	285.46±6.10^{a}
4	1729	29.00±0.18	57.27±0.41	85.22±0.67	152.33±2.38^{a}	216.75±4.10^{ab}	281.53±6.15^{a}
5	1301	29.05±0.18	57.55±0.43	85.83±0.70	154.04±2.45^{a}	217.82±4.22^{ab}	282.00±6.33^{a}
6	935	28.94±0.20	57.27±0.46	85.19±0.75	150.95±2.62^{b}	213.42±4.51^{ab}	275.70±6.77^{a}
7	600	28.82±0.22	56.93±0.52	84.55±0.85	146.31±3.03^{a}	203.33±5.22^{c}	260.74±7.83^{b}
Birth type		NS	NS	NS	NS	NS	NS
Single	6377	29.07±0.07	57.54±0.16	85.47±0.27	153.46±0.94	216.05±1.61	279.83±2.42
Twin	2320	29.23±0.11	57.95±0.25	86.29±0.41	155.20±1.40	216.90±2.42	278.19±3.63
Triplet	96	28.53±0.43	56.23±1.01	83.77±1.64	145.47±5.92	208.80±10.21	273.43±15.30
Month of birth		**	**	**	**	**	**
Jan-Feb	4422	29.21±0.16^{a}	57.97±0.37^{a}	86.64±0.60^{a}	161.19±2.14^{a}	231.50±3.70^{a}	300.92±5.54^{a}
Feb-Mar	3411	29.18±0.16^{a}	57.77±0.38^{a}	86.13±0.62^{a}	155.66±2.19^{b}	221.46±3.78^{b}	286.69±5.67^{b}
Mar-Apr	960	28.44±0.20^{b}	55.97±0.46^{b}	82.77±0.75^{b}	137.28±2.65^{c}	188.79±4.57^{c}	243.84±6.86^{c}
Sex of lamb		**	**	**	**	**	**
Male	4403	28.79±0.16^{b}	56.81±0.37^{b}	84.47±0.60^{b}	144.17±2.17^{b}	200.99±3.75^{b}	258.52±5.62^{b}
Female	4390	29.10±0.16^{a}	57.67±0.38^{a}	85.89±0.63^{a}	158.59±2.21^{a}	226.85±3.81^{a}	295.78±5.71^{a}
LRBW		3.84±0.44**	9.30±1.03**	15.49±1.68**	49.82±5.53**	83.17±9.53**	121.73±14.29**
QRBW		-0.32±0.04**	-0.78±0.10**	-1.29±0.16**	-3.84±0.55**	-6.32±0.94**	-9.19±1.41**
LRDW		-0.02±0.01**	-0.04±0.02^{ns}	-0.07±0.03**	-0.26±0.10**	-0.31±0.17^{ns}	-0.29±0.26^{ns}

LRBW: linear regression of birth weight; QRBW: quadratic regression of birth weight and LRDW: linear regression of dam weight.
The means within the same column with at least one common letter, do not have significant difference (P>0.05).
* (P<0.05) and ** (P<0.01).
NS: non significant.

The positive and negative signs for linear and quadratic regression coefficients showed that longevity increased with lamb birth weight, reached an optimum level and then decreased with increasing lamb birth weight. The negative signs of linear regression coefficients of the mother's body weight showed that lamb's longevity decreased with increasing mother's body weight, however, some regression coefficients were not significant from zero (Table 2).

Genetic parameters of lamb's longevity (days) from birth to yearling age estimated by linear animal model including direct additive genetic effects of animal (autosomal and sex-linked) and maternal additive genetic effects are set out in Table 3. The autosomal heritability estimates for lamb's longevity were low, but these values increased with age of lamb (0.01 to 0.08). The estimates of sex-chromosome heritabilities of lamb's longevity were equal to corresponding autosomal heritabilities up to 3 months of age and constant with increasing age of lamb.

The estimates of maternal heritability of lamb's longevity were constant up to 9 months of age (0.03) and then decreased slightly for 12 months of age (0.02). The autosomal and sex-chromosome genetic correlations between lamb's longevity at different ages were positive and ranged from 0.42 to 0.97, while, maternal genetic correlation were higher relatively (0.85 to 0.99). All genetic correlations between longevity at cumulative time periods close to each other were high and reduced with increasing distance between them.

The estimates of phenotypic correlations between lamb's longevity at different ages were positive, but lower than corresponding genetic correlations relatively. A summary of autosomal and sex-chromosome breeding values of lamb longevity at different ages are shown in Table 4. This Table shows the average breeding values of autosomal and sex-linked effects at different ages, of lamb longevity were greater than zero.

Table 3 Heritability estimates (diagonal), genetic (below diagonal) and phenotypic (above diagonal) correlation estimates for autosomal, sex-linked and maternal effects of longevity in Lori-Bakhtiari lambs

Stay to	1 mo	2 mo	3 mo	6 mo	9 mo	12 mo
Autosomal						
1 mo	**0.01±0.01**	0.95±0.01	0.90±0.01	0.65±0.01	0.55±0.01	0.47±0.01
2 mo	0.89±0.16	**0.01±0.01**	0.98±0.01	0.83±0.01	0.65±0.01	0.58±0.01
3 mo	0.85±0.21	0.99±0.02	**0.01±0.01**	0.88±0.01	0.70±0.01	0.63±0.01
6 mo	0.80±0.25	0.98±0.18	0.96±0.09	**0.06±0.02**	0.94±0.01	0.86±0.01
9 mo	0.60±0.35	0.71±0.18	0.77±0.13	0.91±0.03	**0.07±0.02**	0.97±0.01
12 mo	0.42±0.34	0.55±0.15	0.65±0.13	0.80±0.07	0.97±0.01	**0.08±0.02**
Sex-linked						
1 mo	**0.01±0.01**	-	-	-	-	-
2 mo	0.99±0.13	**0.01±0.01**	-	-	-	-
3 mo	0.99±0.23	0.99±0.07	**0.01±0.01**	-	-	-
6 mo	0.97±0.22	0.98±0.12	0.98±0.10	**0.01±0.01**	-	-
9 mo	0.95±0.20	0.91±0.13	0.97±0.12	0.97±0.21	**0.01±0.01**	-
12 mo	0.90±0.25	0.90±0.12	0.92±0.10	0.91±0.18	0.93±0.20	**0.01±0.01**
Maternal						
1 mo	**0.03±0.01**	-	-	-	-	-
2 mo	0.98±0.01	-	-	-	-	-
3 mo	0.97±0.02	0.99±0.01	**0.03±0.01**	-	-	-
6 mo	0.90±0.10	0.93±0.04	0.94±0.03	**0.03±0.01**	-	-
9 mo	0.85±0.11	0.90±0.15	0.95±0.12	0.99±0.02	**0.03±0.01**	-
12 mo	0.85±0.14	0.86±0.17	0.90±0.13	0.95±0.05	0.97±0.01	**0.02±0.01**

Table 4 Summary of autosomal and sex-linked breeding values of longevity (days) in Lori-Bakhtiari lambs

Stay to	Autosomal				Sex-linked			
	Mean	SD	Min	Max	Mean	SD	Min	Max
1 mo	0.0037	0.0670	-0.3149	0.2291	0.0014	0.0626	-0.3000	0.2667
2 mo	0.0654	0.2545	-1.3905	0.9992	0.0032	0.1448	-0.6936	0.6165
3 mo	0.1409	0.5118	-2.9429	2.0115	0.0062	0.2818	-1.3496	1.9970
6 mo	1.1520	3.6522	-16.9960	10.9801	0.0191	0.0620	-0.3709	0.3000
9 mo	3.6125	7.6173	-36.1396	24.6285	0.0412	0.1340	-0.8009	0.6721
12 mo	6.6690	12.9021	-71.4093	44.9950	0.3545	1.1589	-6.9247	5.8336

SD: standard deviation.

This implies that without any direct selection for lamb longevity during the last years, genetic progress has been made through indirect selection. The mean and standard deviation of autosomal breeding values were higher than sex-linked in all of considered traits. There is significant variation in autosomal and sex-linked breeding values for lamb longevity at all ages in the population (Table 4). The correlation coefficients between breeding values of autosomal (A) and sex-linked (S) are shown in Table 5. Correlation coefficients for A-S, was low to medium. In general, the Spearman correlations were lower than Pearson procedure for all ages and breeding values types.

The overall mean of lamb longevity up to yearling recorded in this study was in range in the literature for different breeds of sheep (Yapi *et al.* 1990; Green and Morgan, 1993; Nash *et al.* 1996; Mukasa-Mugerwa *et al.* 2000; Mandal *et al.* 2007; Sawalha *et al.* 2007). The lamb longevity was higher when compared to values obtained for some sheep breeds such as Menz and Horro (Mukasa-Mugerwa *et al.* 2000).

The highest decrease in lamb longevity occurred during after weaning period (65.74 days), which is in accordance with values reported for Menz and Horro sheep (Mukasa-Mugerwa *et al.* 2000), but higher than values reported for some breeds of sheep such as Scottish Blackface and Muzaffarnagari (Sawalha *et al.* 2007; Mandal *et al.* 2007). This is due to extreme environmental changes in this period. Lambs were weaned at 90 ± 5 days of age and kept on a pasture of cultivated alfalfa which may not cover the lamb's requirements. The significant variations in lamb longevity from birth up to 12 months of age in different years may be attributed to variation in the environmental conditions, feed availability and other management factors. Similarly, Berhan and Van Arendonk (2006) and Mandal *et al.* (2007), observed significant effects of year of birth for lamb mortality rates. The changes in longevity of lamb with age of dam obtained in this study were similar to Smith (1977), who reported that yearling ewes had lambs with smaller birth weight, lower vigor, and higher mortality rates than lambs from older ewes.

Ewe age effects on lamb mortality have been shown to reduce lamb mortality with increasing ewe age (Southey *et al.* 2001; Sawalha *et al.* 2007), although Morris *et al.* (2000), in accordance with the results of this study showed slight decreases in longevity of lambs born to ewes greater than 5 years of age.

Table 5 Correlation coefficients between autosomal (A) and sex-linked (S) breeding values of longevity in Lori-Bakhtiari lambs

Stay to	Pearson correlation	Spearman correlation
1 mo	0.46	0.43
2 mo	0.37	0.37
3 mo	0.36	0.35
6 mo	0.08	0.11
9 mo	0.15	0.20
12 mo	0.17	0.23

The non-significant effect of dam age in lamb longevity up to 6 months of ages could be attributed to the existence of the mother's body weight as a covariate in the model, because some differences are corrected by this covariate. The non-significant effect of the birth type on lamb longevity obtained in this study could be attributed to correction for lamb's birth weight, which included in the model to analysis. The birth weights are normally lower for lambs born in larger litters (Smith, 1977; Morris *et al.* 2000; Sawalha *et al.* 2007) and therefore these lambs may be at greater risk of illnesses. Lambs born in multiple litters may also have higher mortality due to limitation in milk production by the dam, either as a result of low genetic potential for milk production or restricted nutrient intake in limiting environments (Snowder and Knight, 1995). The higher lamb longevity in female lambs compared to males is in agreement with other findings (Nash *et al.* 1996; Mukasa-Mugerwa *et al.* 2000; Mandal *et al.* 2007; Sawalha *et al.* 2007; Vatankhah and Talebi, 2009). Lower longevity in male lambs may be due to sex-linked determinants which have not yet been identified (Mandal *et al.* 2007), but is the main objective in this study. The intermediate optimum range for birth weight obtained in this study has been presented in many evaluations of lamb survival or longevity (Smith, 1977; Lopez-Villalobos and Garrick, 1999; Morris *et al.* 2000; Sawalha *et al.* 2007). Smith (1977) concluded that birth weight had a large influence with most early life mortality occurring in lambs with birth weights below the mean. Morris *et al.* (2000) found similar results with a larger proportion of dead lambs with light birth weights. One possible explanation for the elevated mean mortality rate at birth for lambs with the smallest birth weight is hypothermia, whilst dystocia can be a cause of mortality for lambs with the heaviest birth weights (Sawalha *et al.* 2007). Consequently, selection for optimal birth weight, rather than maximum birth weight, should be practiced when viability and birth weight are to be improved simultaneously.

Autosomal and maternal heritability estimates obtained in this study are in the range of values reported in the literature for direct and maternal heritability. For example, the weighted average heritability of survival rate and longevity to weaning in a review of 24 studies was 0.04 (Fogarty, 1995). Safari *et al.* (2005) in a review of 16 studies reported the range of direct and maternal heritability estimates in lamb survival rate were from zero to 0.11 and zero to 0.19, respectively. Vatankhah (2013) reported low estimates of direct heritabilities (0.01 to 0.09) for survival rate from different linear models, maternal heritabilities ranged from 0.00 to 0.04 and decreased as the age of lambs increased. The sex-linked heritabilities for lamb longevity were similar to the corresponding values of autosomal heritability up to 3 months of age obtained in this study. This implies that the genetic importance of sex-linked effects on longevity of lamb is equal to the genetic autosomal effects. In review the literature, there was not any report on the heritability of sex-linked for traits associated with longevity in sheep. Low estimates of the direct (autosomal and sex-linked) heritability for lamb longevity (<0.10) could be attributed to small additive genetic variance of viability and the impact of non-genetic factors on this trait. According to Riggio *et al.* (2008), one explanation for low heritability for lamb survival or longevity is that it is a composite trait and many factors may lead to death. As improving lamb longevity has a great economic importance, the potential for genetic improvement by within flock selection would be less effective due to low heritability estimates by linear animal models.

According to the estimates of autosomal, sex-liked and maternal genetic correlations, between longevity of lambs before and after weaning, genetic selection on lamb's longevity up to weaning age could improve the longevity after weaning ages.

Due to the lack of direct genetic selection for longevity of lamb and average breeding values greater than zero for autosomal and sex-linked effects obtained in this study, the genetic progress already obtained in this flock can be attributed to indirect selection for traits such as total weight of litter weaned (TWW) per ewe joined and correlated responses. One indirect selection method to improve lamb's longevity is to use composite traits. Selection could be performed on a composite trait that incorporates lamb's longevity or survival rate and is likely to be part of an overall breeding objective, i.e. it has intrinsic economic value to the breeder, as it affects either returns or costs or both. Examples are number of lambs weaned (NLW) or TWW per ewe joined, both relying on a favorable correlated response in lamb's longevity or survival (Brien *et al.* 2014). This seems a reasonable expectation as, for example, favorable genetic correlations of 0.76 (Swan *et al.* 2001) and 0.55

(Hebart *et al.* 2010) have been estimated between NLW and lamb survival from two large Merino flock studies in Australia. A similar, strong genetic correlation of 0.73 has also been reported between TWW and lamb survival (Afolayan *et al.* 2008).

The low to medium estimates of Pearson correlation coefficients and lower estimates of rank correlation coefficients between A-S breeding values of lamb's longevity at all ages indicated that genetic selection on autosomal breeding values solely, cannot choose animals with high breeding values for sex-linked effects.

CONCLUSION

The present investigation revealed that, although the overall lamb's longevity up to yearling observed in this study is not low, but the mortality rate distribution is not suitable, because the highest mortality rate occurred after weaning period (6 to 12 months of age). With regard to non-genetic factors on the longevity of lambs from birth to the yearling age, it could be concluded that in order to improve the longevity of lambs focus should be on improve non-genetic factors and management conditions. Breaking up direct additive genetic variance of lamb's longevity in to autosomal and sex-linked in this study showed that the proportion of the variance related to the sex-linked effects was similar to the proportion of variance attributed to autosomal effects up to weaning. Although, response to selection of lamb's longevity is low due to low estimates of heritability, genetic analysis of longevity using animal models which able to partition total additive genetic variation in to autosomal, sex-linked and maternal effects could make a more effective genetic selection. Thus, genetic selection based on a combination of three sources of BVs with suitable relative economic weights is recommended to improve genetically lamb's longevity.

ACKNOWLEDGEMENT

We are grateful to the head and all of the coworker in Sholi station to collect the data and allowed to use them.

REFERENCES

Afolayan R.A., Fogarty N.M., Gilmour A.R., Ingham V.M., Gaunt G.M. and Cummins L.J. (2008). Reproductive performance and genetic parameters in first cross ewes from different maternal genotypes. *J. Anim. Sci.* **86,** 804-814.

Berhan A. and Van Arendonk J. (2006). Reproductive performance and mortality rate in Menz and Horro sheep following controlled breeding in Ethiopia. *Small Rumin. Res.* **63,** 297-303.

Brien F.D., Cloete S.W.P., Fogarty N.M., Greeff J.C., Hebart M.L., Hiendleder S., Hocking Edwards J.E., Kelly J.M., Kind K.L., Kleemann D.O., Plush K. L. and Miller D.R. (2014). A review of the genetic and epigenetic factors affecting lamb survival. *Anim. Prod. Sci.* **54,** 667-693.

Christley R.M., Morgan K.L., Parkin T.D.H. and French N.P. (2003). Factors related to the risk of neonatal mortality, birthweight and serum immunoglobulin concentration in lambs in the UK. *Prev. Vet. Med.* **57,** 209-226.

Everett-Hincks J.M. and Duncan S.J. (2008). Lamb post-mortem protocol for use on farm: to diagnose primary cause of lamb death from birth to 3 days of age. *Open. Vet. Sci. J.* **2,** 55-62.

Fernando R.L. and Grossman M. (1990). Genetic evaluation with autosomal and X-chromosomal inheritance. *Theor. Appl. Genet.* **80,** 75-80.

Fogarty N.M., Dickerson G.E. and Young L.D. (1985). Lamb production and its components in pure breeds and composite lines. III. Genetic parameters. *J. Anim. Sci.* **60,** 40-57.

Fogarty N.M. (1995). Genetic parameters for live weight, fat and muscle measurements, wool production and reproduction in sheep: a review. *Anim. Breed. Abstr.* **63(3),** 101-143.

Green L.E. and Morgan K.L. (1993). Mortality in early born, housed lambs in south-west England. *Prev. Vet. Med.* **17,** 251-261.

Hebart M.L., Brien F.D., Jaensch K.S., Smith D.H., Walkom S.F. and Grimson R.J. (2010). Genetics of reproductive efficiency: a study of Merino resource flocks in South Australia. Pp. 102-105 in Proc. 9th World Congr. Genet. Appl. Livest. Leipzig, Germany.

Henderson C.R. (1976). A simple method for computing the inverse of a numerator relationship matrix used in prediction of breeding values. *Biometrics.* **32,** 69-83.

Hinch G.N. and Brien F. (2014). Lamb survival in Australian flocks: a review. *Anim. Prod. Sci.* **54,** 656-666.

Lopez-Villalobos N. and Garrick D.J. (1999). Genetic parameter estimates for survival in Romney sheep. *Proc. New Zealand Soc. Anim. Prod.* **58,** 121-124.

Mandal A., Prasad H., Kumar A., Roy R. and Sharma N. (2007). Factors associated with lamb mortalities in Muzaffarnagari sheep. *Small Rumin. Res.* **71,** 273-279.

Meyer K. (2013). WOMBAT- A Program for Mixed Model Analyses by Restricted Maximum Likelihood. User Notes, Animal Genetics and Breeding Unit, Armidale, Australia.

Morris C.A., Hickey S.M. and Clarke J.N. (2000). Genetic and environmental factors affecting lamb survival at birth and through to weaning. *New Zealand J. Agric. Res.* **43,** 515-524.

Mukasa-Mugerwa E., Lahlou-Kassi A., Anindo D., Rege J.E.O., Tembely S., Tobbo M. and Baker R.L. (2000). Between and within breed variation in lamb survival and the risk factors associated with major causes of mortality in indigenous Horro and Menze sheep in Ethiopia. *Small Rumin. Res.* **37,** 1-12.

Nash M.L., Hungerford L.L., Nash T.G. and Zinn G.M. (1996). Risk factors for perinatal and postnatal mortality in lambs. *Vet. Rec.* **139,** 64-67.

Riggio V., Finocchiaro R. and Bishop S.C. (2008). Genetic parameters for early lamb survival and growth in Scottish Blackface sheep. *J. Anim. Sci.* **86,** 1758-1764.

Safari E., Fogarty N.M. and Gilmour A.R. (2005). A review of genetic parameter estimates for wool, growth, meat and reproduction traits in sheep. *Livest. Prod. Sci.* **92,** 271-289.

SAS Institute. (2000). SAS®/STAT Software, Release 9.1. SAS Institute, Inc., Cary, NC. USA.

Sawalha R.M., Conington J., Brotherstone S. and Villanueva B. (2007). Analysis of lamb survival of Scottish Blackface sheep. *Animal.* **1,** 151-157.

Smith G.M. (1977). Factors affecting birth weight, dystocia and preweaning survival in sheep. *J. Anim. Sci.* **44,** 745-753.

Snowder G.D. and Knight A.D. (1995). Breed effects on foster lamb and foster dam on lamb viability and growth. *J. Anim. Sci.* **73,** 1559-1566.

Southey B.R., Rodriguez-Zas S.L. and Leymaster K.A. (2001). Survival analysis of lamb mortality in a terminal sire composite population. *J. Anim. Sci.* **79,** 2298-2306.

Swan A.A., Piper L.R., Brewer H.G. and Purvis I.W. (2001). Genetic variation in reproductive performance of fine wool Merinos. *Proc. Assoc. Adv. Anim. Breed. Genet.* **14,** 417-420.

Vatankhah M. and Talebi M.A. (2009). Genetic and non-genetic factors affecting mortality in Lori-Bakhtiari lambs. *Asian-Australas J. Anim. Sci.* **22(4),** 459-464.

Vatankhah M. (2013). Estimation of genetic parameters for survival rate in Lori-Bakhtiari lambs using linear and Weibull proportional hazard models. *J. Agric. Sci. Technol.* **15(6),** 1133-1143.

Vatankhah M. and Akhoundi A. (2015). The comparison of economic values and relative emphasis of some traits in Lori-Bakhtiari sheep resulted from different ways. *Anim. Sci. J. (Pajouhesh and Sazandegi).* **106,** 71-82.

Yapi C.V., Boylan W.J. and Robinson R.A. (1990). Factors associated with causes of preweaning lamb mortality. *Prev. Vet. Med.* **10,** 145-152.

Young J.M., Trompf J. and Thompson A.N. (2014). The critical control points for increasing reproductive performance can be used to inform research priorities. *Anim. Prod. Sci.* **54,** 645-655.

5

Evaluation of Morphometric and Reproductive Traits of Indigenous Dairy Goat Types in North Western Amhara, Ethiopia

K. Alemayehu[1] and D. Kebede[1*]

[1] Biotechnology Research Institute, Bahir Dar University, Bahir Dar, Ethiopia

*Correspondence E-mail: dakebede10@gmail.com

ABSTRACT

The objective of this study was to evaluate the morphometric and reproductive traits of indigenous dairy goat types. It was conducted in three districts (Ebnat, Gonji-kolela and Farta) based on the agro-ecological differences, goat population potentials and experience of the farmers using goat milk. Semi-structured questionnaires, field observations, morphological and reproductive trait measurements were used. The results revealed that the average age at sexual maturity of male in Farta district, (Mean±SE) was 7.04 ± 0.41 months. However, age at first service of male goats was longer in Gonji kolela (9.84 ± 0.95) and Ebnat (11.54 ± 1.28) districts. The average body length (BL) for dairy, dual and meat type goats were 59.65 ± 0.58, 54.87 ± 0.69 and 55.52 ± 0.74 cm. The mean rear udder length, udder circumference and teat length were 17.14 ± 0.28, 31.39 ± 0.59 and 3.71 ± 0.07 cm, 14.21 ± 0.48, 26.75 ± 1.01 and 3.41 ± 0.12 cm and 13.44 ± 0.38, 24.67 ± 0.81 and 3.01 ± 0.09 cm, respectively. Whereas, the mean rear udder diameter for dairy, dual and meat types for female goats were 10.04 ± 0.22, 8.66 ± 0.37 and 8.01 ± 0.30 cm, respectively. Dairy type female goats had significantly higher ($P<0.001$) values for all udder and teat measurements than dual and meat type female goats. In conclusion, goat breeding and reproduction was not identified as dairy, meat and dual purpose types and there is no specialized dairy goat breeding, reproduction with corresponding production system. Therefore, evaluation of morphometric and reproductive traits of indigenous dairy type goats with specialized and intensive production system are crucial to increase the milk production and productivity.

KEY WORDS goat types, morphometric, reproductive, traits.

INTRODUCTION

Goat is a good source of meat (Chevon), milk, yoghurt, cheese and other by-products such as hide and skin. Goat population has witnessed a positive growth rate in the last 20 years (Morand-Fehr, 2003). The number of goats has increased by almost 50% at world level, cattle increased by just 9%. However, sheep population decreased by 4% (Morand-Fehr and Boyazogly, 1999; Devendra, 2001). In Ethiopia, indigenous Ethiopian goats have been phenotypically classified into 12 types, although genetic characteriza-tion indigenous goats of Ethiopia are grouped into eight distinct genetic entities (Tesfaye, 2004). There are many types of dairy and meat goat breeds introduced to Ethiopia from outside for crossbreeding. The main purpose of crossing with local goats was to improve milk production in areas where goat milk is known to be consumed (Adane and Girma, 2008). Amhara National Regional State also has six goat ecotypes i.e. Gumuz, Awi, Central Abergelle, Abergelle, Begiamedir and Bati (Hassen et al. 2012). The region endowed 5291571 goats; particularly the south Gonder zone and western Gojam have 434230 and 315142 goats, respectively (CSA, 2014), however milk production from goat in this region is insignificant. In Ethiopia, human population increases from time to time. Grazing lands are decreasing and changed to residents and small pieces of farm land. These lead to shortage of grazing and farm land which created difficulty to grow subsistence cash crops and dairy cattle feeds (Alemayehu *et al.* 2015). Many farmers would like to keep dairy cows but do not have sufficient land to grow enough feed. Climate change is also becoming a challenge for dairy cattle producer as it is directly linked with feed shortage and climate induced stress (Alemayehu and Fantahun, 2012). Climate change in Amhara region causes heat stress that cattle and sheep are getting difficulties to survive. However, indigenous goats are able to withstand the stress as goat have low surface to volume ratio. Hence, the number of goats is increasing (Alemayehu and Fantahun, 2012). Besides, dairy goats require less land than dairy cows (CTA, 2007). As a result, keeping dairy goats are becoming more affordable and appropriate option (Legesse *et al.* 2008).

It has been also reported that goat milk as a much less allergenic alternative to cow milk due to its differing protein structure, namely its casein micelle components (Park, 1994). Goat milk has demonstrated significant improvements in colic, minor digestive disorders, asthma and eczema over cow milk, as well as in infants and children with cow milk sensitivities (McCullough, 2003). Haenlein *et al.* (2004) indicated that treatment with goat milk typically resolves between 30 and 40% of problem cases of childhood cow milk allergy, which can be higher in some cases (one study showed improvements in 49 out of 55 children treated with goat milk). Dairy goat farming can contribute to food security of farming families and become an income source through commercialization of raw milk and dairy products. The characteristics of goat milk, both from a nutritional and social standpoint are important and encourage studies to evaluate its production and quality (Fernandes *et al.* 2008). Goat's milk is reported to have higher digestibility and lower allergenic properties compared to cow's milk (Senaka Ranadheera *et al.* 2012). It has also a higher content of short chain fatty acids in milk fat, higher content of zinc, iron, and magnesium, and antibacterial characteristics (Slacanac *et al.* 2010). In addition, these benefits may be further enhanced by using goat's milk as a vehicle for delivering probiotics and prebiotics.

Even though the country as well as the region has large size of goat population, the productivity per unit of animal especially milk production is very low. The contribution of this sector to both the national and the regional economy is relatively low too. Because of less productivity, local goat breeds were subjected to replacement and crossbreeding with imported goat breeds like Anglo-Nubian, Saanen and Toggenburg introduced at different periods by different organization. But this did not improve the productivity except that indiscriminate crossbreeding caused genetic erosion, loss of genetic diversity and reduction of adaptive value for efficient utilization of the existing adapted goat genetic resources (Hassen *et al.* 2012). Therefore, the objective of this paper was to evaluate morphometrically and reproductive traits of indigenous dairy goat type to enhance food security at small holder level.

MATERIALS AND METHODS

Description of the study area

The study was conducted in Ebnat, Farta and Gonji-kolela districts of Amhara region in 2014. These districts were purposively selected based on agro-ecology, goat population potentials and experience of the farmers of using goat milk. Ebnat district considered as lowland, Gonji kolela as midland and Farta as highland. Ebnat and Farta districts are located in the South Gondar administrative zone and Gonji kolela is located south of Bahir Dar in the west Gojjam Zone (Table 1; Figure 1). Ebnat, Farta and Gonhi-kolela districts have 131505 40193 and 52350 indigenous goats respectively (EDARDO, 2013; FDARDO, 2013; GDARDO, 2013).

Data collection methods

The data was collected through individual interviews using questionnaire, focus group discussion and field observations. Focus group discussion and field observation were used to support individual interview to assess and collect information on reproductive performance traits. Focus group discussion were composed of greater than 10 people believed to be knowledgeable about past and present social and economic status of the area, community elders and story tellers and it has to be done in each PA. To supplement the primary data and for background information for the district, secondary data were collected from each district agriculture and rural development offices. About five kebele administrations were selected each. Semi-structured questionnaires, field observations and morphometric trait measurements were conducted based on international dairy goat measurement criteria. A total of 235 dairy (109), dual (50) and meat (76) type goats were identified. The measurements of rear udder diameter (RUD), rear udder length (RUL), udder circumference (UCC), teat length (TTL), Scrotum length (SL), scrotum width (SW) and scrotum circumference (SCC) were highly considered. Besides, morphological quantitative characters such as body length (BL), diagonal body length (DBL), heart girth (HG), pelvic width (PW), pelvic height (PH), rump width (RW) and - rump length (RL) were measured.

Reproductive performance traits such as age at sexual maturity for male and female, age at first kidding, reproductive lifetime of doe's and buck's, number of kidding per doe's life time and kidding interval were evaluated. Mean and standard error were used for reproductive performance analysis.

Data analysis

Multivariate analysis of SPSS version 20 was used for quantitative data ascertain the goat type, district, dentition (1PPI, 2PPI, 3PPI 4PPI and broken teeth) and sex. General linear model (GLM) of multivariate analysis of SAS (2000) was used to quantify fixed effects such as the goat type, agro-ecology and sex. Pair wise comparison was employed when it was significant to reveal the difference between means. The model used for GLM analysis was in the following formula:

$$Y_{ijkl} = \mu + A_i + S_j + D_k + B_{1 + AiBi + BiDk} + e_{ijkl}$$

Where:

Y_{ijk}: observed measurement.

μ: overall mean.

A_i: effect of ith age group.

S_j: effect of jth sex (male and female).

D_k: effect of kth district or agro-ecology.

B_1: effect of lth breed (dairy type goat, dual purpose goat and meat type goat).

AiBi: effect of age by breed interaction.

BiDk: effect of breed by district interaction.

e_{ijkl}: random residual error.

RESULTS AND DISCUSSION

Flock structure

The proportion of goats at different sex and age classes follow similar trends, where breeding did represent the largest class, followed by kids (Table 2). These results were in line with Tesfaye, (2009); Dhaba *et al.* (2012) and Solomon (2014). The sampled goat population structure by age and sex. About 22.2% older females' ages structured with 3PPI and 47.8% with 4PPI. The younger males structured (39.3%) with 1PPI and 39.3% with 2PPI. The younger females (1PPI and 2PPI) and extreme older females accounted 9.7% 1PPI, 18.8% 2PPI and 1.4% broken teeth, respectively. The older male goats 3PPI, 4PPI and broken teeth accounted 14.3%, 3.6% and 3.6%, respectively (Table 3). The proportions of breeding were higher in a sampled population and followed by kids. Similar results were reported in Ethiopia indicating that herders keep more proportion of females than males (Dhaba *et al.* 2013; Dereje *et al.* 2013).

Reproductive performances of goats

In Farta district, the average age at sexual maturity of male (Mean±SE) was 7.04 ± 0.41 months. This was in line with the result of Solomon (2014) who reported 7.4 ± 2.01 months for western Lowland male goats. However, age at first service of male goats was longer in Gonji kolela (9.84±0.95) and Ebnat (11.54±1.28) districts (Table 4).

The two districts show delayed age at sexual maturity of males compared to Farta district but younger than male Abergelle goats which were reported 12.3 months (Solomon, 2014).

The overall average age at sexual maturity of female goats was 7.36 ± 0.26 months. This was closer to the report of Tesfaye (2009).

Goat types characterization based on morphometric traits

Average body length (BL) for dairy, dual and meat type goats were 59.65 ± 0.58, 54.87 ± 0.69 and 55.52 ± 0.74 cm, respectively (Table 5). There were significantly higher (P<0.05) values of average body length for dairy type goats than dual and meat type goats. When compared from the exotic goats, the body length of indigenous dairy goat type was lower than Alpine and Saanen goats which have 74.33 ± 3.93 cm and 74.28 ± 4.78 cm, respectively (Nemeth *et al.* 2010).

About 25-36 months and above female goats have 69.31 ± 1.85 cm body lengths in Peshawar, Pakistan which was also higher than this result (Hamayun *et al.* 2006). However significance difference observed among indigenous goat types in the study population.

Average diagonal body length (DBL) for dairy, dual and meat type goats were 64.52 ± 0.62, 59.93 ± 0.73 and 61.22 ± 0.78, respectively. Average heart girth (HG) in these breeds was 79.27 ± 0.69, 75.44 ± 0.81 and 75.31 ± 0.87 cm, respectively. There were significantly higher (P<0.05) values of average diagonal body length and heart girth for dairy type goats than dual and meat type goats. Heart girth of the present dairy goats was lower than Beetal goats in Pakistan which have 82.0 ± 3.49 cm heart girth (Abdul, 2011).

Heart girth of Jamnapari female goats in India was 76.1 ± 0.38 cm, which was lower than the result of dairy type goats (FAO, 1982). About 25-36 months and above Beetal goats have 74.89 ± 0.93 cm heart girth which was lower than the present dairy type goats (Hamayun *et al.* 2006).

Dairy type goats was significantly higher (P<0.05) average pelvic width (15.05±0.20 cm) than dual (13.09±0.24 cm) and meat (12.70±0.25 cm) type goats (Table 5). Alpine and Saanen goats have 17.21 ± 1.07 and 18.49 ± 2.29 cm, respectively pelvic width in Hungary, which is higher than the present finding (Nemeth *et al.* 2010).

Table 1 Description of the study areas

Study area characteristics	Agro-ecology		
	Lowland (Ebnat)	Midland (Gonji-kolela)	Highland (Farta)
Distance from Regional city (Bahir Dar)	122	70	97
Altitude a.s.l (m)	< 2216	2216	1920-4135
Rainfall (mm)	500-900	1338	900-1099
Temperature (co)	25-30	21	9-25
Area (ha)	249837	64186	103457
Human population (No)	122514	21333	281280
Livestock population (No)	562040	260685	432822
Indigenous goat population	131505	52350	40193

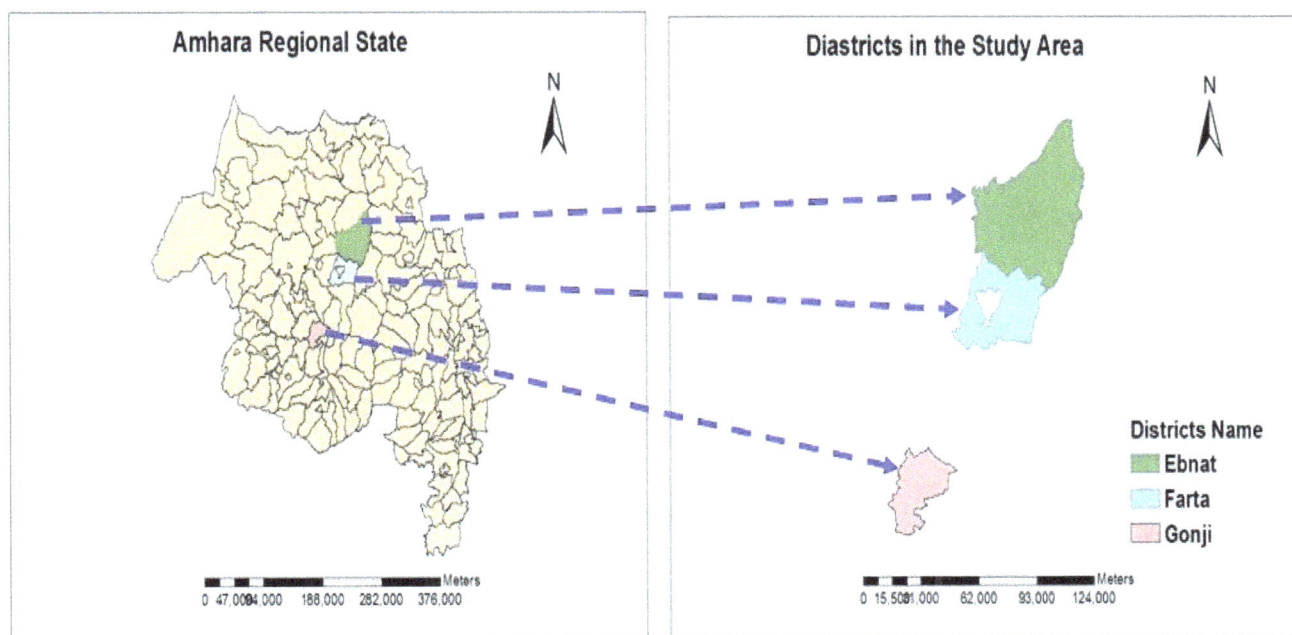

Figure 1 Map of the study areas

Similarly, pelvic height (PH) of dairy type goats (73.37±0.54 cm) was significantly higher (P<0.05) than dual (71.42±0.64 cm) and meat (71.44±0.68 cm) type goats.

Average rump width (RW) and rump length (RL) were 19.31 ± 0.19 cm and 16.61 ± 0.16 cm for dairy type goat, 17.61 ± 0.23 cm and 15.29 ± 0.19 cm for dual type goat and 17.35 ± 0.25 cm and 15.37 ± 0.21 cm for meat type goats (Table 6). Dairy type goats have significantly higher (P<0.001) values for rump width and rump length than dual

and meat type goats. The present result of rump width for dairy type goats was higher than Beetal goats which have a rump width of 15.4 ± 2.60 and 17.6 ± 3.55 cm in the front and 14.7 ± 1.15 and 16.5 ± 1.68 cm towards back in females and males, respectively (Abdul, 2011). However, the finding of rump length of dairy type goatwas comparable to Beetal goats which have a rump length of 15.8 ± 1.12 and 17.3 ± 1.12 cm, for females and males, respectively (Abdul, 2011).

Table 2 Flock structure of goats by age and sex group (Mean±SE)

Goat categories	Ebnat		Farta		Gonji-kolela	
	Mean±SE	%	Mean±SE	%	Mean±SE	%
Kids < 6 months	11.62±4.06[a]	31.86	1.77±0.32[b]	25.84	2.68±0.25[b]	27.42
Male kids 6-12 months	2.38±0.83[a]	6.54	0.38±0.14[b]	5.62	1.05±0.23[b]	10.75
Female kids 6-12 months	3.69±0.91[a]	10.13	0.31±0.17[b]	4.49	1.21±0.34[b]	12.37
Bucks	3.85±0.68[a]	10.55	0.77±0.39[b]	11.24	0.95±0.19[b]	9.68
Does	14.08±4.97[a]	38.61	2.92±0.40[b]	42.7	3.00±0.26[b]	30.65
Castrates	0.85±0.34[a]	2.32	0.69±0.40[a]	10.11	0.89±0.21[a]	9.14

The means within the same row with at least one common letter, do not have significant difference (P>0.05).
SE: standard error.

Table 3 10 goat flock proportion / structure by age and sex groups in a sampled population

Age	Ebnat			Farta			Gonji-kolela			Over all		
	F (%)	M (%)	Total (%)	F (%)	M (%)	Total (%)	F (%)	M (%)	Total (%)	F (%)	M (%)	Total (%)
1PPI	2 (2.8)	4 (26.7)	6 (6.9)	11 (16.7)	7 (53.8)	18 (22.8)	7 (10.1)	0 (0.0)	7 (10.1)	20 (9.7)	11 (39.3)	31 (13.2)
2PPI	15 (20.8)	5 (33.3)	20 (23.0)	12 (18.2)	6 (46.2)	18 (22.8)	12 (17.4)	0 (0.0)	12 (17.4)	39 (18.8)	11 (39.3)	50 (21.3)
3PPI	13 (18.1)	4 (26.7)	17 (19.5)	16 (24.2)	0 (0.0)	16 (20.3)	17 (24.6)	0 (0.0)	17 (24.6)	46 (22.2)	4 (14.3)	50 (21.3)
4PPI	42 (58.3)	1 (6.7)	43 (49.4)	26 (39.4)	0 (0.0)	26 (32.9)	31 (44.9)	0 (0.0)	31 (44.9)	99 (47.8)	1 (3.6)	100 (42.6)
Broken	0 (0.0)	1 (6.7)	1 (1.1)	1 (1.5)	0 (0.0)	1 (1.3)	2 (2.9)	0 (0.0)	2 (2.9)	3 (1.4)	1 (3.6)	4 (1.7)
Total	72 (82.8)	15 (17.2)	87 (37.0)	66 (83.6)	13 (16.4)	79 (33.6)	69 (100.0)	0 (0.0)	69 (29.4)	207 (88.1)	28 (11.9)	235 (100.0)

PPI: pair of permanent incisors.

Table 4 Reproductive performances of goats

Parameters estimated	Ebnat	Farta	Gonji-kolela	Overall
Age at sexual maturity for male (months)	11.54±1.28[a]	7.04±0.41[b]	9.84±0.95[a]	9.52±0.60
Age at sexual maturity for female (months)	7.54±0.50[a]	7.46±0.53[a]	7.16±0.40[a]	7.36±0.26
Age at first kidding (months)	12.54±0.47[a]	12.42±0.46[a]	12.16±0.40[a]	12.34±0.25
Reproductive lifetime of doe's (year)	8.69±0.62[a]	7.38±0.62[a]	8.53±0.50[a]	8.24±0.33
Reproductive lifetime of buck's (year)	4.23±0.44[a]	3.73±0.56[a]	4.13±0.24[a]	4.04±0.22
Number of kidding per doe's life time	17.77±1.45[a]	12.00±0.95[b]	15.47±1.58[ab]	15.13±0.88
Kidding interval (months)	5.80±0.13[a]	8.61±0.42[b]	6.21±0.14[a]	6.79±0.22

The means within the same row with at least one common letter, do not have significant difference (P>0.05).

Table 5 LSM ± SE body length, neck length, diagonal body length, heart girth and height at wither for goat type, district, age, sex and age by goat type interaction

Effect and level	BL LSM±SE	NL LSM±SE	DBL LSM±SE	HG LSM±SE	HW LSM±SE
Grand	57.08±0.42	30.58±0.26	62.23±0.43	77.04±0.47	70.80±0.41
Goat type	***	***	***	***	**
Dairy	59.65±0.58[a]	32.26±0.37[a]	64.52±0.62[a]	79.27±0.69[a]	72.16±0.61[a]
Dual	54.87±0.69[b]	29.62±0.43[b]	59.93±0.73[b]	75.44±0.81[b]	69.05±0.72[b]
Meat	55.52±0.74[b]	29.01±0.46[b]	61.22±0.78[b]	75.31±0.87[b]	70.66±0.77[ab]

BL: body length; NL: neck length; DBL: diagonal body length; HG: heart girth and WH: wither height.
The means within the same column with at least one common letter, do not have significant difference (P>0.05).
* (P≤0.05) ** (P≤0.01) and *** (P≤0.001).
LSM: least square mean; Nm: not measured; NS: non significant and SE: standard error.
PPI: pair of permanent incisors.

Dairy specific reproductive trait characterization

The mean rear udder length (RUL), udder circumference (UCC) and teat length (TTL) in these breeds averaged were 17.14 ± 0.28, 31.39 ± 0.59 and 3.71 ± 0.07 cm, 14.21 ± 0.48, 26.75 ± 1.01 and 3.41 ± 0.12 cm and 13.44 ± 0.38, 24.67 ± 0.81 and 3.01 ± 0.09 cm, respectively.

Whereas, the mean rear udder diameter (RUD), for dairy, dual and meat type female beetal goats were 10.04 ± 0.22, 8.66 ± 0.37 and 8.01 ± 0.30 cm, respectively (Table 7). Dairy type female goats had significantly higher (P<0.001) values for all udder and teat measurements than dual and meat type female goats.

Table 6 LSM ± SE pelvic width, pelvic height, rump width, rump length, for goat type, district, by goat type interaction

Effect and level	PW	PH	RW	RL
	LSM±SE	LSM±SE	LSM±SE	LSM±SE
Grand	13.82±0.15	72.26±0.36	18.27±0.14	15.88±0.12
Goat type	***	*	***	***
Dairy	15.05±0.20[a]	73.37±0.54[a]	19.31±0.19[a]	16.61±0.16[a]
Dual	13.09±0.24[b]	71.42±0.64[b]	17.61±0.23[b]	15.29±0.19[b]
Meat	12.70±0.25[b]	71.44±0.68[b]	17.35±0.25[b]	15.37±0.21[b]
Disrtict	*	**	NS	***
Ebnat	13.69±0.20[a]	72.68±0.46[a]	18.24±0.21	15.39±0.16[a]
Farta	13.88±0.21[a]	71.68±0.49[a]	18.20±0.22	16.36±0.17[b]
Gonji-kolela	14.47±0.23[ab]	74.07±0.52[b]	18.35±0.24	16.33±0.18[b]

PW: pelvic width; PH: pelvic height; RW: rump width and RL: rump length.
The means within the same column with at least one common letter, do not have significant difference (P>0.05).
* (P≤0.05) ** (P≤0.01) and *** (P≤0.001).
LSM: least square mean; NS: non significant and SE: standard error.

Table 7 LSM ± SE rear udder diameter, rear udder length, udder circumference, teat length, scrotum length, scrotum width and scrotum circumference for goat type, district, age, sex and age by goat type interaction

Effects and level	RUD	RUL	UCC	TTL	SL	SW	SCC
	LSM±SE	LSM±SE	LSM±SE	LSM±SE	LSM±SE	LSM±SE	LSM±SE
Grand	9.21±0.17	15.55±0.24	28.63±0.48	3.46±0.05	12.69±0.42	8.88±0.35	23.56±0.66
Goat type	***	***	***	***	NS	NS	NS
Dairy	10.04±0.22[a]	17.14±0.28[a]	31.39±0.59[a]	3.71±0.07[a]	13.75±1.09	9.37±0.89	23.75±1.74
Dual	8.66±0.37[b]	14.21±0.48[b]	26.75±1.01[b]	3.41±0.12[b]	12.85±0.69	9.20±0.56	24.26±1.10
Meat	8.01±0.30[b]	13.44±0.38[b]	24.67±0.81[b]	3.01±0.09[c]	12.21±0.63	8.44±0.52	22.92±1.00
District	NS	NS	NS	NS	NS	*	*
Ebnat	9.07±0.29	15.78±0.40	28.52±0.82	3.42±0.09	13.03±0.56	9.45±0.43[a]	24.71±0.82[a]
Farta	9.10±0.31	15.24±0.43	28.14±0.88	3.52±0.10	12.23±0.65	8.09±0.50[b]	22.00±0.95[b]
Gonji-kolela	9.45±0.29	15.60±0.40	29.16±0.82	3.44±0.09	Nm	Nm	Nm

RUD: rear udder diameter; RUL: rear udder length; UCC: udder circumference; TTL: teat length; SL: scrotum length; SW: scrotum width and SCC: scrotum circumference.
The means within the same column with at least one common letter, do not have significant difference (P>0.05).
* (P≤0.05) ** (P≤0.01) and *** (P≤0.001).
LSM: least square mean; Na: not applicable; Nm: not measured; NS: non significant and SE: standard error.

West African Dwarf (WAD) goat has 12.00 ± 0.34 cm udder widths in semi intensive management system in Nigeria, which was higher than the present rear udder diameter (Abu *et al.* 2013). However, udder circumference (25.44±0.58 cm) and teat length (2.40±0.11 cm) of greater than two years WAD goat in Nigeria were lower than the present dairy type goats (Abu *et al.* 2013). Scrotum length, scrotum width and scrotum circumference were 13.75 ± 1.09, 9.37 ± 0.89 and 23.75 ± 1.74 cm in dairy type male goats, 12.85 ± 0.69, 9.20 ± 0.56 and 24.26 ± 1.10 cm in dual male goats and 12.21 ± 0.63, 8.44 ± 0.52 and 22.92 ± 1.00 cm in meat type male goats, respectively.

CONCLUSION

From this study, it was possible to see that goats' milk provides more nutritional and therapeutic value than dairy cows, which can satisfy the nutritional requirement with small amount. Goats' milk is easy to drink and is a richer food because it has more calcium, phosphorous and chlorine than cow's milk. But the dairy goat industry is not yet started in Ethiopia in general and in Amhara region in particular as the goat breeding and reproduction was not identified as dairy, meat and dual purpose types. There were significant differences among indigenous dairy type goats from dual and meat type goats in almost all morphological quantitative traits but there is no specialized production and system in the area to enhance production and productivity. The results of dairy specific trait morphological characterization such as the mean rear udder length, udder circumference and teat length and rear udder diameter for dairy goat types were comparable with other internationally known dairy goat breeds. However, there is no specialized dairy goat breeding, reproduction and corresponding production system in the region. Therefore, morphological traits identification, breeding and multiplication of superior dairy type goats with specialized and intensive production system are crucial to increase the milk productivity of their goats and thereby improving the livelihood of the farmers.

ACKNOWLEDGEMENT

We would like to thank the farmers, development agents, agricultural office workers of Ebinat, Farta, and Gonjji Kollela Districts for their cooperation during data collection. We would like to thank also Bahir Dar University, Biotechnology Research Institute for funding this research.

REFERENCES

Abdul W. (2011). Characterization of Goats for linear type traits in Pakistan. Ph D. Thesis. Faisalabad University of Agriculture, Pakistan.

Abu A.H., Mhomga L.I. and Akogwu E.I. (2013). Assessment of udder characteristics of West African Dwarf (WAD) goats reared under different management systems in Makurdi, Benue State, Nigeria. *African J. Agric. Res.* **8(25)**, 3255-3258.

Adane H. and Girma A. (2008). Economic significance of sheep and goats. Pp. 2-24 in Sheep and Goat Production Handbook for Ethiopia. A. Yami and R.C. Markel, Eds. Addis Ababa, Ethiopia.

Alemayehu K. and Fantahun T. (2012). The effect of climate change on ruminant livestock population dynamics in Ethiopia. *Livest. Res. Rural Dev.* Available at: http://www.lrrd.org/lrrd24/10/kefy24185.htm.

Alemayehu K., Kebede D., Melese A. and Andualem S. (2015). Farmers perception on utilization of milk products and stress tolerance capabilities of goats in north-western Amhara, Ethiopia. *J. Harmon. Res. Appl. Sci.* **3(4)**, 193-200.

CSA. (2014). Central Statistical Agency. Agricultural sample survey. Report on livestock and livestock characteristics. The Federal Democratic Republic of Ethiopia. Private Peasant Holdings. Statistical Bulletin 570, Addis Ababa, Ethiopia.

CTA. (2007). The ACP-EU Technical Centre for Agricultural and Rural Cooperation CTA Practical Guide Series, Wageningen, Netherlands.

Dereje T., Berhanu B. and Aynalem H. (2013). Morphological characterization of indigenous Hararghe Highland Goat Breed in their native environment, west Hararghe, Ethiopia. *American-Eurasian J. Sci. Res.* **8(2)**, 72-79.

Devendra C. (2001). Small ruminants' imperatives for productivity enhancement, improved livelihoods and rural growth- a review. *Asian-Australas J. Anim. Sci.* **14(10)**, 1483-1496.

Dhaba U., Belay D., Solomon D. and Taye T. (2012). Sheep and goat production systems in Ilu Abba Bora Zone of Oromia Regional State, Ethiopia: feeding and management strategies. *Glob. Vet.* **9(4)**, 421-429.

Dhaba U., Belay D., Solomon D. and Taye T. (2013). Breeding practices and reproductive performance of traditionally managed indigenous sheep and goat in Ilu Abba Bora Zone of Oromia Regional State, Ethiopia. *Glob. Vet.* **10(6)**, 676-680.

Ebnat District Agriculture and Rural Development Office. (2013). Socio-Economic survey report (un-published).

FAO (1982). Sheep and Goat Breeds of India. Animal Production and Health Paper. Rome, Italy.

Farta District Agriculture and Rural Development Office. (2013). Socio-Economic survey report (un-published).

Fernandes M.F., Queiroga R.C.R.E., Medeiros A.N., Costa R.G., Bomfim M.A.D. and Braga A.A. (2008). Características físico-químicas e perfillipídico do leite de cabras mestiças Moxo-tó alimentadas com dietas suplementadas com óleo de semente de algodão ou de girassol. *Revis. Bras. Zootec.* **37(4)**, 703-710.

Gonji kolela District Agriculture and Rural Development Office (2013). Socio-Economic survey report (un-published).

Hassen H., Baum M., Rischkowsky B. and Markos T. (2012). Phenotypic characterization of Ethiopian indigenous goat populations. *African J. Biotechnol.* **11(73)**, 13838-13846.

Hamayun K., Fida M., Riaz A., Gul Nawaz R. and Muhammad Z. (2006). Relationship of body weight with linear body measurements in goats. *J. Agric. Biol. Sci.* **1(3)**, 51-54.

Legesse G., Abebe G., Siegmund-Schultze M. and Valle Zárate A. (2008). Small ruminant production in two mixed-farming systems of southern Ethiopia: status and prospects for improvement. *Exp. Agric.* **44(3)**, 399-412.

McCullough F. (2003). Nutritional evaluation of goat's milk. *Health Food J.* **105(45)**, 239-251.

Morand-Fehr P. and Boyazogly J. (1999). Present status and future outlook of the small ruminant sector. *Small Rumin. Res.* **34**, 259-269.

Morand-Fehr P. (2003). Strategy for goat farming in the 2st Century. *Small Rumin. Res.* **51(2)**, 175-183.

Nemeth T., Molnar A., Baranyai G. and Kukovics S. (2010). Morphologic characterization and body measurement of Hungarian goats. Pp. 271 in Proc. 56th Ann. Meet. European Assoc. Anim. Prod. Uppsala, Sweeden.

Park Y. (1994). Hypo allergenic and therapeutic significance of goat milk. *Small Rumin. Res.* **14**, 151-159.

SAS Institute. (2000). SAS®/STAT Software, Release 9.1. SAS Institute, Inc., Cary, NC. USA.

Senaka Ranadheera C., Evans C.A., Adams M.C. and Baines S. K. (2012). Probiotic viability and physico-chemical and sensory properties of plain and stirred fruit yogurts made from goat's milk. *Food Chem.* **135(3)**, 1411-1418.

Slacanac V., Bozanic R., Hardi J., Rezessyne Szabo J., Lucan M. and Krstanovic V. (2010). Nutritional and therapeutic value of fermented caprine milk. *Int. J. Dairy Technol.* **63(2)**, 171-189.

Solomon A. (2014). Design of community based breeding programs for two indigenous goat breeds of Ethiopia. Ph D. Thesis. Boku University of Natural Resources and Life science, Vienna, Austria.

Tesfaye A. (2004). Genetic characterization of indigenous goat population of Ethiopia using microsatellite markers. Ph D. Thesis. National Dairy Research Institute, Karnal, India.

Tesfaye T. (2009). Characterization of goat production systems and on- farm evaluation of the growth performance of grazing goats supplemented with different protein sources in Metema Woreda, Amhara Region, Ethiopia. MS Thesis. Haramaya Univ., Haramaya, Ethiopia.

6

Effect of Essential Oils of *Zataria multiflora*, *Eucalyptus globolus* and their Combination on Fermentation Parameters using Merghoz Goat Rumen Liquor

S. Mirzaei Cheshmehgachi[1], M.M. Moeini[1], F. Hozhabri[1*] and M.E. Nooryan Soroor[1]

[1] Department of Animal Science, Faculty of Agriculture, Razi University, Kermanshah, Iran

*Correspondence E-mail: hozhabri@razi.ac.ir

ABSTRACT

In this experiment, different doses of *Zataria multiflora* (ZM) and *Eucalyptus globolus* (EG) essential oils (EOs) (0, 100, 250, 500, 750 and 1000 µg/mL) and a mixture of them were added to alfalfa hay incubated with buffered rumen liquor of Merghoz goat to assess *in vitro* gas production, rumen fermentation and protozoa population. In all treatments the asymptotic gas volume (B) and the rate constant (c) were decreased while the lag phase was increased. There were no effects of ZM and EG on pH, but in high doses of ZE pH was higher than that of control (P<0.001). The ammonia-N concentration was decreased due to addition of combination of EOs (P<0.01) and volatile fatty acid concentration was reduced (P<0.01) following incorporation of EOs. Gas production and organic matter digestibility were decreased (P<0.01) 24 h after incubation, whereas the partitioning factor was increased. Metabolizable energy was decreased, (P<0.05). By inclusion of EOs, total protozoa population and individual genera reduced (P<0.001). The results revealed that EOs of ZM and EG could be potentially used to modulate rumen fermentation, but using them at high level doses have anti-protozoal effects.

KEY WORDS methane production, organic matter digestibility, partitioning factor, protozoa population.

INTRODUCTION

Improving the protein and energy efficiency in ruminant nutrition is a major concern. Essential oils (EOs) have strong antimicrobial properties and can modulate ruminal fermentation to improve nutrient utilization in ruminants by decreasing deamination, methanogenesis activity and methane production in the rumen (Benchaar *et al.* 2008). *Zataria multiflora* (ZM) is a medicinal plant which belongs to the family Labiatae. The essential oils of ZM have strong inhibitory effects against some bacteria. Carvacrol, a monoterpenoid phenol, is the main constituent of ZM essential oil (Talebzadeh *et al.* 2012). The other major constituents were p-cymene, thymol, p-pinene and carvacrol methyl ether. Eucalyptus globolus (EG) is a tall evergreen tree and produce a wide variety of oils. The main active ingredient of EG essential oil is 1, 8- cineole. Pinene, o-cymene and limonene are the other components of EG (Maciel *et al.* 2010). Based upon these characteristics, an *in vitro* experiment was conducted to study the addition effect of ZM and ZM essential oils, and their combination to alfalfa hay incubated with buffered rumen liquor of Merghoz goat on *in vitro* gas production, rumen fermentation and protozoa population.

MATERIALS AND METHODS

Essential oils

Air-dried aerial parts of ZM at full flowering stage (collected from Shiraz Province, Iran) and Eucalyptus leaves (collected from, Kermanshah Province, Iran), were hydro-distillated for 2.5 h, using Clevenger-type apparatus, according to the method described by the British Pharmacopoeia (1988). The amount of oil that was obtained from ZM and EG were 2.24% and 2.60%, respectively. Essential oils were dried over anhydrous sodium sulphate and stored in sealed glass vials at 4 °C. Stock solutions were prepared by dissolving the essential oils in absolute ethanol (mg/mol). For control bottles also equal volumes of ethanol (1% vol/vol) were added as a positive control.

Rumen inoculum

Rumen inoculums were collected from six Merghoz goats using esophageal tube before morning feeding of a diet containing alfalfa hay. The chemical composition of the diet was organic matter (OM), 930 g/kg; crude protein (CP), 138.6 g/kg; neutral detergent fiber (NDF), 544 g/kg; ether extract, 15.6 g/k on dry matter (DM) basis. Rumen contents were strained through four layers of cheese cloth and were continuously purged with CO_2 to stabilize anaerobic condition and kept at 39 °C in a water bath before use.

In vitro gas production (IVGP)

For measuring the kinetics of gas production, 200 mg of alfalfa hay was weighed into a 120 mL Wheaton vial. The vials were subsequently filled with 30 mL of inoculation medium consisting of 10 mL of rumen fluid and 20 mL of buffer solution as described by Menke and Steingass (1988). ZM or EG essential oils (0, 150, 300, 450 and 600 µg/mL) and a combination of them (0, 250, 500, 750 and 1000 µg/mL) were added to the vials, subsequently. Three bottles as blanks containing 30 mL of inoculation medium were also included. The vials were sealed (under CO_2) and placed in a rotor inside incubator (39 °C). The gas pressure was recorded at 0, 2, 4, 6, 8, 10, 12, 24, 32, 48, 72, 80 and 96 h after incubation. The pressure of gas produced in each vial was recorded using a pressure transducer (Testo 512; Testo Inc. Germany). These recorded pressures were used to estimate the generated gas volumes (Lopez *et al.* 2010). After subtraction of gas production from blank bottles, data were fitted to exponential model (Ørskov and McDonald, 1979):

$$y= B [1 - exp -c \times (t-lag)]$$

Where:

y: cumulative volume of gas produced at time t (h).

B: asymptotic gas volume.

c: rate constant.

lag: time (h) between inoculation and commencement of gas production.

Chemical analysis

Alfalfa sample was oven-dried and ground through a 1 mm screen mill (Foss, model CyclotecTM 1093). 500 mg of the substrate and 40 mL of buffered rumen fluid were added to the bottles (Makkar, 2010) and different doses of EOs were included, subsequently. After 24 h incubation, the pressure of gas produced in the headspace of each bottle was recorded using a pressure transducer (Testo 512; Testo Inc. Germany). Then bottles were respectively transferred to an ice bath to stop fermentation and then opened to measure medium pH using a pH meter (Inolab level 2, Germany). Supernatants were collected and frozen at −20 °C until ammonia and total volatile fatty acids (TVFA) analysis. NH_3-N concentration of the bottle content was determined by spectrophotometer (CARY100, VARIAN) according to Broderick and Kang (1980). Total VFAs concentration was measured by Markham apparatus according to the method described by Barnett and Reid (1957) and methane content of the produced gas was determined according to Demeyer *et al.* (1988) and Fievez *et al.* (2005). The metabolizable energy (ME) of substrate was calculated on the basis of the formula proposed by Menke and Steingass (1988), as follows:

$$ME (MJ/kg DM)= 2.20 + 0.136 \times GP + 0.0057 \times CP + 0.00029 \times EE^2$$

Where:
ME: metabolizable energy (MJ/kg DM).
EE: ether extract.
GP: cumulative gas production after 24 h incubation.

In a separate run, *in vitro* organic matter digestibility (OMD) after 24 h incubation was calculated using method described by Makkar (2010). The ratio of substrate truly degraded (mg) to gas volume (mL) at different incubation times was expressed as the partitioning factor (PF) which was determined according to Blümmel *et al.* (1997). Also microbial mass was calculated as mg substrate truly degraded − (mL gas volume×stoichiometrical factor) as described by Blümmel *et al.* (1997).

Protozoa enumeration

For counting protozoa population, whole contents of vials were sustained by diluting with an equal volume formalin solution.

Total numbers and three subfamilies of Entodiniinae, Ophryscolecinae, Diplodiniinae and family Isotrichdae of ciliate protozoa were identified according to the procedures described by Dehority (1993).

Statistical analyses

The observations of experiment were subjected to statistical analysis of variance using the following model to examine the effects of different doses of ZM, EG or their combination on all parameters in three replicates:

$$Y_{ij} = \mu + T_i + e_{ij}$$

Where:

Y_{ij}: observation.

μ: overall mean for each parameter.

T_i: effect of doses.

e_{ij}: residual error.

Data were analyzed using the procedure of SPSS 23.0 software (SPSS, 2015). For all analyses, specific orthogonal contrasts were used to test 1) control *vs.* the average of EOs doses and 2) linear (L), quadratic (Q) and cubic (C) effects of EOs doses on parameters. For protozoal count data, normality assumptions of residuals were tested using Proc Univariate (SPSS 23.0) with the Kolmogorov–Smirnov test. For all statistical analyses, significance was declared at (P<0.05) and trends at (P<0.1). The data for kinetics were processed with the y= B [1 − exp −c × (t−lag)] using the Prism 3.0 software. The results were subjected to one-way variance analysis and compared by using the Duncan test with 5% probability.

RESULTS AND DISCUSSION

Effects of essential oils on gas production

Results showed that control group had the higher (P<0.001) 'B' and 'c' values and the lower (P<0.001) lag time (L) than those of other treatments (Table 1). A large increase in lag time was observed in high doses of essential oils for all treatments.

Similar to the present study, Taghavi-Nezhad *et al.* (2014) found that asymptotic gas production and rate of gas production decreased with the addition of *Zataria multiflora* essential oil to a concentrate-based substrate and Talebzadeh *et al.* (2012) reported comparable results with the incorporation of 150-600 mg/mL of *Zataria multiflora* essential oil to the incubation medium. This reduction can be due to decreased fermentation activity of microorganisms. Gallucci *et al.* (2009) reported that carvacrol and thymol (the main constituents of ZM) are known to have bactericidal or bacteriostatic effects.

The eucalyptus is also a rich source of an antiseptic component (cineole) and contains substances with strong antibacterial properties (Sallam *et al.* 2009). Results showed that ZM and EG were more effective than their combination in reducing gas production.

A noticeable increase in the 'L' value was observed at high doses of EOs. This is due to the fact that essential oils decrease colonization and digestion of readily fermentable substrates without effect on fibrous substrates (Wallace *et al.* 2002). Others have also shown that phenolic compounds inhibit digestion of soluble fractions of feeds as well as the attachment of bacteria to insoluble components of feeds (McAllister *et al.* 1994).

The gas production after 24 h (GP$_{24}$) was decreased by different levels of essential oils of ZM, EgG (P<0.001) and their combination (P<0.05). This finding is in agreement with observations of Macheboeuf *et al.* (2008) reporting decrease in gas production up to 83% after addition of oregano to the incubation media. Carvacerol and thymol caused a reduction in gas production (Benchaar *et al.* 2007). Reduction in gas production may due to decline in TVFA (Table 2), methane productions (Table 1) and fermentable organic matter (Table 2).

Methane production decreased (P<0.001) with increased level of essential oils in all the treatments and it might be due to decreased gas production which represent reduction in fermentation of incubated material. Sallam *et al.* (2009) also observed the linear reduction in methane emission due to the Eucalyptus essential oil supplementation. They emphasized that the reduction in methane production was attributed to a decrease in the fermentable substrate rather than to a direct effect on methanogenesis.

Garcia-González *et al.* (2008) in their study also showed that plant active compounds can reduce methane production by affecting protozoa population. Methane production decreased in batch culture when essential oils were added at 1 μL/mL or at 70, 140 and 280 ppm (Jahani-Azizabadi *et al.* 2014).

Fermentation parameters and digestibility

Treatment with Zataria and Eucalyptus did not affect pH of media and it was in normal range but the combination of EOs increased the value of pH (P<0.05). Supplementation of diet with cinnamaldehyde (the active compound of cinnamon) in dairy cows (Benchaar *et al.* 2008) and beef cattle (Yang *et al.* 2010) did not alter ruminal pH. The ammonia nitrogen (NH$_3$-N) was not affected by Eucalyptus or low doses of Zataria but it decreased due to incorporation of the combination of EOs (P<0.05) and high doses of Zataria (P<0.001). At all doses of Eucalyptus and low doses of Zataria the concentration of NH$_3$-N remained unchanged. At high doses of Zataria decline in NH$_3$-N was observed.

Table 1 Effect of different doses of essential oils on kinetics of gas production

Parameters	Essential oil (EO, µg/mL)						SEM	Contrasts[2]			
	0	100	250	500	750	1000		Control *vs.* EO	L	Q	C
Zataria multiflora											
B	281.56[d]	330.73[e]	290.06[d]	213.63[c]	124.26[b]	52.09[a]	23.96	***	***	***	***
C	0.051[c]	0.029[abc]	0.017[ab]	0.017[ab]	0.041[bc]	0.009[a]	0.004	**	*	NS	*
L	1.30[a]	4.24[ab]	10.67[bc]	15.89[c]	10.61[bc]	13.84[c]	1.44	***	***	*	NS
GP$_{24}$ mL/g OMD	374.43[e]	325.40[d]	249.46[c]	178.86[b]	156.06[b]	64.13[a]	25.61	***	***	NS	NS
Methane % of GP$_{24}$	29.97[d]	26.95[d]	18.41[c]	17.38[c]	11.96[b]	2.47[a]	2.24	***	***	NS	NS
Methane mL/g OMD	112.36[f]	87.66[e]	45.93[d]	31.06[c]	18.60[b]	1.80[a]	9.45	***	***	***	NS
Eucalyptus globolus											
B	281.56[c]	298.36[cd]	311.50[d]	272.23[c]	191.46[b]	100.99[a]	18.1	***	***	***	NS
C	0.051[b]	0.031[a]	0.025[a]	0.016[a]	0.018[a]	0.032[b]	0.003	**	*	**	NS
L	1.30[a]	3.23[ab]	3.22[ab]	10.24[bc]	10.56[bc]	13.84[c]	1.36	*	***	NS	NS
GP$_{24}$ mL/g OMD	374.43[d]	369.2[d]	355.7[d]	306.16[c]	262.8[b]	193.86[a]	16.10	***	***	***	NS
Methane % of GP$_{24}$	29.97[e]	26.60[cd]	23.96[cd]	20.48[bc]	16.58[b]	3.42[a]	2.22	***	***	*	NS
Methane mL/g OMD	112.36[d]	94.56[cd]	88.26[c]	63.23[b]	43.66[b]	7.16[a]	8.85	***	***	*	NS
Zataria and Eucalyptus combination											
B	281.56[c]	-	320.16[d]	274.56[c]	176.66[b]	113.90[a]	20.63	***	***	***	**
C	0.051[b]	-	0.020[a]	0.013[a]	0.013[a]	0.023[a]	0.004	***	**	**	NS
L	1.30[a]	-	10.13[ab]	17.04[bc]	23.08[c]	17.30[bc]	2.33	**	**	*	NS
GP$_{24}$ mL/g OMD	374.43[c]	-	348.2[bc]	308.46[b]	257.06[a]	230.03[a]	15.66	*	***	NS	NS
Methane % of GP$_{24}$	29.97[c]	-	25.27[c]	18.57[b]	11.24[a]	11.03[a]	2.13	***	***	NS	NS
Methane mL/g OMD	112.36[d]	-	88.03[c]	57.13[b]	29.36[a]	25.23[a]	9.32	**	***	NS	NS

[1] B: the asymptotic gas volume; c: the rate constant and L: lag time.
[2] L: linear; Q: quadratic and C: cubic.
GP: gas production and OMD: organic matter digestibility.
The means within the same column with at least one common letter, do not have significant difference (P>0.05; P>0.01 and P>0.001).
SEM: standard error of means.
NS: non significant.
* (P<0.05); ** (P<0.01) and *** (P<0.001).

However, in this study the NH$_3$-N in all levels of essential oils was in normal rang (85-300 mg/L, McDonald *et al.* 2010). A reduction in ammonia concentration reflects an inhibitory effect of EOs on proteolytic activity of rumen microorganisms. As reported essential oils inhibit amino acid deamination by ruminal microbes (Mcintosh *et al.* 2003) and lead to a reduction in protozoal population (Newbold *et al.* 2004).

The levels of 750 and 1000 µg/mL of EOs, decreased the concentration of TVFA. It might be a result of inhibited protozoa activity in the rumen (Williams and Coleman, 1992; Table 3).

The inconsistency of VFAs concentration because of essential oils was observed in the literatures. With the use of plant secondary metabolites, Spanghero *et al.* (2008) found decrease in VFAs, but Newbold *et al.* (2004) reported that essential oils tended to stimulate VFA production and Talebzadeh *et al.* (2012) observed an increase in TVFA by adding low level (150 µg/mL) of ZM to the fermentation media. In contrast, Beauchemin and McGinn (2006) reported no changes in VFAs production, and Castillejos *et al.* (2007) observed different responses to EOs concerning VFAs production depending on the type and dose of EOs and experimental conditions.

These differences may be due to the synergistic effects of cineole, carvacrol and other secondary metabolites in eucalyptus and ZM essential oil (Joch *et al.* 2016). Organic matter digestibility was influenced by incorporation of EOs and the effect was more significant in EG than other treatments. Reduction in OMD might be a consequence of decrease in fermentation of substrate due to EOs as can been seen from reduced gas production (Table 1). Same to this result, cinnamon oil (Fraser *et al.* 2007) and thymol (Castillejos *et al.* 2007) caused decline in dry matter digestibility, however, addition of eugenol had no significant effect (Castillejos *et al.* 2007). Higher values for PF were obtained at levels > 500 µg/mL of EOs. Microbial biomass (MB) and efficiency of microbial biomass by adding ZM (at all levels) or EG and their combination (at high levels) were increased (Table 2). Similar to these results, other investigators reported an increase in PF and MB by supplementation of *Thymus kotschyanus* (Mirzaei *et al.* 2016) and *Zingiber multifloria* (Talebzadeh *et al.* 2012) essential oils. As partitioning factor (PF) represents the efficiency of fermentation and microbial protein production (Blümmel *et al.* 1997), probably digested organic matter by addition of EOs resulted in greater microbial biomass growth rather than VFA production (Taghavi-Nezhad *et al.* 2011).

Table 2 Effect of different doses of essential oils on *in vitro* fermentation parameters

Parameters	Essential oil (EO, µg/mL)						SEM	Contrasts[2]			
	0	100	250	500	750	1000		Control *vs.* EO	L	Q	C
Zataria multiflora											
pH	6.65	6.61	6.67	6.74	6.75	6.72	0.231	NS	NS	NS	NS
NH₃-N (mg/L)	162.29[bc]	171.83[c]	182.50[c]	147.96[bc]	126.87[ab]	94.5[a]	11.1	NS	**	*	NS
TVFA (mmol/L)	51.66[bc]	61.66[c]	53.33[bc]	47.58[b]	22.00[a]	21.01[a]	3.98	*	***	**	*
OMD (mg)	179.43[cd]	189.43[d]	187.56[d]	170.06[c]	139.43[b]	127.03[a]	5.94	**	***	***	**
PF	3.05[a]	3.23[ab]	3.35[ab]	3.46[b]	3.34[c]	6.49[d]	0.202	***	***	***	***
MB (mg)	50.13[a]	60.38[ab]	64.36[b]	61.90[ab]	68.71[b]	84.01[c]	2.85	**	***	NS	*
EMB (%)	27.77[a]	31.82[ab]	34.26[b]	36.31[b]	49.28[c]	66.07[d]	3.21	***	***	***	*
ME (MJ/kg DM)	10.27[e]	10.26[e]	9.94[d]	9.05[c]	6.74[b]	6.94[a]	0.488	***	***	***	*
Eucalyptus globolus											
pH	6.65	6.71	6.64	6.73	6.80	6.76	0.215	NS	NS	NS	NS
NH₃-N (mg/L)	162.29[a]	146.37[a]	156.63[a]	164.25[a]	165.50[a]	165.67[a]	0.854	NS	NS	NS	NS
TVFA (mmol/L)	51.66[b]	43.33[ab]	65.00[c]	68.33[c]	30.83[a]	31.66[a]	3.82	NS	**	***	NS
OMD (mg)	179.43[b]	153.90[a]	146.8[a]	159.7[a]	160.66[a]	159.4[a]	2.77	***	NS	**	**
PF	3.05[c]	2.64[ab]	2.53[a]	2.90[bc]	3.38[d]	5.08[e]	0.210	*	***	***	*
MB (mg)	50.13[c]	25.99[ab]	19.25[a]	38.69[bc]	56.17[c]	90.42[d]	5.18	NS	***	***	NS
EMB (%)	27.77[b]	16.83[a]	13.03[a]	24.15[b]	34.78[c]	56.71[d]	3.52	NS	***	***	NS
ME (MJ/kg DM)	10.27[d]	10.19[d]	10.17[d]	9.76[c]	8.47[b]	6.65[a]	0.339	***	***	***	***
Zataria and Eucalyptus combination											
pH	6.65[a]	-	6.59[a]	6.73[ab]	6.86[bc]	6.92[c]	0.037	*	***	NS	NS
NH₃-N (mg/L)	162.29[b]	-	124.39[a]	115.65[a]	120.99[a]	109.33[a]	5.56	***	***	*	NS
TVFA (mmol/L)	51.66[b]	-	42.66[ab]	42.65[ab]	33.33[a]	36.00[a]	2.03	**	**	NS	NS
OMD (mg)	179.43[d]	-	176.78[d]	150.70[c]	130.43[b]	117.41[a]	6.66	***	***	NS	*
PF	3.05[a]	-	3.02[a]	3.61[b]	4.20[c]	4.96[d]	0.20	***	***	**	NS
MB (mg)	50.13[a]	-	48.13[a]	58.81[ab]	61.97[b]	65.28[b]	2.21	**	***	*	*
EMB (%)	27.77[a]	-	27.22[a]	39.01[b]	47.46[c]	55.57[d]	3.02	***	***	*	*
ME (MJ/kg DM)	10.27[c]	-	10.23[c]	7.96[b]	6.06[a]	5.50[a]	0.544	***	***	NS	**

[1] NH₃-N: ammonia nitrogen; TVFA: total volatile fatty acids; OMD: organic matter digestibility; PF: partitioning factor; MB: microbial biomass; EMB: efficiency of microbial biomass and ME: metabolizable energy.

[2] L: linear; Q: quadratic and C: cubic.

The means within the same column with at least one common letter, do not have significant difference (P>0.05; P>0.01 and P>0.001).

SEM: standard error of means.

NS: non significant.

* (P<0.05); ** (P<0.01) and *** (P<0.001).

Table 3 Effect of different doses of essential oils on protozoa population (×10⁴/mL)

Parameters	Essential oil (EO, µg/mL)						SEM	Contrasts			
	0	100	250	500	750	1000		Control *vs.* EO	L	Q	C
Zataria multiflora											
Total protozoa	15.00[c]	9.44[b]	8.05[ab]	4.72[a]	5.27[a]	5.83[ab]	0.93	***	***	**	NS
Entodinium spp.	11.39[c]	8.33[bc]	7.50[ab]	4.44[a]	5.27[ab]	5.83[ab]	0.65	***	***	*	NS
Isotricha spp.	0.833[b]	0.28[ab]	0.00[a]	0.00[a]	0.00[a]	0.00[a]	0.11	*	*	NS	NS
Diplodiniinae	0.833[b]	0.28[ab]	0.00[a]	0.00[a]	0.00[a]	0.00[a]	0.11	*	*	NS	NS
Ophryoscolecinae	1.94[b]	0.56[a]	0.28[a]	0.28[a]	0.00[a]	0.00[a]	0.22	**	**	NS	NS
Eucalyptus globolus											
Total protozoa	15.00[c]	9.16[b]	6.11[ab]	3.05[a]	4.45[a]	3.05[ab]	1.11	***	***	**	NS
Entodinium spp.	11.39[c]	6.66[b]	5.55[ab]	3.05[a]	3.89[ab]	3.05[a]	0.77	***	***	**	NS
Isotricha spp.	0.833[a]	0.56[a]	0.28[a]	0.00[a]	0.00[a]	0.00[a]	0.13	NS	NS	NS	NS
Diplodiniinae	0.833[a]	0.833[a]	0.28[a]	0.00[a]	0.56[a]	0.00[a]	0.18	NS	NS	NS	NS
phryoscolecinae	1.94[b]	1.11[a]	0.56[a]	0.00[a]	0.55[a]	0.00[a]	0.26	NS	NS	NS	NS
Zataria and Eucalyptus combination											
Total protozoa	15.00[b]	-	3.88[a]	4.72[a]	3.05[a]	2.77[a]	1.30	***	***	**	*
Entodinium spp.	11.39[b]	-	3.05[a]	4.16[a]	3.05[a]	2.77[a]	0.94	***	***	**	*
Isotricha spp.	0.833[b]	-	0.00[a]	0.00[a]	0.00[a]	0.00[a]	0.12	**	*	NS	NS
Diplodiniinae	0.833[a]	-	0.28[a]	0.28[a]	0.00[a]	0.00[a]	0.13	*	*	NS	NS
Phryoscolecinae	1.94[b]	-	0.56[ab]	0.28[ab]	0.00[a]	0.00[a]	0.25	*	*	NS	NS

[1] L: linear; Q: quadratic and C: cubic.

The means within the same column with at least one common letter, do not have significant difference (P>0.05; P>0.01 and P>0.001).

SEM: standard error of means.

NS: non significant.

* (P<0.05); ** (P<0.01) and *** (P<0.001).

Addition of essential oils into substrate caused a reduction in metabolizeable energy (ME). This result may be related to reduction in gas production, VFA concentration and OMD in the fermentation medium especially in high doses.

Effects of essential oils on protozoa concentration

The results showed that essential oils decreased total protozoal count (P<0.001). The concentration of *Entodinium* spp. (P<0.01), *Isotricha* spp., *Diplodiniinae* and *Ophryoscolecinae* also reduced (P<0.001). The antiprotozoal effect of EOs was most likely due to the phenolic structure of its main active compounds (Talebzadeh *et al.* 2012). Such a structure can lead to demolition of cell membrane, inhibition of enzymes and lack of substrates which are essential for cell metabolism (Goel *et al.* 2005) and it may be related to the lipophilic nature of compounds such as anethol which facilitates permeation of EO across the protozoal membrane (Cardozo *et al.* 2006). It appeared that population of protozoa was more sensitive to combination of Zataria and Eucalyptus essential oils than each of them (Table 3).

CONCLUSION

The results of this experiment indicate that EOs of *Zataria multiflora*, *Eucalyptus globolus* have a potential to manipulate rumen fermentation favorably with antimethanogenic and defaunating properties. As regards to the essential oils combination, there is a need to identify the suitable doses without adverse effect on feed digestibility.

ACKNOWLEDGEMENT

This study was financed by Razi University, Iran. Authors would like to thank Dr Hadi Hajarian, the Head of Animal Science Department of Agriculture Faculty, for kindly collaboration.

REFERENCES

Barnett A.J.G. and Reid R.L. (1957). Studies on production of volatile fatty acids from grass by rumen liquid in an artificial rumen. *J. Agric. Sci.* **48**, 315-321.

Beauchemin K.A. and Mcginn S.M. (2006). Methane emissions from beef cattle: effects of fumaric acid, essential oil and canola oil. *J. Anim. Sci.* **84**, 1489-1496.

Benchaar C., Calsamiglia S., Chaves A.V., Fraser G.R., Colombatto D., McAllister T.A. and Beauchemin K.A. (2008). A review of plant-derived essential oils in ruminant nutrition and production. *Anim. Feed Sci. Technol.* **145**, 209-228.

Benchaar C., Chaves A.V., Fraser G.R., Wang Y., Beauchemin K.A. and McAllister T.A. (2007). Effects of essential oils and their components on *in vitro* rumen microbial fermentation. *C-anadian J. Anim. Sci.* **87**, 413-419.

Blümmel M., Makkar H.P.S. and Becker K. (1997). *In vitro* gas production, a technique revisited. *J. Anim. Physiol. Anim. Nutr.* **77**, 24-34.

British Pharmacopoeia. (1988). Her Majesty's Stationery Office. HMSO, London, UK.

Broderick G.A. and Kang J.H. (1980). Automated simultaneous determination of ammonia and total amino acids in ruminal fluid and *in vitro* media. *J. Dairy Sci.* **63**, 64-75.

Cardozo P.W., Calsamiglia S., Ferret A. and Kamel C. (2006). Effects of alfalfa extract, anise, capsicum and a mixture of cinnamaldehyde and eugenol on ruminal fermentation and protein degradation in beef heifers fed a high concentrate diet. *J. Anim. Sci.* **84**, 2801-2808.

Castillejos L., Calsamiglia S., Ferret A. and Losa R. (2007). Effects of dose and adaptation time of a specific blend of essential oil compounds on rumen fermentation. *Anim. Feed Sci. Technol.* **132**, 186-201.

Dehority B.A. (1993). Laboratory Manual for Classification and Morphology of Rumen Ciliate Protozoa. CRC Press, Boca Raton, Florida, USA.

Demeyer D.I., Meulemeester M., de Graeve K. and Gupta B.W. (1988). Effect of fungal treatment of nutritive value of straw. *Med. Fac. Landbouww. Rijksuniv. Gent.* **53**, 1811-1819.

Fievez V., Babayemi O.J. and Demeyer D. (2005). Estimation of direct and indirect gas production in syringes: a tool to estimate short chain fatty acid production that requires minimal laboratory facilities. *Anim. Feed Sci. Technol.* **123**, 197-210.

Fraser G.R., Chaves A.V., Wang Y., McAllister T.A., Beauchemin K.A. and Benchaar C. (2007). Assessment of the effects of cinnamon leaf oil on rumen microbial fermentation using two continuous culture systems. *J. Dairy Sci.* **90**, 2315-2328.

Gallucci M.N., Oliva M., Casero C., Dambolena J., Luna A. and Zygadlo J. (2009). Antimicrobial combined action of terpenes against the food-borne microorganisms *Escherichia coli*, *Staphylococcus aureus* and *Bacillus cereus*. *Flavour Frag. J.* **24**, 348-354.

Garcia-González R., Lopez S., Fernandez M. and Gonzalez J.S. (2008). Dose response effects of *Rheum officinale* root and *Frangula alnus* bark on ruminal methane production *in vitro*. *Anim. Feed Sci. Technol.* **145**, 319-334.

Goel G., Puniya A.K., Aguliar C.N. and Singh K. (2005). Interaction of gut microflora with tannins in feeds. *Naturwissenschaften.* **92**, 497-503.

Jahani-Azizabadi H., Danesh Mesgaran M., Vakili A. and Rezayazdi K. (2014). Effect of some plant essential oils on *in vitro* ruminal methane production and on fermentation characteristics of a mid-forage diet. *J. Agric. Sci. Technol.* **16**, 1543-1554.

Joch M., Cermak L., Hakl J., Hucko B., Duskova D. and Marounek M. (2016). *In vitro* screening of essential oil active compounds for manipulation of rumen fermentation and methane mitigation. *Asian-Australas J. Anim. Sci.* **29(7)**, 952-959.

Lopez S., Makkar H.P.S. and Soliva C.R. (2010). Screening plants and plant products for methane inhibitors. Pp. 191-231 in *In vitro* Screening of Plant Resources for Extra-Nutritional Attributes in Ruminants: Nuclear and Related Methodologies. P.E. Vercoe, H.P.S. Makkar and A.C. Schlink, Eds. Springer, Dordrecht, the Netherlands.

Macheboeuf D., Morgavi D.P., Papon Y., Mousset J.L. and Arturo-Schaan M. (2008). Dose-response effects of essential oils on *in vitro* fermentation activity of the rumen microbial population. *Anim. Feed Sci. Technol.* **145**, 335-350.

Maciel M.V., Morais S.M., Bevilaqua C.M.L., Silva R.A., Barros R.S., Sousa R.N., Sousa L.C., Brito E.S. and Souza-Neto M.A. (2010). Chemical composition of *Eucalyptus* spp. essential oils and their insecticidal effects on *Lutzomyia longipalpis*. *Vet. Parasitol.* **167**, 1-7.

Makkar H.P.S. (2010). *In vitro* screening of feed resources for efficiency of microbial protein synthesis. Pp. 107-144 in *In vitro* Screening of Plant Resources for Extra-Nutritional Attributes in Ruminants: Nuclear and Related Methodologies. P.E. Vercoe, H.P.S. Makkar and A.C. Schlink Eds. Springer, Dordrecht, Netherlands.

McAllister T.A., Bae H.D., Jones G.A. and Cheng K.J. (1994). Microbial attachment and feed digestion in the rumen. *J. Anim. Sci.* **72**, 3004-3018.

Mcdonald P., Edwards R.A., Greenhalgh J.F.D., Morgan C.A., Sinclair L.A. and Wilkinson R.G. (2010). Animal Nutrition. Prentice Hall, London, UK.

Mcintosh F.M., Williams P., Losa R., Wallace R.J., Beever D.A. and Newbold C.J. (2003). Effects of essential oils on ruminal microorganisms and their protein metabolism. *Appl. Environ. Microbiol.* **69**, 5011-5014.

Menke K.H. and Steingass H. (1988). Estimation of the energetic feed value obtained from chemical analysis and *in vitro* gas production using rumen fluid. *Anim. Res. Dev.* **28**, 7-55.

Mirzaei Z., Hozhabri F. and Alipour D. (2016). Thymus kotschyanus essential oil component and their effects on *in vitro* rumen fermentation, protozoal population and acidosis parameters. *Iranian J. Appl. Anim. Sci.* **6(1)**, 53-58.

Newbold C.J., McIntosh F.M., Williams P., Losa R. and Wallace R.J. (2004). Effects of a specific blend of essential oil compounds on rumen fermentation. *Anim. Feed Sci. Technol.* **114**, 105-112.

Ørskov E.R. and McDonald I. (1979). The estimation of protein degradability in the rumen from incubation measurements weighted according to rate of passage. *J. Agric. Sci.* **92**, 499-503.

Sallam S.M.A., Bueno I.C.S., Brigide P., Godoy P.B., Vitti D.M.S.S. and Abdalla A.L. (2009). Efficacy of eucalyptus oil on *in vitro* rumen fermentation and methane production. *Options Mediterraneennes*. **85**, 267-272.

Spanghero M., Zanfi C., Fabbro E., Scicutella N. and Camellini C. (2008). Effects of a blend of essential oils on some end products of *in vitro* rumen fermentation. *Anim. Feed Sci. Technol.* **145**, 364-374.

SPSS Inc. (2015). Statistical Package for Social Sciences Study. SPSS for Windows, Version 23.0. Chicago SPSS Inc.

Taghavi-Nezhad M., Alipour D., Flythe M.D., Zamani P. and Khodakaramian G. (2014). The effect of essential oils of *Zataria multiflora* and *Mentha spicata* on the *in vitro* rumen fermentation, and growth and deaminative activity of amino acid-fermenting bacteria isolated from Mehraban sheep. *J. Anim. Prod. Sci.* **54**, 299-307.

Taghavi-Nezhad M., Alipour D., TorabiGoudarzi M., Zamani P. and Khodakaramian G. (2011). Dose response to carvone rich essential oils of spearmint (*Mentha spicata*): *in vitro* ruminal fermentation kinetics and digestibility. *J. Agric. Sci. Technol.* **13**, 1013-1020.

Talebzadeh R., Alipour D., Saharkhiz M.J., Azarfar A. and Malecky M. (2012). Effect of essential oils of *Zataria multiflora* on *in vitro* rumen fermentation, protozoal population, growth and enzyme activity of anaerobic fungus isolated from Mehraban sheep. *Anim. Feed Sci. Technol.* **172**, 115-124.

Wallace R.J., McEwan N.R., McIntosh F.M., Teferedegne B. and Newbold C.J. (2002). Natural products as manipulators of rumen fermentation. *Asian-Australas J. Anim. Sci.* **15**, 1458-1468.

Williams A.G. and Coleman G.S. (1992). The Rumen Protozoa. Springer-Verlag, London, UK.

Yang W.Z., Ametaj B.N., Benchaar C. and Beauchemin K.A. (2010). Dose response to cinnamaldehyde supplementation in growing beef heifers: ruminal and intestinal digestion. *J. Anim. Sci.* **88**, 680-688.

7

Effects of Two Sources of Tannins on Performance, Nitrogen Utilization and Efficiency of Microbial Nitrogen Synthesis in Dairy Goats

A. Mokhtarpour[1*], A.A. Naserian[2], F. Pourmollae[3] and S. Safa[4]

[1] Research Center of Special Domestic Animals, Research Institute at University of Zabol, Zabol, Iran
[2] Department of Animal Science, Faculty of Agriculture, Ferdowsi University of Mashhad, Mashhad, Iran
[3] Agricultural Jihad Organization of Khorasan-Razavi, Mashhad, Iran
[4] Department of Animal Science, Faculty of Agriculture, University of Tabriz, Tabriz, Iran

*Correspondence E-mail: am.mokhtarpour@uoz.ac.ir

ABSTRACT

The aim of this study was to evaluate the effects of tannic acid (TA) and pistachio by-product tannin extract (PBE) on N utilization and efficiency of microbial nitrogen synthesis (EMNS) in Saanen dairy goats. The experimental design was a change over design with three treatments and three periods using six mid lactation multiparous dairy goats. Tannic acid solution and PBE were added to fresh alfalfa at ensiling to get the final concentration of about 1% tannin as dry matter (DM) basis. Dietary treatments were as follows: 1) untreated alfalfa silage (AS), 2) AS treated with TA (AS+TA) and 3) AS treated with PBE (AS+PBE). Intake of DM, organic matter (OM) and N and also milk yield and composition (as percent or as g/d) were not affected by tannin additions. A considerable shift occurred in the pattern of N excretion in goats fed PBE diet compared to control silage diet. Urinary N losses had a tendency (P<0.10) to lower (17.4 *vs.* 19.2 g/d) and fecal N losses was higher for goats fed AS + PBE compared with those fed the AS silage diet (14.6 *vs.* 12.6 g/d). As a proportion of N intake or as g N loss per d, retained N was higher (P<0.05) in goats fed AS + PBE compared to those fed the control diet. Microbial nitrogen synthesis (MN) was not affected by tannin additions. However, a trend (P<0.10) for less efficiency of microbial nitrogen synthesis (EMN) was observed in tannin fed goats compared to those fed the control diet. Although, tannins added to AS may exert some beneficial effects on N utilization and also environmental N outputs, they may interfere with rumen fermentation lead to decrease in EMN in goats especially those fed with PBE.

KEY WORDS dairy goats, nitrogen utilization, pistachio by-product extract, tannic acid.

INTRODUCTION

Extensive alfalfa protein degradation to non protein nitrogen (NPN) during ensiling and in the rumen (Muck *et al.* 2003) will lead to inefficient utilization of N by ruminants and increase environmental N excretion (Deaville *et al.* 2010). It also causes the need of expensive supplemental protein in silage based diets (Givens and Rulquin, 2004). Considerable attention has recently been focused on using natural plant extract containing polyphenolic compounds (tannins) in order to decrease protein degradation. Tannins are plant secondary metabolites that can bind with macromolecules such as protein and carbohydrates, hence retard rate and extent of their degradations (McSweeny *et al.* 2001; Frutos *et al.* 2004). One of the sources of tannins which produce in Iran with an annual production of about 765500 tonnes is pistachio by-product (PB). PB is obtained after de-hulling of fresh pistachio (*Pistacia vera*) and contains a relatively high amounts of phenolic compounds (9.06-14.57% of DM) and tannins (4.97-8.67% of DM) depending on extraction methods, particle size of samples, extraction time and type of solvent (Mokhtarpour *et al.* 2014). Previous studies demonstrated that addition of tannins or tannin rich extract to legumes or grasses can exert some beneficial effects on protein degradability and N utilization both *in vitro* and *in vivo* (Tabacco *et al.* 2006; Deaville *et al.* 2010; Mokhtarpour *et al.* 2015). Recently, it has been reported that treating soybean meal by PB extract (BPE) in the diets of Holstein bulls (Jolazadeh *et al.* 2015) and also incorporation of PBE in the diets of lambs (Rajaei-Sharifaadi and Naserian, 2014) decreased ruminal degradation of protein, increased retained N and resulted in better animal performance. We hypothesized that addition of PBE to alfalfa silage not only may decrease protein degradability but also may improve milk production and N retention of dairy goats. Furthermore, since tannin concentration in PBE was expressed as tannic acid equivalent (Makkar, 2000) and tannins in PBE were mostly hydrolysable, tannic acid was also used as a source of commercial hydrolysable tannin for treating fresh alfalfa to compare these two sources of tannins. Therefore, the purpose of this study was to determine whether addition of TA and PBE to fresh alfalfa at ensiling can affect Saanen dairy goat's performance, nitrogen balance and microbial nitrogen synthesis.

MATERIALS AND METHODS

Preparation of alfalfa silages

It has been reported that more than 70% of TT in PB can be extracted by water (Mokhtarpour *et al.* 2014) and thus eliminating costly solvents such as acetone, methanol or ethanol which is commonly used for tannin extraction (Makkar, 2000). Preparations of PB extract and alfalfa silages for this experiment are described in our previous study (Mokhtarpour *et al.* 2105). Briefly, sun-dried PB was collected from pistachio de-hulling factory in Feizabad (Khorasan-e-Razavi Province, Iran), ground to pass a 2-mm screen and then was soaked in water with a ratio of 1:5 (w/v) at room temperature for 12 h. The contents were filtered through cheesecloth and then filtrated extract was sprayed on chopped alfalfa ($\sim$ 25% DM) with a ratio of 500 ml/kg DM to get the final concentration of about 1% tannin as tannic acid equivalent on DM basis (AS+PBE). To the second silo, tannic acid was first dissolved in the same amount of water and then sprayed on alfalfa to get the final concentration of about 1% tannin. The same amount of water was also applied to the third silo as control silage. All silos were wilted for 20 h to increase DM content and then were ensiled for 60 d.

Animals, experimental diets and management

Six multiparous Saanen dairy goats (110 ± 13 days in milk, 1.42 ± 0.29 kg milk production) were used in a change over design with three treatments and three periods. Animals were kept in individual cage to facilitate separate collection of feces and urine. Guidelines for the care and use of animals were approved by the Animal Care Committee of the Ferdowsi University of Mashhad. Dietary treatments were: 1) untreated alfalfa silage (AS), 2) AS treated with TA (AS+TA) and 3) AS treated with PBE (AS+PBE). Each experimental period lasted 21 d with the first 14 d used for diet adaptation and 7 d for sample collection. Experimental diets were formulated to meet the nutrient requirements of lactating goats using the Small Ruminant Nutrition System (SRNS; version 1.9.4468; Tedeschi *et al.* 2010). Diets were fed as a total mixed ration (TMR) with forage to concentrate ratio of 50:50. Diets were offered twice daily *ad libitum* at 08:00 and 16:00 h and goats had free access to water. Chemical composition, phenolic compounds and crude protein (CP) fractions of silages and the ingredients and chemical composition of the diets are presented in Tables 1 and 2, respectively.

Sampling and data collection

Feeds and orts were collected before the morning feeding and weighed daily during the measurement period and were composited for each goat. Dry matter intake was calculated by difference between total amount of DM offered and refused. Complete fecal collection of each goat was done through the 5-day collection periods and then dried in an oven. Daily dried samples were ground and later composited for each 5-day periods. Composite samples of the TMR, feed refusal and feces were dried in an oven, then ground to pass through a 2-mm screen and stored for later analysis. Urine from each goat was collected daily in plastic vessels containing 100 mL 4 N H_2SO_4 to maintain the pH level below 3 to prevent bacterial destruction of total purine derivatives (TPD) (Chen and Gomes, 1995). Urine was weighed once a day and mixed well and 10% of daily aliquot was pooled over the 5-day collection period per animal and stored at $-20°C$ for later TPD (i.e., allantoin, uric acid and xanthine hypoxanthine) analysis and N content.

Goats were milked twice daily at 07:00 and 15:00 h and individual milk yields were recorded at each milking. A daily composite milk sample from the morning and afternoon milking was taken during the collection period and fresh subsamples were analyzed daily for chemical composition. Animals were weighed on day 22 before feeding, and then the diet given to each cow was changed.

Chemical analyses

Dry matter content of silages, feeds, orts and feces was determined by drying in an oven at 100 °C to a constant weight (method 934.01). Ash (method 942.05), and acid detergent fiber (ADF) (method 973.18) were determined according to AOAC (2005) procedures. Crude protein (Kjeldahl N×6.25) was determined by the block digestion method using copper catalyst and steam distillation into boric acid (method 2001.11) on 2100 Kjeltec distillation unit as described in AOAC (2005). Neutral detergent fiber (NDF) was determined by the method of Van Soest *et al.* (1991). The sodium sulphite and α-amylase were not used and both NDF and ADF were expressed exclusive of residual ash. Crude protein fractions were determined by Licitra *et al.* (1996) procedure. Tannin assay was conducted according to Makkar (2000) procedure. Briefly, after drying samples in an oven at 40 °C to constant weight in order to minimize changes in tannin content and activity, they were ground to pass a 2 mm sieve and then 0.5 mm sieve. Approximately, 200 mg samples were extracted in four replicates in 70% aqueous acetone (v/v) by using an ultrasonic bath for 20 min. After centrifugation (3000 g/min, 4 °C, 10 min), the supernatant was collected and kept in refrigerator (4 °C). Total phenolic compounds (TP) and total tannins (TT) were determined by Folin-Ciocalteu reagent using tannic acid (Merck GmbH, Darmstadt, Germany) as a standard (Makkar, 2000). The values of TP and TT were expressed as tannic acid equivalent.

Microbial nitrogen synthesis (MNS) was estimated on urinary purine derivatives (PD) excretion based on the relationship derived by Chen and Gomes (1995). The amounts of allantoin, uric acid and xanthine plus hypoxanthine were determined by spectrophotometric method. The urinary PD excreted in a day was used in the iteration process to calculate the microbial nitrogen supply as described by Chen and Gomes (1995). Efficiency of microbial nitrogen synthesis (EMNS) was calculated as:

$$EMNS = MN (g/d) / DOMR$$

Where:
DOMR: apparently digested OM in the rumen (assuming that rumen digestion was 65 g/100 g OM of digestion in total tract.
$$DOMR = DOMI \times 0.65$$
DOMI: digestible organic matter intake according to ARC (1984).

Milk samples were analyzed for fat, protein, lactose, solids-not fat, and total solids content by Milk-O-Scan 605 analyser (Foss Electric, Hillerød, Denmark). The gross energy content in milk was calculated according to Tyrrell and Reid (1965) as:

Milk energy content (MJ/kg) = $4.184 \times 2.204 \times [41.63 \times$ fat (%) $+ 24.13 \times$ protein (%) $+ 21.60 \times$ lactose (%) $- 117.2] / 1000$.

Statistical analyses

Mixed procedure of SAS (2001) was used to analyze data as a following model for a change over design:

$$Y_{ijk} = \mu + T_i + P_j + C_k + \varepsilon_{ijk}$$

Where:
Y_{ijk}: dependent variable.
μ: overall mean.
T_i: fixed effect of treatment (i=1, 2, 3).
P_j: fixed effect of period (j= 1, 2, 3).
C_K: random effect of goat.
ε_{ijk}: random residual error.

Least squares means procedure (LSMEANS) was used to test the differences among means if a value of P < 0.05 was detected. Trends were discussed at P < 0.10. All results are reported as least squares means.

RESULTS AND DISCUSSION

The concentration of total tannins in untreated and treated silages were, 0.5 and 1.2% of DM, respectively. The pH values decreased in silages treated with TA and PBE (Table 1). A substantial decreased in ammonia N concentration occurred in PBE treated silage compared to control and TA silages (6.5 *vs.* 14.7 and 13.5 mg/dL, respectively). Soluble protein significantly decreased in AS + PBE compared to control (44.8 *vs.* 49.6). Intake of DM, OM and N and also milk yield and composition were not affected by addition of tannins (Table 3). Intake, losses and retention of N are shown in Table 4. Addition of PBE to alfalfa silage increased fecal N (P<0.05) and lowered urinary N excretions (P<0.10) when expressed as g N per day. However, as proportion of N intake, excretion of fecal N was not affected by addition of PBE, but urinary N excretion decreased significantly. A tendency for lower urinary N excretion (P<0.10) was also observed by addition of TA. Milk N as g per day or as percentage of N intake was not affected by addition of tannins. Retention of N was significantly higher in goats fed PBE compared to other treatments when expressed either as g per day or percentage of N intake. Microbial N synthesis was not affected by addition of tannins, however, EMNS had a tendency to decrease in goats fed PBE compared to those fed control diet. In spite of the same concentration of tannin in AS + TA and AS + PBE, addition of TA to the silage had no effect on any of the parameters mentioned above.

Lower pH value in AS + PBE may be due to a relatively low pH of 4.95 in PBE (data not shown). Another reason could be possibly due to organic acids production during fermentation of water soluble carbohydrates in PBE.

Table 1 Chemical composition and phenolic compounds of silages

Item	Treatment			SEM	P-value
	AS	AS + TA	AS + PBE		
Chemical composition, % of DM					
Dry matter	27.7	27.7	28.3	0.30	0.72
Crude protein	17.8	17.9	17.5	0.10	0.08
Neutral detergent fiber	42.8	42.7	43.7	0.39	0.61
Acid detergent fiber	31.5	27.2	29.3	0.95	0.20
Ash	9.9	10.5	10.0	0.15	0.31
Total phenolic compounds[†]	1.1[b]	1.4[ab]	1.7[a]	0.11	0.04
Total tannins[†]	0.5[b]	1.2[a]	1.2[a]	0.13	< 0.01
pH	5.44[a]	5.09[b]	4.68[c]	0.139	< 0.01
NH_3-N (mg/dL)	14.7[a]	13.5[a]	6.5[b]	1.30	
CP fractions, % of CP[‡]					
Soluble protein (A+B1)	49.6[a]	47.3[ab]	44.8[b]	0.91	0.01
A	46.7	44.3	42.3	1.03	0.25
B1	3.0	2.9	2.5	0.59	0.96
B2	37.2	41.0	42.5	1.20	0.19
B3	8.2	7.7	8.2	0.45	0.90
C	4.9	4.0	4.5	0.24	0.37

AS: alfalfa silage (control); AS + TA: alfalfa silage with tannic acid and AS + PBE: alfalfa silage with pistachio by-product extract.
[†] Expressed as grams of tannic acid equivalent.
[‡] A= non protein nitrogen; B1, B2 and B3= true protein based on decreasing solubility and C= acid detergent insoluble protein.
The means within the same row with at least one common letter, do not have significant difference (P>0.05).
SEM: standard error of the means.

The action of tannins in AS + PBE to bind proteins and reduced proteolysis might cause a decreased in NH_3-N and soluble protein concentrations (Table 1).

The lack of a significant effect on dry matter intake (DMI) as a result of feeding TA or PBE to dairy goats in our trial may have been due to the low concentration of TT in silages (Table 1) and in diets (Table 2). The similar milk yield and composition in goats fed treated or untreated silages (Table 3) can be due to similar DMI (Table 3). However, a slight increase in DMI and organic matter intake (OMI) (about 9%) was observed in goats fed AS + PBE. Dry matter intake in goats fed with PB at 32% of diet DM in replacement of alfalfa hay in the ration (2.12% dietary tannin) of dairy goats had no effect on DM intake (Sedighi-Vesagh et al. 2015). Use of PB extract as a source of tannin in treating alfalfa silage had no effect on DM intake of goats in early lactation (Mokhtarpour et al. 2015). Naserian et al. (2015) reported that DM intake of dairy goats was increased by increasing pistachio skins (soft external hulls) in replacement of wheat bran up to 21% of diet DM (0.45-1.9% dietary tannin). Higher DM intake in goats fed pistachio skins may be due to the positive effect of secondary metabolites on ruminal fermentation and higher nutrient digestibility observed in that study.

Similar to our results, feeding PB to dairy goats at 30 and 32% of diet DM containing 1.81 and 2.12% tannin respectively had no effect on milk yield and composition (Ghaffari et al. 2014; Sedighi-Vesagh et al. 2015). In contrast, Mokhtarpour et al. (2015) observed an increase in milk fat and protein yield by treating alfalfa silage with 1% PBE in goats at early lactation.

A linear increase in milk fat content also occurred by increasing pistachio skins in dairy goats ration, whereas concentrations of other milk components were not affected by experimental treatments (Naserian et al. 2015). They stated that the higher milk fat content in goats fed pistachio skins may be partly associated with an increased intake of ether extracts (EE) due to higher EE content in pistachio skins than wheat bran.

Regarding similar dietary ingredients and similar N intake, changes in N utilization among treatments can be attributed to the tannin added to the silages (Table 4). The higher loss of fecal N (g/d) in PBE fed goats suggests that already formed tannin–protein complexes may incompletely dissociate post ruminally. This reversibility depends on the affinity of tannins to bind proteins (Makkar, 2003) and the larger the molecular size of tannins, the higher the binding affinity (Ozdal et al. 2013). Thus, our results indicated that PB tannins may be generating some indigestible rumen escape protein. According to Butter (1999) an increment in fecal N excretion as a result of tannin addition could be due to protein bound to tannin, reduced digestive enzymes activity, impaired intestinal function and increase in secretion of endogenous proteins. Patra (2010) also reviewed that fecal N as percentage of N intake increased in sheep fed high foliage diets containing high concentrations of tannins. In spite of the same concentration of tannins in alfalfa silages treated with TA and PBE, lack of effect on N utilization pattern following feeding dairy goats with AS + TA revealed that the nature of tannin (chemical structures and molecular weight) can influence the animal response to dietary tannins.

Table 2 Feed ingredients and chemical composition of experimental diets

Item	Treatment		
	AS	AS + TA	AS + PBE
Diet ingredients, %			
Alfalfa silage	50.0	50.0	50.0
Barley grain	26.0	26.0	26.0
Wheat bran	23.0	23.0	23.0
Limestone	0.35	0.35	0.35
Salt	0.15	0.15	0.15
Vitamin-mineral mix[†]	0.50	0.50	0.50
Chemical composition, % of DM			
Dry matter	41.8	42.5	42.8
Organic matter	91.8	92.0	91.3
Crude protein	14.2	14.2	14.2
Neutral detergent fiber	41.3	41.3	40.9
Acid detergent fiber	24.0	23.3	23.4
Non fiber carbohydrate[‡]	33.2	33.3	33.0
Ether extract	3.2	3.2	3.2
Total phenolic compounds[§]	0.68	1.28	1.37
Total tannins[§]	0.29	0.65	0.67

AS: alfalfa silage (control); AS + TA: alfalfa silage with tannic acid and AS + PBE: alfalfa silage with pistachio by-product extract.
[†] Contained (g/kg premix; DM basis): vitamin A: 330000 IU; vitamin D: 60000 IU; vitamin E: 1000 IU; Ca: 160 g; P: 85 g; Na: 63 g; Mg: 45 g; Zn: 2100 mg; Mn: 1500 mg; Cu: 535 mg; Se: 12 mg and I: 45 mg.
[‡] Non-fibre carbohydrates calculated as: 100 – (neutral detergent fiber+crude protein+ether extract+ash) (NRC 2001).
[§] Expressed as grams of tannic acid equivalent.

Table 3 Effect of treatment on dry matter intake, milk yield and composition

Item	Treatment			SEM	P-value
	AS	AS + TA	AS + PBE		
Intake, kg/d					
Dry matter	1.91	1.90	2.09	0.056	0.22
Organic matter	1.75	1.74	1.90	0.049	0.25
Milk production, kg/d	1.40	1.37	1.37	0.067	0.71
Milk Composition, %					
Fat	4.47	4.62	4.55	0.181	0.61
Protein	3.57	3.53	3.62	0.031	0.13
Lactose	4.38	4.39	4.39	0.049	0.88
Solid not fat	9.13	9.02	9.09	0.092	0.32
Total solids	12.72	12.80	12.81	0.221	0.68
Milk component yield, g/d					
Fat	62.80	64.68	62.97	4.628	0.43
Protein	49.90	48.60	51.37	1.090	0.47
Lactose	61.15	60.42	60.38	3.181	0.88
Milk energy, MJ/kg[†]	3.27	3.33	3.32	0.081	0.55
Milk energy output, MJ/d	4.59	4.62	4.58	0.284	0.86

AS: alfalfa silage (control); AS + TA: alfalfa silage with tannic acid and AS + PBE: alfalfa silage with pistachio by-product extract.
[†] Estimated according to the equation of Tyrrell and Reid (1965).
SEM: standard error of the means.

Similar to our results, Getachew *et al.* (2008a) reported that the pattern of N utilization by sheep was not affected by alfalfa hay supplemented with different levels of TA (3 to 9% DM).

However, inclusion of PBE shifted partitioning in N from urine to feces which is consistent with the results of the most studies on tannins effects (Deaville *et al.* 2010; Ahnert *et al.* 2015).

Reduction in urinary N excretion as a proportion of N intake (P<0.05) following PBE supply and also a trend (P<0.10) for lower urinary N losses (g/d) in goats fed AS + TA or AS + PBE can be attributed to the action of tannins in binding proteins and consequently lower ruminal degradation of N and ammonia-N losses (Patra and Saxena, 2011). This is confirmed by the strong protein protection from *in vitro* ruminal degradation of alfalfa hay by addition of TA and quebracho tannins (Getachew *et al.* 2008b).

Increased in retained N as a result of supplementing PBE to alfalfa silage is consistent with recent report for tannin action in N retention (Ahnert *et al.* 2015).

Table 4 Effect of treatment on nitrogen utilization and microbial nitrogen synthesis

Item	Treatment			SEM	P-value
	AS	AS + TA	AS + PBE		
N intake, g/d	43.5	43.4	47.3	1.30	0.23
N losses					
Feces	12.6[b]	12.7[b]	14.6[a]	0.35	0.02
Urine	19.2	17.5	17.4	0.29	0.08
Milk	7.8	7.6	8.1	0.16	0.47
Retained	4.0[b]	5.6[ab]	7.2[a]	0.48	0.04
N losses and retention as percentage of N intake (%)					
Feces	28.8	29.3	30.9	0.66	0.47
Urine	44.3[a]	41.0[ab]	37.2[b]	1.14	0.04
Milk	18.1	17.9	17.1	0.38	0.52
Retained	8.9[b]	12.8[ab]	15.1[a]	0.92	0.03
Microbial N, g/d	22.7	21.9	20.9	0.87	0.76
Efficiency of Microbial N, g N/kg DOMR[†]	25.8	23.4	22.5	0.63	0.07

AS: alfalfa silage (control); AS + TA: alfalfa silage with tannic acid and AS + PBE: alfalfa silage with pistachio by-product extract.
[†] DOMR= apparently digested organic matter in the rumen (65% of apparently digested organic matter in total tract) according to ARC (1984).
The means within the same row with at least one common letter, do not have significant difference (P>0.05).
SEM: standard error of the means.

In contrast, it was noted a decline in urinary N excretion counteracting the elevated fecal N excretion in goats (Al-Dobaib *et al.* 2009) and sheep (Deaville *et al.* 2010) fed a diet supplemented with hydrolysable (HT) or condensed tannin (CT) which resulted in similar retained N at both experiments. However, effect of tannins on N retention is still contradictory.

Researches carried out in the last decade have highlighted the risk of large losses of N compounds to the environment. A shift in N excretion from urine to feces in goats fed AS + PBE can exert beneficial effects environmentally as urinary N is mainly in the form of urea, which is rapidly convert into ammonia and then to volatile nitrous oxides (green house gases) (Patra and Saxena, 2011). Moreover, nitrate produced by oxidation of ammonia (ammonium) causes water pollution (Eckard *et al.* 2010). Therefore, lower urinary N content means less volatile ammonia and nitrous oxides losses into the environment. In our study, MNS was not affected by treatments, however numerically higher OM intake in goats fed AS + PBE compared with untreated silage (1.90 *vs.* 1.75 kg/d) may lead to a tendency (P<0.10) for decrease in MNS efficiency. Furthermore, free phenolic compounds in crude PBE may negatively affect the activity of ruminal microorganisms and / or interfere with their enzyme secretion (McSweeny *et al.* 2001). Ghasemi *et al.* (2012) reported that substitution of PB at 25% of DM (1.92% dietary tannin) with alfalfa hay in sheep ration had no effect on MNS and efficiency of MNS. However, they observed a significant decrease in both MNS and efficiency of MNS at 50% of DM PB (3.07% tannin). Inclusion of PEG to the goats fed a high tannin containing diet increased urinary excretion of allatoin, thus MNS which was likely the result of improved N bioavailability in the rumen (Salem *et al.* 2005).

Al-Dobaib *et al.* (2009) reported that MNS decreased in goats fed alfalfa hay treated with 3% quebracho tannin, however, supplementing with 1 and 2% tannin slightly increased MNS. They also stated that efficiency of MNS significantly enhanced at all levels of tannin (1, 2 and 3%). Makkar (2003) concluded that low levels of tannins can decrease the rate of digestion of feeds, hence, synchronizing the release of various nutrients resulting in higher microbial protein efficiency.

However, studies are required to know the levels of tannins in order to have this positive response. Different responses of tannins among different studies can be ascribed to tannin concentration, type of tannin, chemical structure and type of diet besides other factors such as animal species and physiological state of the animal (Makkar, 2003; Patra and Saxena, 2011).

CONCLUSION

Presence of TA and PBE at 0.7% DM in diet had no effect on milk yield and composition. Addition of PB tannin extract increased fecal N losses and reduced urinary N and also increased retained N. This shift from urine to feces may have some beneficial effects on environment in the case of reducing N pollution. However, a tendency to lower EMN as a result of PBE addition may indicate that PB tannins can bind strongly with proteins which would need to be counteracted with a competitive agent such as PEG. However, at the same level of tannin, TA had no significant effect on pattern of N utilization, MNS and EMNS in dairy goats compared to PBE indicating the different biological effect of tannins. However, it is suggested to identify the effect of PB on N utilization and EMNS by using purified PB tannins.

ACKNOWLEDGEMENT

Authors would like to thank H. Ghorbani and M. Hosseini Ghaffari for their assistance.

REFERENCES

Ahnert S., Dickhoefer U., Schulz F. and Susenbeth A. (2015). Influence of ruminal Quebracho tannin extract infusion on apparent nutrient digestibility, nitrogen balance and urinary purine derivatives excretion in heifers. *Livest. Sci.* **177**, 63-70.

Al-Dobaib S.N. (2009). Effect of different levels of quebracho tannin on nitrogen utilization and growth performance of Najdi sheep fed alfalfa (*Medicago sativa*) hay as a sole diet. *Anim. Sci. J.* **80**, 532-541.

AOAC. (2005). Official Methods of Analysis. Vol. I. 18th Ed. Association of Official Analytical Chemists, Arlington, VA, USA.

ARC. (1984). The nutrient requirement of ruminant livestock. Supplement No. 1. Commonwealth Agricultural Bureaux, Farnham Royal, England.

Butter N.L. (1999). The effect of condensed tannins and dietary protein on ruminant intestinal nematode infections. Ph D. Thesis. University of Nottingham, Nottingham, UK.

Chen X.B. and Gomes M.J. (1995). Estimation of microbial protein supply to sheep and cattle based on urinary excretion of purine derivatives. Pp. 1-8 in An Overview of the Technical Details. Rowett Research Institute, Occasional Publication, Bucksburn, UK.

Deaville E.R., Givens D.I. and Mueller-Harvey I. (2010). Chestnut and mimosa tannin silages: Effects in sheep differ for apparent digestibility, nitrogen utilisation and losses. *Anim. Feed Sci. Technol.* **157**, 129-138.

Eckard R.J., Grainger C. and de Klein C.A.M. (2010). Options for the abatement of methane and nitrous oxide from ruminant production. *Livest. Sci.* **130**, 47-56.

Frutos P., Hervás G., Giráldez F.J. and Mantecón A.R. (2004). Tannins and ruminant nutrition. *Spanish J. Agric. Res.* **2**, 191-202.

Getachew G., Pittroff W., DePeters E.J., Putnam D.H., Dandekar A. and Goyal S. (2008a). Influence of tannic acid application on alfalfa hay: *in vitro* rumen fermentation, serum metabolites and nitrogen balance in sheep. *Animal.* **2**, 381-390.

Getachew G., Pittroff W., Putnam D.H., Dandekar A., Goyal S. and DePeters E.J. (2008b). The influence of addition of gallic acid, tannic acid, or quebracho tannins to alfalfa hay on *in vitro* rumen fermentation and microbial protein synthesis. *Anim. Feed Sci. Technol.* **140**, 444-461.

Ghaffari M.H., Tahmasbi A.M., Khorvash M., Naserian A.A. and Vakili A.R. (2014). Effects of pistachio by-products in replacement of alfalfa hay on ruminal fermentation, blood metabolites and milk fatty acid composition in Saanen dairy goats fed a diet containing fish oil. *J. Appl. Anim. Res.* **42**, 186-193.

Ghasemi S., Naserian A.A., Valizadeh R., Tahmasebi A.M., Vakili A.R. and Behgar M. (2012). Effects of pistachio by-product in replacement of lucerne hay on microbial protein synthesis and fermentative parameters in the rumen of sheep. *Anim. Prod. Sci.* **52**, 1052-1057.

Givens D.I. and Rulquin H. (2004). Utilisation by ruminants of nitrogen compounds in silage-based diets. *Anim. Feed Sci. Technol.* **114**, 1-18.

Jolazadeh A.R., Dehghan-banadaky M. and Rezayazdi K. (2015). Effects of soybean meal treated with tannins extracted from pistachio hulls on performance, ruminal fermentation, blood metabolites and nutrient digestion of Holstein bulls. *Anim. Feed Sci. Technol.* **203**, 33-40.

Licitra G, Hernandez T.M. and Van Soest P.J. (1996). Standardization of procedures for nitrogen fractionation of ruminant feeds. *Anim. Feed Sci. Technol.* **57**, 347-358.

Makkar H.P.S. (2000). Quantification of Tannins in Tree Foliage. A Laboratory Manual for the FAO/IAEA Co-ordinated Research Project on Use of Nuclear and Related technique to Develop Simple Tannin Assays for Predicting and Improving the safety and Efficiency of Feeding Ruminants on Tanniniferous Tree Foliage. Joint FAO/IAEA, FAO/IAEA of Nuclear Techniques in Food and Agriculture. Animal Production and Health Sub-62 program, FAO/IAEA Working Document. IAEA, Vienna, Austria.

Makkar H.P.S. (2003). Effects and fate of tannins in ruminant animals, adaptation to tannins, and strategies to overcome detrimental effects of feeding tannin-rich feeds. *Small Rumin. Res.* **49**, 241-256.

McSweeny C.S., Palmer B., McNeill D.M. and Krause D.O. (2001). Microbial interaction with tannin: nutritional consequences for ruminants. *Anim. Feed Sci. Technol.* **91**, 83-93.

Mokhtarpour A., Naserian A.A., Valizadeh R., Danesh-Mesgaran M. and Pourmollae F. (2014). Extraction of phenolic compounds and tannins from pistachio by-products. *Ann. Res. Rev. Biol.* **4**, 1330-1338.

Mokhtarpour A., Naserian A.A., Pourmollae F. and Ghaffari M.H. (2015). Effect of treating alfalfa silage with pistachio by-products extract on Saanen dairy goats performance and microbial nitrogen synthesis. *J. Anim. Physiol. Anim. Nutr.* **100**, 758-767.

Muck R.E., Moser L.E. and Pitt R.E. (2003). Postharvest factors affecting ensiling. Pp. 251-304 in Silage Science and Technoogy. D. Buxton, R. Muck and J. Harrison, Eds. ASA-CSSA-SSSA, Madison, Wisconsin.

Naserian A.A., Staples C.R. and Ghaffari M.H. (2015). Effects of replacing wheat bran by pistachio skins on feed intake, nutrient digestibility, milk yield, milk composition and blood metabolites of dairy Saanen goats. *J. Anim. Physiol. Anim. Nutr.* **100**, 259-263.

NRC. (2001). Nutrient Requirements of Dairy Cattle. 7th Ed. National Academy Press, Washington, DC, USA.

Ozdal T., Capanoglu E. and Altay F. (2013). A review on protein-phenolic interactions and associated changes. *Food Res. Int.* **51**, 954-970.

Patra A.K. (2010). Aspects of nitrogen metabolism in sheep-fed mixed diets containing tree and shrub foliages. *Br. J. Nutr.* **103**, 1319-1330.

Patra A.K. and Saxena J. (2011). Exploitation of dietary tannins to improve rumen metabolism and ruminant nutrition. *J. Sci. Food Agric*. **91,** 24-37.

Rajaei-Sharifabadi H. and Naserian A.A. (2014). Growth performance and nitrogen retention in lambs fed diets containing two different levels of crude protein supplemented with pistachio by-product extract as a source of tannins. *Res. Opin. Anim. Vet. Sci.* **4,** 273-280.

Salem H.B., Salem I.B. and Said M.B. (2005). Effect of the level and frequency of PEG supply on intake, digestion, biochemical and clinical parameters by goats given kermes oak (*Quercus coccifera*)-based diets. *Small Rumin. Res.* **56,** 127-137.

SAS Institute. (2001). SAS®/STAT Software, Release 9.1. SAS Institute, Inc., Cary, NC. USA.

Sedighi-Vesagh R., Naserian A.A., Ghaffari M.H. and Petit H.V. (2015). Effects of pistachio by-products on digestibility, milk production, milk fatty acid profile and blood metabolites in Saanen dairy goats. *Anim. Physiol. Anim. Nutr.* **99,** 777-787.

Tabacco E., Borreani G., Crovetto G.M., Galassi G., Colombo D. and Cavallarin L. (2006). Effect of chestnut tannin on fermentation quality, proteolysis, and protein rumen degradability of alfalfa silage. *J. Dairy Sci.* **89,** 4736-4746.

Tedeschi L.O., Cannas A. and Fox D.G. (2010). A nutrition mathematical model to account for dietary supply and requirements of energy and other nutrients for domesticated small ruminants: the development and evaluation of the small ruminant nutrition system. *Small Rumin. Res.* **89,** 174-184.

Tyrrell H.F. and Reid J.T. (1965). Prediction of the energy value of cow's milk. *J. Dairy Sci.* **48,** 1215-1223.

Van Soest P.J., Robertson J.B. and Lewis B.A. (1991). Methods for dietary fiber, neutral detergent fiber and non starch polysaccharides in relation to animal nutrition. *J. Dairy Sci.* **74,** 3583-3597.

8

Influence of Wet and Dry Season on Milk Composition of Dromedary Camels (*Camelus dromedarius*) from Tunisia

H. Hamed[1*], A. El Feki[1] and A. Gargouri[1]

[1] Departement des Science de la Vie, Faculte des Science de Sfax, Unite de Physiopathologie Environnementale, Valorisation des Molecules Bioactives et Modelisation Mathematique, Sfax, Tunisia

*Correspondence E-mail: houdahamed1@yahoo.fr

ABSTRACT

This study investigated the effect of season (wet and dry) on milk composition of the dromedary camel (*Camelus dromedarius*). Milk samples representing the wet and dry seasons were analyzed for total solids (TS), fat (F), solids-non-fat (SNF), ash, crude protein (CP), nitrogen soluble at pH 4.6 (SN), mineral content (MC) and casein fraction (CnF). Camel milk had higher (P<0.05) fat content in the wet season. Crude protein, TS, SNF and SN contents (%) were also higher (P<0.01) during the wet season. Ash content was similar (P>0.05) in the two seasons. The concentrations of Mg, Na and K were affected by season while the concentrations of Ca and Cl were not. The results from the present study indicate that the increase of milk CnF observed in the wet season was not due to the increase in β-casein and α1-casein contents. Among caseins, the κ-casein and α2-casein fractions did not differ between the two seasons. Conversely, milk from summer camels showed higher contents of αs1-casein and β-casein. It was concluded that the effects of season on the chemical composition of camel milk may impact the processing characteristics of camel milk.

KEY WORDS camel milk, casein fractions, chemical composition, season.

INTRODUCTION

Dromedary camel (*Camelus dromedarius*) can survive and produce considerable amount of milk in hot and dry environments (Bekele *et al.* 2011). Thus, camel milk is considered one the most valuable food sources for nomadic people in the arid and semi-arid areas and has been consumed for centuries due to its nutritional values (Kenzhebulat *et al.* 2000; Mal *et al.* 2006; Lorenzen *et al.* 2011). Many factors influence the concentration of major constituents (fat, protein, lactose and minerals) in milk within species. Previous finding pointed out that the variation in camel milk composition could be attributed mainly to geographical origin and seasonal variations (Nagy *et al.* 2013; Konuspayeva *et al.* 2009; Sallam *et al.* 2008).

Some authors have reported seasonal changes (Todorova, 1998; Bertoni *et al.* 2005) on milk protein fractions, but these data are not conclusive in camel milk. The relative proportions of individual components of casein are subject to considerable variation, which can have an effect on to properties of milk during technical processe (Pabst, 1994; Remeuf, 1994). However, limited information is available on camel milk composition under pastoral systems in Tunisia. The objective of this study was to determine the effects of season (wet and dry) on camel milk composition.

MATERIALS AND METHODS

Milk sampling
The study was carried out using individual milk samples

from 36 dromedary camels (*Camelus dromedarius*) of Maghrabi breed from the south and the center of Tunisia. The dromedaries were fed throughout the year exclusively by grazing. Individual samples during early morning milking were collected into sterile bottles between May 2008 and March 2009. The first few streams of milk from each quarter were discarded. Milk samples were obtained by hand milking. Milk samples (about 1000 mL each) were collected and chilled (4 °C) before transferred to the laboratory. At the laboratory, each milk sample was sub-sampled and aliquots were taken for analyses. All physicochemical parameters were determined on the day of sampling. In addition, another aliquot of about 100 mL of milk sample was taken and stored at -20 °C for further analysis on protein fraction. For each milk sample, all of the analytical assessments were carried out in duplicate.

Chemical composition determinations

Milk fat (F) content was determined using Gerber method (International Dairy Federation, 1981). Total solid (TS) contents were determined using the forced draft oven method (Marshal, 1993). Ash content was determined burning away all the organic matter at 550 °C in a muffle furnace (Marshall, 1993). SNF % was calculated by subtracting the fat % from TS % and calculated according to the following equation:

$$SNF \% = TS \% - fat \%$$

The mineral content was determined in an autoanalyzer (SYNCHRON CX9 ALX system, Beckman Coulter Inc (ref:442790).

Separation of milk nitrogen and protein fractions

The pH 4.6-insoluble fraction containing the isoelectric caseins was prepared by precipitation of milk with 1 M sodium acetate buffer (pH 4.6) followed by centrifugation at 6000 rpm and 5 °C for 15 min. The casein pellet recovered was first washed three times with 1 M sodium acetate buffer (pH 4.6) and then, to remove the remaining fat and other low density components, they were washed twice with a mixture of sodium acetate buffer and dichloromethane (1:1, v/v). The final protein precipitate was then lyophilized before analysis.

The pH 4.6-soluble nitrogen fraction (SN) was filtered through filter paper (Whatman No. 1) and kept frozen until used. TN and SN fractions were determined in triplicate by the Dumas method (International Dairy Federation, 2002). Casein nitrogen (CnN), crude protein (CP) and Casein (Cn), expressed as g per 100 mL of milk, were calculated as follows according to Ribadeau-Dumas and Grappin (1989):

$$CnN = TN - SN$$
$$CP = TN \times 6.38$$
$$Cn = CnN \times 6.36$$

Casein as percentage of CP was calculated as: (Cn/CP) × 100. The proportion Cn/CP was used as an index of proteolysis (Ma *et al*. 2003).

Reversed phase-HPLC casein analysis

Whole casein from individual dromedary milks was separated in duplicate by reverse phase HPLC according to the procedure developed by Alim *et al*. (2005) with a Summit × 2 dual gradient HPLC system (Dionex, Indstein, Germany). Casein samples were reduced for 1 h at room temperature in a sample buffer containing 1 ml of 8 M urea, 0.1 M Bis-Tris, 0.3% mercaptoethanol and 1.3% sodium citrate. Reduced samples were diluted (1:5, v/v) with 6 M urea and 0.1% trifluoroacetic acid. Samples (20 µL) previously filtered through a 0.45 µm filter were injected into a C18-bonded silica gel (250 mm×4.6 mm) with a particle diameter of 5 µm and pore width of 300 nm (Europa Protein, Teknokroma, St Cugat, Spain), at a constant temperature of 46 °C. The mobile phase consisted in 0.1% trifluoroacetic acid in ultrapure water (solvent A) and 0.1% trifluoroacetic acid in acetonitrile (solvent B). For casein separation, elution was achieved with a linear gradient from 33% to 49% of solvent B in 35 min at a flow rate of 1 mL/min, and the eluted peaks were detected by UV absorbance at 220 nm. Data were processed with the chromatographic system"s software ChromoLion (Dionex) and the percentage of each casein fraction was determined.

Hydrolysis of dromedary casein by chymosin

Solutions of dromedary casein (1%, w/v) in 50 m*M* sodium acetate buffer at pH 6.6 containing 0.02% thimerosal to prevent microbial activity were treated with chymosin (180 International Milk Clotting Units mL−1, Maxiren 180, DSM Food Specialties, Seclin Cedex, France) at a level of 0.1% (v/v). The solutions were rotated (13 rpm) at 30 °C during 30 and 60 min. At the end of each period, chymosin was inactivated by heating (90 °C, 5 min) and the pH was lowered to 4.6. After centrifugation (4500 g, 15 min), the pellets were redissolved in the chromatographic sample buffer and aliquots were taken for chromatographic separation.

Statistical analysis

Statistical treatments of data were performed using SPSS software (version 13). Data were arranged according to two seasons; dry season (Aug-Oct) and wet season (Nov-Jan). Data were analyzed by one-way analysis of variance

(ANOVA). The differences among the means of the analysis data were compared at a significance level of (P<0.05).

RESULTS AND DISCUSSION

The fat content in camel milk was lower (P<0.05) in the dry season (Table 1).

Table 1 Composition of camel milk in the wet and dry seasons (Mean±SE)

Milk constituents	Wet season (n=16)	Dry season (n=20)	P-value
Fat (%)	4.64±0.31	3.25±0.20	0.029
TS (%)	12.52±0.64	10.24±0.32	0.002
Ash (%)	0.50±0.05	0.60±0.24	0.213
SNF (%)	7.88±0.33	6.99±0.12	0.011
CP (g/100 mL)	2.61±0.06	2.36±0.03	0.004
TN (g/100 mL)	0.41±0.01	0.37±0.006	0.721
SN (g/100 mL)	0.14±0.01	0.11±0.002	0.003
CnN (g/100 mL)	0.27±0.00	0.26±0.004	0.045

TS: total solids; SNF: solids-non-fat; CP: crude protein (N×6.38); TN: total nitrogen; SN: nitrogen soluble at pH 4.6 and CnN: casein nitrogen.
SE: standard error.

This might be due to nutritional status of the animals during the wet season, where feeds are more easily available and are richer in crude protein, carbohydrates, minerals and vitamins. Our results are consistent with those reported by Sevi *et al.* (2004), suggesting that reduction in fat content of milk, probably is a consequence of a greater secretion of prolactin whose concentration in plasma is higher in the summer than in the winter.

Total solid content was highest (P<0.05) in the wet season and decreased significantly during the dry season, which is in agreement with Elvan and Sebnem (2008). This might be attributed to the reason that camels during hot seasons provides milk with lower total solid because the calves needs more fluids (Shuiep *et al.* 2008). The SNF content in camel milk was lower (P<0.05) in the dry season. Sharma *et al.* (2002) confirmed the effect of seasons on SNF content, who found that SNF content varied among seasons being highest in winter (8.98%) followed by summer (8.84%).

There was no significant correlation between ash content and season (Table 1). Similar results were expressed by Biye *et al.* (2014) and Rao and Mishra (2010), who reported that ash content was not significantly influenced by season. The mean CP content of the camel milk was 2.93 (g/100 mL) for the wet season and 2.36 (g/100 mL) for the dry season. Similarly, Haddadin *et al.* (2008) found that the CP content is lowest in August (2.48%) and highest in December and January (2.9%). The variation in fat and protein correspond to the data given in the literature. Seasonal patterns in the fat and protein have been described in different

countries and under different management practices. In the present study, number of important camel milk constituents (i.e. F, CP, TS, SNF, SN and CnN) showed the highest mean values in wet season and the lowest in dry season. It is attributed to the fact that the green fodder is available in the south and the center of Tunisia during rainy season. Higher Mg was obtained during dry season compared to the wet season (4.92 *vs.* 2.26 mmol/L, Table 2).

Table 2 Variation in mineral content (mmol/L) of camel milk during the wet and dry seasons (Mean±SE)

Mineral	Wet season (n=16)	Dry season (n=20)	P-value
Mg	2.26±0.33	4.92±0.41	0.000
Cl	61.12±0.09	61.93±3.82	0.891
K	63.59±2.99	52.4±2.55	0.010
Na	30.57±0.66	33.53±1.18	0.028
Ca	10.47±0.20	10.15±0.65	0.125

Mg: magnesium; Cl: chloride; K: potassium; Na: sodium and Ca: calcium.
SE: standard error.

During the dry season, the average of Na content was found to be significantly higher (P<0.05) than that of the wet season while K was the opposite being significantly lower (P<0.01) in the dry season. The difference in milk mineral concentrations between seasons is due to a "dilution effect" as reported by Guler (2007), which is related to animal feeding behaviour and changes in pasture composition.

In the present study, the reduction in CP contents of camel milk during the dry season was mainly due to the reduction in casein content (Table 1). The reduction of casein content in summer milk has also been reported by other authors (Hermansen *et al.* 1999; Mackle *et al.* 1999). β-casein and α1-casein content were lower (P<0.05) in the wet season, while no difference was found for κ-Cn and α2-Cn between seasons (Figure 1).

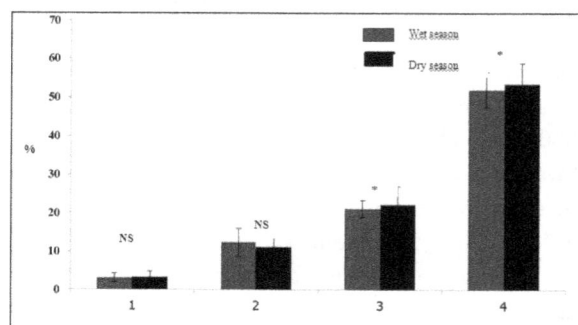

Figure 1 Mean±SD of casein fraction in the wet and dry season in camel milk (1: κ-Cn; 2: α2-Cn; 3: α1-Cn and 4: β-Cn)
Values are the mean ± SE
NS: non significant
* (P<0.05)

However, Kroeker *et al.* (1985) observed no definitive seasonal trend for the relative percentage of casein fractions.

However others authors (Bernabucci *et al.* 2002) showed that β-CN and α1-CN decreased during summer in cow's milk. Our results suggest that decrease in αS1 and β-casein contents may cause the poor cheese making properties of wet season. The results of the present study clearly demonstrated that chemical composition of camel milk is affected by the season. It could, therefore, be concluded that camel milk composition is a reflection of seasonal changes in quality and availability of feed as well as parity differences. However, more work is needed to verify these effects and also to study the effects of management and breed differences on milk composition. Our study showed that the casein fractions were affected by season. For this reason, during the rainy season, it is necessary to prevent deterioration of the quality of milk that can affect the yield and quality of cheese.

CONCLUSION

The results of the present study clearly demonstrated that chemical composition of camel milk is affected by the season. It could, therefore, be concluded that camel milk composition is a reflection of seasonal changes in quality and availability of feed as well as parity differences. However, more work is needed to verify these effects and also to study the effects of management and breed differences on milk composition. Our study showed that the casein fractions were affected by season. For this reason, during the rainy season, it is necessary to prevent deterioration of the quality of milk that can affect the yield and quality of cheese.

ACKNOWLEDGEMENT

This work was supported by Grants from the Tunisian Ministry of Higher Education, Scientific Research and Technology.

REFERENCES

Alim N., Fondrini F., Bounizi I., Felligini M. and Enne G. (2005). Characterisation of casein fractions from Algerian dromedary (*Camelus dromedarius*) milk. *Pakistan J. Nutr.* **4**, 112-116.

Bekele T., Lunderheim N. and Dahlbron K. (2011). Milk feeding and feeding behaviour in the camel (*Camelus dromedarius*) during 4 watering regimens. *J. Dairy Sci.* **94**, 1310-1317.

Bernabucci U., Lacetera N., Ronchi B. and Nardone A. (2002). Effects of the hot season on milk protein fractions in Holstein cows. *Anim. Res.* **51**, 25-33.

Bertoni G., Calamari L., Maianti G. and Battistotti B. (2005). Milk for protected denomination of origin (PDO) cheeses: I. The main required features. Pp. 217-228 in Indicators of Milk and Beef Quality. J. F. Hocquette and S. Gigli, Ed. EAAP Pubblication No. 112, Wageningen Academic Publishers, Wageningen, the Netherlands.

Biye C., Michael J. and Lewis Alistair S. (2014). Effect of seasonal variation on the composition and properties of raw milk destined for processing in the UK. *Food Chem.* **158**, 216-223.

Elvan O. and Sebnem S. (2008). The effect of seasonal variation on the composition of cow milk in Van province. *Pakistan J. Nutr.* **7**, 161-164.

Guler Z. (2007). Levels of 24 minerals in local goat milk, its strained yoghurt and salted yoghurt (tuzlu yogurt). *Small Rumin. Res.* **71**, 130-137.

Haddadin M.S.Y., Gammoh S.I. and Robinson R.K. (2008). Seasonal variations in the chemical composition of camel milk in Jordan. *J. Dairy Res.* **75**, 8-12.

Hermansen J.E., Ostersen S., Justesen N.C. and Aaes O. (1999). Effects of dietary protein supply on caseins, whey proteins, proteolysis and renneting properties in milk from cows grazing clover or N-fertilized grass, *J. Dairy Res.* **66**, 193-205.

International Dairy Federation. (2002). Milk and Milk Products. Determination of Nitrogen Content. Routine Method Using Combustion According to the Dumas Principle. Geneva (Switzerland), IDF Standard Publisher, Brussels, Belgium.

International Dairy Federation. (1981). Milk Determination of Fat Content. Gerber Butyrometers, Brussels, Belgium.

Kenzhebulat S., Ermuhan B. and Tleuov A. (2000). Composition of camel milk and its use in the treatment of infectious diseases in human. Pp. 101 in Proc. 2nd Camelid Conf. Agroecon. Camelid Farming, Almaty, Kazakhstan.

Konuspayeva G., Faye B. and Loiseau G. (2009). The composition of camel milk: A meta-analysis of the literature data. *J. Food Comp. Anal.* **22**, 95-101.

Kroeker E.M., Ng-Kwai-Hang K.F., Hayes J.F. and Moxley J.E. (1985). Effects of environmental factors and milk protein polymorphism on composition of casein fraction in bovine milk. *J. Dairy Sci.* **68**, 1752-1757.

Lorenzen P., Wernery R., Johnson B., Jose S. and Wernery U. (2011). Evaluation of indigenous enzyme activities in raw pasteurized camel milk. *Small Rumin. Res.* **97**, 79-82.

Ma Y., Barbano D. and Santos M. (2003). Effect of CO_2 addition to raw milk on proteolysis and lipolysis at 4 ˚C. *J. Dairy Sci.* **86**, 1616-1631.

Mackle T.R., Bryant A.M., Petch S.F., Hill J.P. and Auldist M.J. (1999). Nutritional influences on the composition of milk from cows of different protein phenotypes in New Zealand. *J. Dairy Sci.* **82**, 172-180.

Mal G., Sena D.S., Jain V.K. and Sahani M.S. (2006). Therapeutic value of camel milk as a nutritional supplement for multiple drug resistant (MDR) tuberculosis patients. *Israel J. Vet. Med.* **61**, 88-91.

Marshall R.T. (1993). Standard Methods for the Examination of Dairy Products. American Public Health Association, Washinton, DC.

Nagy P., Faye B., Marko O., Thomas S., Wernery U. and Juhasz J. (2013). Microbiological quality and somatic cell count in bulk milk of dromedary camels (*Camelus dromedarius*) descriptive statics, correlations and factors of variation. *J. Dairy Sci.* **96**, 1-5.

Pabst K. (1994). Die bedeuntung von milchprotein-varianten für die herstellung von käse. *J. Milch. Forsch.* **46**, 263-274.

Rao A. and Mishra S. (2010). An assessment of the nutritional profile of milk in different seasons and locations in Varanasi through modern laboratory techniques. *Indian J. Prev. Soc. Med.* **41**, 237-239.

Remeuf F. (1994). Relation entre caractéristiques physico-chimiques et aptitudes fromagères des laits. *Rec. Med. Vet.* **170**, 359-365.

Ribadeau-Dumas B. and Grappin R. (1989). Milk protein analysis. *Lait.* **69**, 357-416.

Sallam A.B., Ali M.A.M. and Abdel Moneim M.A.N. (2008). Camels (*Camelus dromedarius*) under pastoral systems in North Kordofan, Sudan: Seasonal and parity effects on milk composition. *J. Camel. Sci.* **1**, 32-36.

Sevi A., Albenzio M., Marino R., Santillo A. and Muscio A. (2004). Effects of lambing season and stage of lactation on ewe milk quality. *Small Rum. Res.* **51**, 251-259.

Sharma R.B., Manish K. and Pathak V. (2002). Effect of differents seasons on Cross bred cow milk composition and Paneer yield in Sub-Himalayan Region. *Asian-Australas J. Anim. Sci.* **15**, 528-530.

Shuiep E.S., El Zubeir I.E.M., El Owni O.A.O. and Musa H.H. (2008). Influence of season and management on composition of raw cow camel (*Camelus dromedarius*). *Trop. Subtrop. Agroecosys.* **8**, 101-106.

Todorova D. (1998). Influence of the diet and season alteration on the cow's milk composition and properties. *Belgium J. Agric. Sci.* **4**, 525-530.

9

An Evolutionary and Phylogenetic Study of the *BMP15* Gene

M. Mahdavi[1,2] **and G.R. Dashab**[1*]

[1] Department of Animal Science, Faculty of Agriculture, University of Zabol, Zabol, Iran
[2] Razi Vaccine and Serum Research Institute, Karaj, Iran

*Correspondence E-mail: dashab@uoz.ac.ir

ABSTRACT

DNA sequence data contains a wealth of biologically useful information. Recent innovations in DNA sequencing technology have greatly increased our capacity to determine massive amounts of nucleotide sequences. These sequences can be used to specify the characteristics of different regions, interpret the evolutionary relationships between categorized groups, likelihood of performing multiple comparisons between an unidentified sequence and millions of specified sequences existing in DNA information gene banks. This study focuses on bioinformatics analysis of molecular genetic variation of the *BMP15* gene of some selected mammalian species with a view to providing relevant genetic information for breeding and selection programs in the studied species using computational methods. A total of twenty-three *BMP15* nucleotide sequences and amino acids sequences were retrieved from the *NCBI* gene bank. Sequence alignment, pair and multi-alignment comparison of the *BMP15* gene of the various species were done with *MEGA6*. High degree of polymorphism of the *BMP15* gene was observed among the studied species. Despite high polymorphism in *BMP15* similarity between sequences was greater than 98% and differences existing in fertility rates among species may be species-specific due to differences in the processing of *BMP15*. The mean diversity among populations and the scattering coefficient of evolution in the *BMP15* position were calculated to be 2.15 and 1.25 bp, respectively and the mean divergence was calculated to be 1.7 bp. Also, the mean divergence within the species was 4.1, 0.4 and 2.1 bp in sheep, cows and goats, respectively. Bases substitution rate and pattern in *BMP15* showed that cows and goats have the minimum genetic distance and human has the maximum genetic distance with sheep. In addition to existing variation in DNA and amino acids sequences between *BMP15* in different species, this study highlighted specific segments significantly more conserved in *BMP15* from mono-ovulating as compared to super-ovulating species.

KEY WORDS *BMP15*, phylogenetic, transition, transversion.

INTRODUCTION

In recent years, with the development of genetic engineering technology, rapid DNA sequencing has been made possible and a massive amount of nucleotide sequence information from different parts of the genome in different species is growing at a rapid pace. Deoxyribonucleic acid (DNA) sequence data contains a wealth of biologically useful information. The analysis of these sequences can be used by comparing the sequences of the genes with identified functions to specify the characteristics of different regions, such as active sites and functional regions of the DNA to interpret the evolutionary relationships between categorized groups (phylogenetic) and comparing an unidentified sequence with millions of sequences available in DNA information gene banks. Also, new and increasing reports on the sequences of the DNA have provided and created conditions as well as a motivation for evolutionary investigations using comparative studies. Therefore, it would be very desirable to use the statistical methods to estimate the evolutionary distance between similar sequences and the number of nucleotide substitutes. Estimating evolutionary distances between protein and DNA sequences is very important to form a phylogenetic tree, knowing the time and origin of the divergence and branching of species, and understanding the mechanism of the evolution of genes, proteins, and populations.

To date, three categories of genes affecting the growth of follicles and ovulation rate have been identified, including activin receptor-like kinase 6 (ALK6), growth and differentiation factor-9 (GDF9) and bone morphogenetic proteins (BMP) category, which is the most famous of bone morphogenetic protein (*BMP15*) gene. All these genes are among the transforming growth factor β (TGF-β) large family and they affect the regulation of the expression and secretion of the hormones affecting the follicular growth and ovulation rate. The growth factors *BMP15*, *BMP6* and growth differentiation factor 9 (*GDF9*) are produced by ova (McNatty *et al.* 2005), while the receptors BMP type A1, A2 and type 2 are located on ova and somatic cells of ovarian follicles (Souza *et al.* 2002) and also on the cells of the pituitary gland of sheep.

The *BMP15* gene is located on chromosome X and consists of two exons, that are separated from each other by an intron with a length of 5.4 kb. Their full transcription product is a sequence of 1179 nucleotides which encodes a prepeptide with a length of 393 amino acids and its full peptide has a length of 125 amino acids (Galloway *et al.* 2002). The study of the above positions in rodents has shown that mRNA transcribed from the position of *BMP15* and the protein resulting from its expression are found in the early stages of ovulation in the oocyte cell (Dube *et al.*1998; Otsuka *et al.* 2001). Also, the *BMP15* mutations cause a reduction in the amount of mature protein or a change in binding to the receptors on the cell surface. Thus, investigating the phenotype resulting from the mutation of the *BMP15* locus shows that the presence of certain haplotypes is necessary for ovarian follicular evolution, normal ovulation, and formation of the corpus luteum in sheep. Besides, reports have shown that mRNA and the *BMP15* protein exist at all stages of ovarian follicles of goats (Silva *et al.* 2004).

To date, eight mutations have been identified in the gene encoding this factor in sheep, namely, Inverdal, Hanna, Belclare and Galway, Lacaune, Rasa Aragonesa, Grivette and Olkuska according to the name of the breed of the sheep in which these mutations are identified for the first time. The FecX[I] allele in sheep is related to a thymine to adenine change at position 896 cDNA encoding the *BMP-15* factor. The FecX[H] allele is related to a cytosine to thymine change at position 871, which leads to the creation of a stop codon in amino acid 23 of the complete protein and this stop codon also leads to the loss of the biological activity of the *BMP15* factor (Fabre *et al.* 2006). Additionally, the FecX[G] allele results from a thymine to adenine change at position 718. The FecX[G] mutation leads to the creation of a stop codon in amino acid 239 of the protein.

Thus, the processed protein is not produced. The FecX[B] allele results from changing guanine to thymine at nucleotide 1100 and in this case, the change of amino acid serine to isoleucine occurs at position 99 of the protein sequence (Table 1) (Fabre *et al.* 2006). In Lacaune breed, the mutant allele (FecX[L]) associated with high prolificacy and was identified as a Cys321Tyr substitution altering the *BMP15* protein function (Drouilhet *et al.* 2013). Both FecX[Gr] and FecX[O] mutations are closely located into two very well conserved domains of the sheep, cow, pig, human and mouse *BMP15* proteins, FecX[Gr], which corresponds to a substitution of a threonine to an isoleucine, clearly affected the hydrophobicity of the protein while FecX[O] altered the polarity and the molecular weight of the protein by replacing an asparagine to a histidine. These two mutations clearly affect he intrinsic properties of the *BMP15* protein since they correspond to substitutions of polar amino acids by non-polar and basic amino acids suspected to modify consequently its three-dimensional structure (Demars *et al.* 2013). The recombinant *BMP15* gene increases the proliferation of granulosa cells in mice and humans. Moreover, *BMP15* in granulosa cells potentially stimulates the mRNA encoding the kit messenger (a factor necessary for the growth of ovum in primary follicles). Therefore, both *BMP15* and kit play an important role in the early growth of follicles. Also, *BMP15* is able to control the production of steroids. In fact, *BMP15* in mice selectively controls the biological effects of follicle-stimulating hormone (FSH) on granulosa cells, by inhibiting the production of the FSH resulting from the production of progesterone and with no effects on the FSH resulting from estradiol synthesis. The fundamental mechanism points to the control of the negative feedback of the FSH receptor which leads to the prevention of the accumulation of the mRNA resulting from multiple genes related to FSH, such as genes steroidogenic acute regulatory protein (SARP), 3ß-hydroxysteroid dehydrogenase (P450scc), P450 side chain cleavage enzyme - (3β-HSD) the luteinizing hormone (LH) receptor and inhibitor/activator subunits.

BMP15 in sheep increases the proliferation rate of granulosa cells and prevents the secretion of the baseline FSH and the FSH resulting from the progesterone of the granulosa cells of small anterior follicles (Fabre *et al.* 2006). Totally the effect of sheep variants seems tightly related to the kind of mutations described.

Table 1 The mutations reported in the locus of the *BMP15* gene

Mutation	Locus (chromosome)	Allele	Position of change in nucleotide sequence	Change in protein structure	Breed
Inverdale	BMP-15 (X)	FecX^I	T 896 A	V299D / V31D	Romney
Hanna	BMP-15 (X)	FecX^H	T 871 C	Q291stop / Q23stop	Romney
Belclare	BMP-15 (X)	FecX^B	T1100G	S3671 / S991	Belclare
Galway	BMP-15 (X)	FecX^G	T 718 C	T239stop / no	Belclare, Cambridge
Lacaune	BMP-15 (X)	FecX^L	G 803 A	C321Y / C53Y	Laucane
Rasa aragonesa	BMP-15 (X)	FecX^R	17 nt deletion (525-541)	-	Rasa aragonesa
Grivette	BMP-15 (X)	FecX^{Gr}	C 950 T	T317I	Grivette
Olkuska	BMP-15 (X)	FecX^O	A 1009 C	N337H	Olkuska

Indeed, 3 out of the 8 mutations identified so far are amino acids deletion FecX^R or premature stop codon (FecX^G and FecX^H) in the *BMP15* sequence impairing consequently the production of the *BMP15* active form (Monteagudo *et al.* 2009).

Thus, the objective of the present study is the bioinformatics analysis in order to analyze sequences, including searching in existing information gene banks, matching sequences, and estimating evolutionary distances and structure of the phylogenetic tree in species, such as sheep, mice, cows, goats, guinea pigs, humans, pigs, and other species, using partial and total sequences of the *BMP15* gene available in *NCBI* gene bank.

MATERIALS AND METHODS

The analysis method and DNA information gene banks

One of the most informative methods used in sequence data analysis is similarity searching. For DNAs, similarity at the sequence level implies some structural or functional similarity between the protein products or regulatory elements of gene expression.

Searching a database with an uncharacterized gene sequence can identify homologues in other species or sequence elements that encode structural domains within the protein. Searches can be conducted with either nucleotide or peptide sequences. However, detection of similarity at the nucleotide level is difficult unless the sequences are closely related. For analysis of coding DNAs, similarity searching with the translated protein sequence is more informative.

A commonly used tool for similarity searching is BLAST (Basic Local Alignment Search Tool) because of its practical balance of speed, sensitivity and selectivity. In the present study, 23 sequences of the *BMP15* gene, including mRNA and DNA, were taken from the *NCBI* information gene bank, and using the tool BLAST on the website http://ncbi.nlm.nih.gov, similar sequences, their similarities (both nucleotide and protein) and possible mutations were investigated (Table 2).

Comparison among sequences and knowing the genetic parameters, such as the number of mutations, nucleotide diversity, the number of positions in which similar substitutions have taken place as well as their diversity, were determined using Dnasp v5 software program (Librado and Rozas, 2009) and determining the alignment of the sequence of the *BMP15* gene with sequences of other organisms was done using MEGA6 (Tamura *et al.* 2013) software program.

Phylogeny and determination of evolutionary direction
In order to draw the phylogenetic tree, the protein sequence of the *BMP15* gene was predicted for the species being studied using MEGA6 software program (Tamura *et al.* 2013). After editing the sequences and deleting the noncoding regions, the phylogenetic tree was drawn using neighbor-joining (NJ) method. In this method, a matrix (Q) was used, so that in this matrix, all the branches are used and the lowest value which represents high similarity between two branches will be selected and employed in a branching of the phylogenetic tree. Bootstrap values were obtained through 100 times of re-sampling. The phylogenetic tree was drawn using NJ method and equation 1.

1) $Q(i, j) = (r - 2)d(i, j) - \Sigma d(i, k) - \Sigma d(j, k)$

Where:
d(i, j): distance between branches i and j.
k: k-th branch of the tree.
r: total number of the branches.
Q(i, j): numerical value of branches i and j.

Also, the maximum composite likelihood method was used to obtain the succession to substitution rate of nucleotides in purine and pyrimidine bases. Investigating the nucleotide changes that have changed amino acids (d_N) in relation to the nucleotide changes that have not affected the resulting amino acid (d_S), is a highly efficient and useful method for detecting the trend of natural selection for genes during evolution.

Table 2 Characteristics of DNA sequences used for the bioinformatics analysis of the *BMP15* gene

Number	Accession number	Spices/genus	Length (bp)
1	AY572412.1	*Bos taurus* (*BMP15*) mRNA	1185
2	AY572412.1	*Bos taurus* (*BMP15*) mRNA complete cds	1185
3	XM_010821881.1	*Bos taurus* (*BMP15*) transcript variant X1 mRNA	3066
4	DQ463368.1	*Bos taurus BMP15* mRNA complete cds	1189
5	JN655670.1	*Capra hircus* (*BMP15*) gene complete cds	6648
6	NM_001285588.1	*Capra hircus* (*BMP15*) mRNA	1230
7	JN655670	*Capra hircus* breed Beetal (*BMP15*) gene complete cds	6648
8	EU888137.1	*Capra hircus* breed Black Bengal (*BMP15*) mRNA complete cds	1182
9	EU847289.1	*Capra hircus* breed Boer *BMP 15* mRNA complete cds	1185
10	FJ429281.1	*Capra hircus* breed Guizhou White *BMP 15* precursor (*Bmp15*) gene complete cds	1436
11	JF824149.1	*Capra hircus* breed Lezhi black goat *BMP15* mRNA complete cds	1230
12	JN655669.1	*Capra hircus* breed Teddy *BMP15* gene complete cds	6648
13	XM_004999940.1	*Cavia porcellus Bmp15* mRNA	1224
14	NM_005448.2	*Homo sapiens BMP15* mRNA	1262
15	NM_009757.4	*Mus musculus BMP15* mRNA	3086
16	BC055363.1	*Mus musculus BMP15* mRNA (cDNA clone MGC:60557 IMAGE:30051425) complete cds	3059
17	AF082348.1	*Mus musculus BMP15* precursor (*Bmp15*) mRNA complete cds	1603
18	NM_001114767.1	*Ovis aries BMP15* mRNA	1182
19	EU402923.1	*Ovis aries BMP15* mRNA partial cds	260
20	JN655671.1	*Ovis aries* breed Lohi bone morphogenetic protein 15 (*BMP15*) gene complete cds	6638
21	KF114861.1	*Sus scrofa BMP15* gene promoter region and 5 UTR	2162
22	NM_001005155.1	*Sus scrofa BMP15*mRNA	1194
23	HQ450759.1	*Sus scrofa BMP15* mRNA complete cds alternatively spliced	1298

Thus, using the numerical value of this ratio (d_N/d_S), the trend of natural selection was identified for the *BMP15* gene. d_N and d_S values can be calculated using equations 2 and 3, respectively.

2) $d_N = -3/4Ln(1-4/3P_N)$
3) $d_S = -3/4Ln(1-4/3P_S)$

Where:

P_N: ratio of the positions with non-similar substitutions.
P_S: ratio of the positions with similar substitutions.

The significance of the numerical value of d_N/d_S was investigated using Fisher's test at 5% likelihood level.

RESULTS AND DISCUSSION

The mean inter-population genetic diversity was calculated to be 2.15 bp using the maximum composite likelihood method. Also, the scattering coefficient of evolution, which is another indicator for measuring the inter-population diversity, was estimated to be 1.25 bp using the number of base pair of nucleotides, Nei and Kumar (2000) and Tamura (2004) methods. The mean divergence between all sequence pairs was calculated to be 1.7, which represents the number of base substitutions at each site and an average of all sequence pairs. Divergence is defined using the average number of base substitutions at each site, between all

the base pairs within the groups. This divergence was estimated to be 0.4 bp in cows, 2.1 bp in goats, zero in mice, 1 in pigs and 4.1 in sheep. Also, the mean distance between species, was estimated according to the sequence data taken from the *NCBI* gene bank, using the maximum composite likelihood method, which is based on the number of base pairs (Table 3).

Bases substitution rate and pattern were estimated using tamura-nei model (Table3). These presented rates express the likelihood of the replacement of each base with another one. These likelihoods have been estimated according to the sequences being analyzed and changes in the sequence of bases (Table 4).

As shown in Table 3, the maximum distance was observed between sheep and human and mouse while there was the lowest value between sheep and cow and goat. When each of the bases is evaluated, the probability of base substitution (r values) related to each of them should be considered. To simplify this issue, the sum of the likelihood values has been considered to be 100 (Tamura *et al.* 2004; Yang and Kumar, 1996).

The bioinformatics analysis with 23 nucleotide sequences at 221 positions, after excluding missing data and checking for deleted distance showed 22.27%, 25.85%, as 24.61%, and 27.27% nucleotide frequency for adenine, thymine/uracil, cytosine and guanine, respectively.

The divergence between species was estimated using pair comparison method and the number of bases.

Table 3 The distance between species using the number of base pairs in the locus of the *BMP15* gene

Spices/genus	Bos taurus	Capra hircus	Cavia porcellus	Homo sapiens	Mus musculus	Ovis aries	Sus scrofa
Bos taurus	-	-	-	-	-	-	-
Capra hircus	0.1	-	-	-	-	-	-
Cavia porcellus	0.3	0.6	-	-	-	-	-
Homo sapiens	0.2	0.6	0.3	-	-	-	-
Mus musculus	0.3	0.5	0.5	0.3	-	-	-
Ovis aries	0.4	-0.4	0.6	0.9	0.9	-	-
Sus scrofa	0.1	0.3	0.4	0.2	0.3	0.4	-

Table 4 The probability of substitution (*r*) from one base (row) to another base (column)[1,2]

Wild allele mutant allele	A	T/U	C	G
A	-	*6.21*	*5.91*	**13.79**
T/U	*5.35*	-	**13.12**	*6.55*
C	*5.35*	**13.78**	-	*6.55*
G	**11.27**	*6.21*	*5.91*	-

[1] Each number in the table, the likelihood of replacement (r) of one base (row) with another base (columns).
[2] The substitution rates of the bases from the same family (the replacement of a purine with a purine, or a pyrimidine with a pyrimidine) are shown in a diagonal (**BOLD**), and the transition rates of the bases from different families, are shown in *Italics*.

The numbers presented in detail in Table 5 are the number of base substitutions in the each site between sequences. The maximum divergence is 5 bp, in related to the comparison between sheep and cows.

MEGA6 (Tamura *et al.* 2013) is a commonly used program for multiple sequence alignment. It uses a progressive algorithm to align sequences in successively larger groups, beginning with the most closely related sequences. Using MEGA6, 23 sequences being studied are compared and a tentative measure of similarity is derived, represented by a distance matrix. This is used to produce a phylogenetic guide tree (Figure 1), using the neighbour-joining (NJ) method (Saitou and Nei, 1987). The branching pattern of the tree is used to determine the most closely related pair of the sequences. A final alignment is obtained by repeating this procedure until it reaches the root of a tree. The resulting molecular phylogenetic tree represents two main branches of the phylogenetic relationships between the sequences. The final nodes (leaves) of the tree represent the existing sequences and refer to the practical units of classification, while the internal nodes represent hypothetical ancestor sequences. Presented tree is a branching, that is, each node creates two branches, each of which represents the occurrence of a specific event or differences between *BMP15* in studied sequences. Comparative studies of sequences were used in a wide range of taxonomic levels, to evaluate phylogenetic relationships. Results showed different regions and intragenic distances of the DNA varied among species within a *BMP15* sequences. Despite some similarity between sequences, phylogenetic tree and genetic scattering rate showed distances among species in *BMP15s*. The phylogeny results of a recent study (Bwaseh *et al.* 2016) based on nucleotide and amino acid sequences of

BMP15 showed a similar clustering of sequences among the various species with those obtained in this study, although there was some intermingling between the species. An investigation on the sequences of the *BMP15* gene in the species being studied carried out using MEGA6 (Tamura *et al.* 2013) and BioEdit software programs as well as basic local alignment search tool (BLAST) showed a lot of similarity among the species being studied in the locus of the *BMP15* gene. One of the *BMP15* gene sequences belong to sheep was set as query sequence and then in BLAST output, results of other 22 sequences were compared with query sequence. In Figures 2 to 5, most similar sequences were shown. In many cases, the similarity was 100 and in all cases, this similarity was greater than 98%. The result of comparing similarity, were rank first for Max Score/Total Score and the least E-values respectively.

Score of the pairwise comparison between query DNA sequence and the desired DNA sequence in the NCBI database was calculated as fellow: +2 for each match; -1 for each mismatch, and -2 for a gap). Higher scores mean better alignments. In Figure 3, the sequence with the maximum similarity and the relevant E-values are shown in order, from top to bottom.

The results of some two by two comparisons at some sites, using the BLAST tool are shown in Figures 6 to 8. Despite many similarities, the sequences at some sites had base variation and replacement too.

In Figure 4, there are some various deletions in some regions of the gene in some sequences. As shown, most of mRNA sequences containing a deletion and sequences without deletion are DNA sequences. These deletions are mainly related to introns which were usually happened by splicing in RNA processing.

Table 5 The estimate of the divergence between the sequences of the *BMP15* gene in domesticated species of animals

Spices/genus[1]	1	2	3	4	5	6	7	8	9	10	11	12	13	14	15	16	17	18	19	20	21	22	23
1 *Bos taurus* (*BMP15*) mRNA	-	-	-	-	-	-	-	-	-	-	-	-	-	-	-	-	-	-	-	-	-	-	-
2 **Bos taurus (BMP15) mRNA complete cds**	0.0	-	-	-	-	-	-	-	-	-	-	-	-	-	-	-	-	-	-	-	-	-	-
3 *Bos taurus* (*BMP15*) transcript variant X1 mRNA	0.6	0.6	-	-	-	-	-	-	-	-	-	-	-	-	-	-	-	-	-	-	-	-	-
4 **Bos taurus BMP15 mRNA complete cds**	0.0	0.0	0.6	-	-	-	-	-	-	-	-	-	-	-	-	-	-	-	-	-	-	-	-
5 *Capra hircus* (*BMP15*) gene complete cds	3.8	3.8	2.9	3.8	-	-	-	-	-	-	-	-	-	-	-	-	-	-	-	-	-	-	-
6 **Capra hircus (BMP15) mRNA**	0.0	0.0	0.6	0.0	4.0	-	-	-	-	-	-	-	-	-	-	-	-	-	-	-	-	-	-
7 *Capra hircus* breed Beetal (*BMP15*) gene complete cds	3.1	3.1	3.0	3.1	0.0	3.9	-	-	-	-	-	-	-	-	-	-	-	-	-	-	-	-	-
8 **Capra hircus breed Black Bengal (BMP15) mRNA complete cds**	0.0	0.0	0.6	0.0	4.0	0.0	3.9	-	-	-	-	-	-	-	-	-	-	-	-	-	-	-	-
9 *Capra hircus* breed Boer BMP15 mRNA complete cds	0.0	0.0	0.6	0.0	4.0	0.0	3.9	0.0	-	-	-	-	-	-	-	-	-	-	-	-	-	-	-
10 **Capra hircus breed Guizhou White BMP15 precursor (Bmp15) gene complete cds**	0.0	0.0	0.6	0.0	4.0	0.0	3.9	0.0	0.0	-	-	-	-	-	-	-	-	-	-	-	-	-	-
11 *Capra hircus* breed Lezhi black goat BMP15 mRNA complete cds	0.0	0.0	0.6	0.0	4.0	0.0	3.9	0.0	0.0	0.0	-	-	-	-	-	-	-	-	-	-	-	-	-
12 **Capra hircus breed Teddy BMP15 gene complete cds**	3.1	3.1	3.0	3.1	0.0	3.9	0.0	3.9	3.9	3.9	3.9	-	-	-	-	-	-	-	-	-	-	-	-
13 *Cavia porcellus* Bmp15 mRNA	0.3	0.3	0.9	0.3	3.8	0.3	3.7	0.3	0.3	0.3	0.3	3.7	-	-	-	-	-	-	-	-	-	-	-
14 **Homo sapiens BMP15 mRNA**	0.2	0.2	0.7	0.2	4.1	0.2	4.1	0.2	0.2	0.2	0.2	4.1	0.3	-	-	-	-	-	-	-	-	-	-
15 *Mus musculus* BMP15 mRNA	0.3	0.3	1.0	0.3	3.8	0.3	3.7	0.3	0.3	0.3	0.3	3.7	0.5	0.3	-	-	-	-	-	-	-	-	-
16 **Mus musculus BMP15 mRNA (cDNA clone MGC:60557 IMAGE:30051425) complete cds**	0.3	0.3	1.0	0.3	3.8	0.3	3.7	0.3	0.3	0.3	0.3	3.7	0.5	0.3	0.0	-	-	-	-	-	-	-	-
17 *Mus musculus* BMP 15 precursor (*Bmp15*) mRNA complete cds	0.3	0.3	1.0	0.3	3.8	0.3	3.7	0.3	0.3	0.3	0.3	3.7	0.5	0.3	0.0	0.0	-	-	-	-	-	-	-
18 **Ovis aries BMP15 mRNA**	0.1	0.1	0.5	0.1	4.1	0.1	4.0	0.1	0.1	0.1	0.1	4.0	0.4	0.3	0.4	0.4	0.4	-	-	-	-	-	-
19 *Ovis aries* BMP15 mRNA partial cds	4.4	4.4	5.0	4.4	2.9	4.5	2.9	4.5	4.5	4.5	4.5	2.9	3.8	4.4	4.7	4.7	4.7	4.7	-	-	-	-	-
20 **Ovis aries breed Lohi bone morphogenetic protein 15 (BMP15) gene complete cds**	3.1	3.1	2.9	3.1	0.0	3.9	0.0	3.9	3.9	3.9	3.9	0.0	3.7	4.1	3.7	3.7	3.7	4.0	3.4	-	-	-	-
21 *Sus scrofa* BMP15 gene promoter region and 5 UTR	2.3	2.3	2.8	2.3	4.1	2.2	4.2	2.2	2.2	2.2	2.2	4.2	2.5	2.2	2.2	2.2	2.2	2.4	4.1	4.2	-	-	-
22 **Sus scrofa BMP15 mRNA**	0.1	0.1	0.7	0.1	4.0	0.1	3.9	0.1	0.1	0.1	0.1	3.9	0.4	0.2	0.4	0.4	0.4	0.2	4.4	3.9	1.8	-	-
23 *Sus scrofa* BMP15 mRNA complete cds alternatively spliced	0.1	0.1	0.7	0.1	4.0	0.1	3.9	0.1	0.1	0.1	0.1	3.9	0.4	0.2	0.4	0.4	0.4	0.2	4.4	3.9	1.8	0.0	-

[1] Column numbers are same as row numbers (spices/genus).

This is the reason why there are differences in different species and even in different organs and parts of the body of a species despite the same gene sequence.

When DNA sequences are aligned with each other, identification of the appropriate corresponding nucleotides is quite difficult because there are only four types of nucleotides. Alignment of amino acid sequences is easier and can be more meaningful if the intent is to compare a group of related sequences for potential functional characteristics. In order to do this, amino acid sequences of *BMP15* related to four most used species such as mouse, pigs, sheep, and humans were first retrieved from *NCBI* and then have been compared using BLAST.

The *BMP15* protein sequences was from *Ovis aries* (GenBank AAF81688.1), *Sus scrofa* (GenBank NP_001005155.1), *Mus musculus* (GenBank NP_033887.1) and *Homo sapiens* (GeneBank NP_005439.2) were aligned and compared.

Results showed that differences in nucleotide sequences leading to changes in the protein sequence (Figure 9).

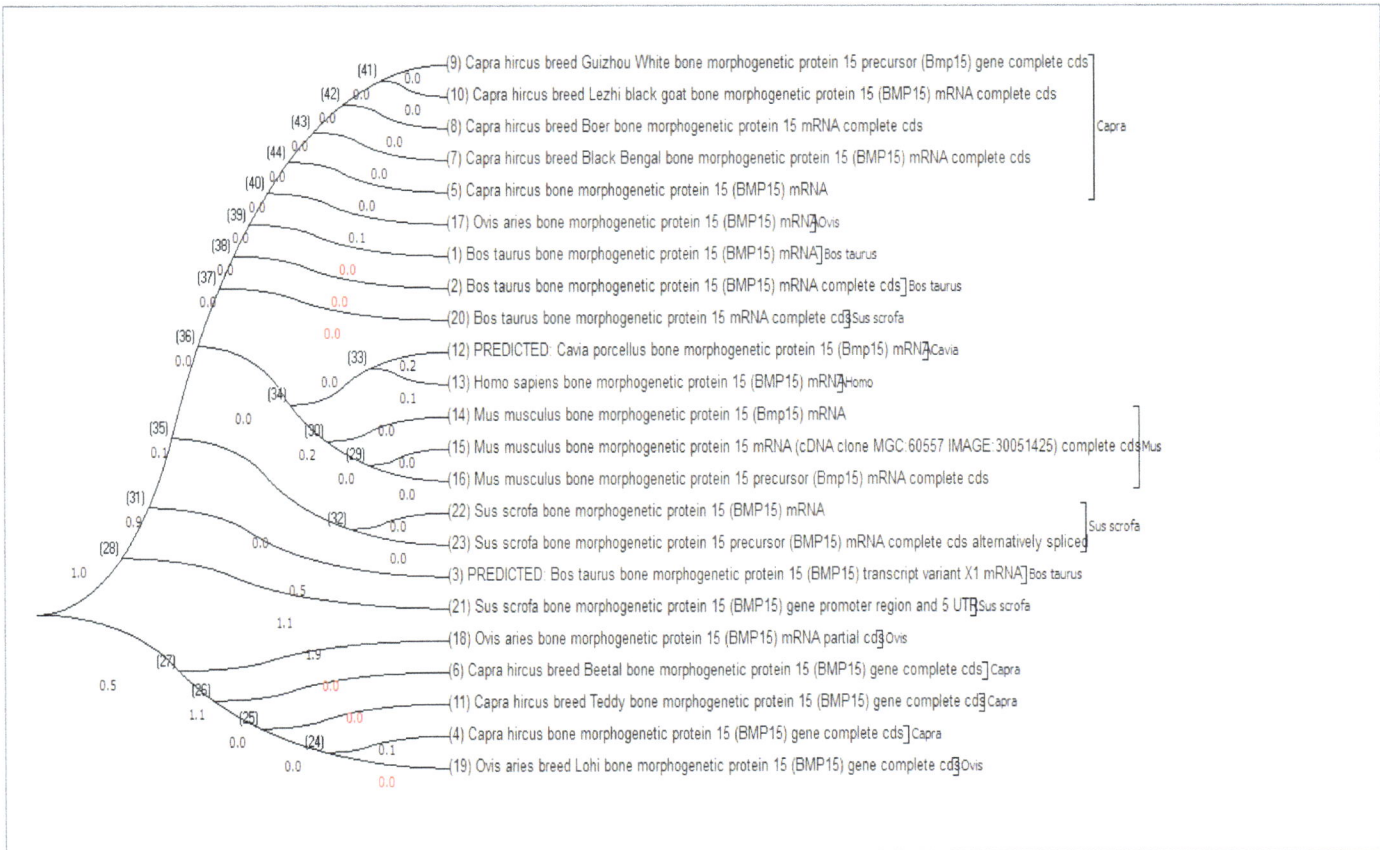

Figure 1 Phylogenetic tree drawn based on 23 nucleotide sequences of the *BMP15* gene

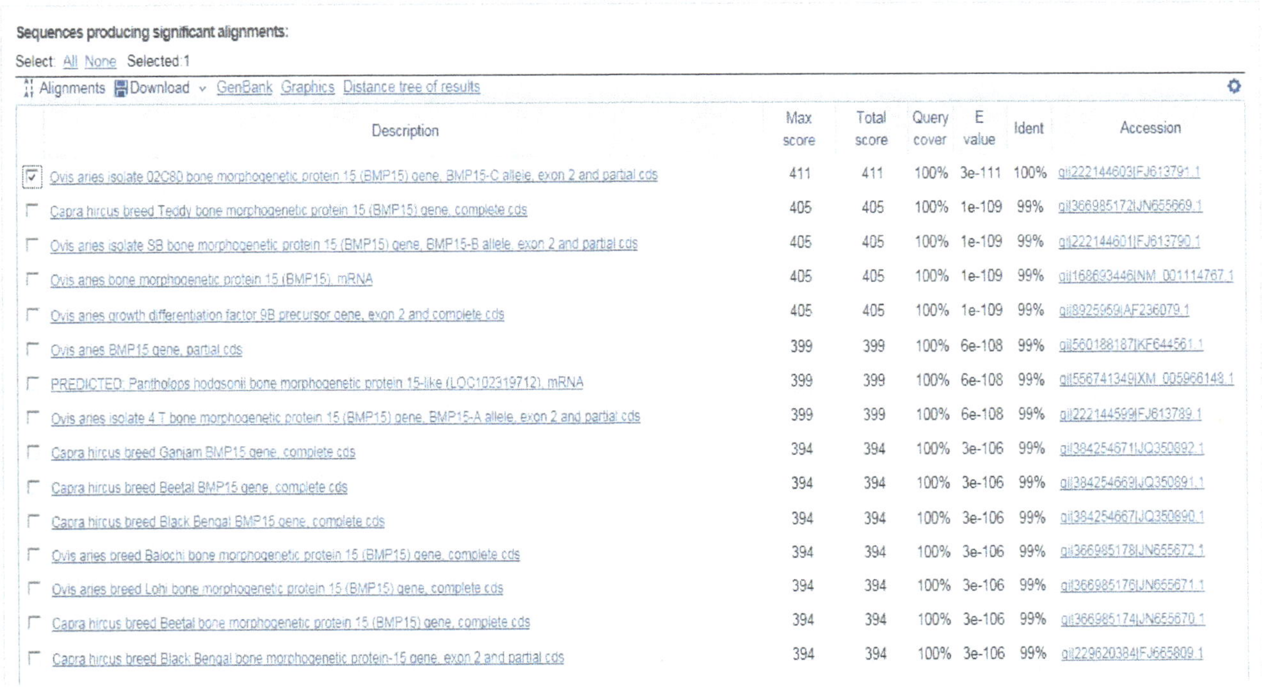

Figure 2 The scoring of the similarity and matching rate of sequences using the tool "BLAST"

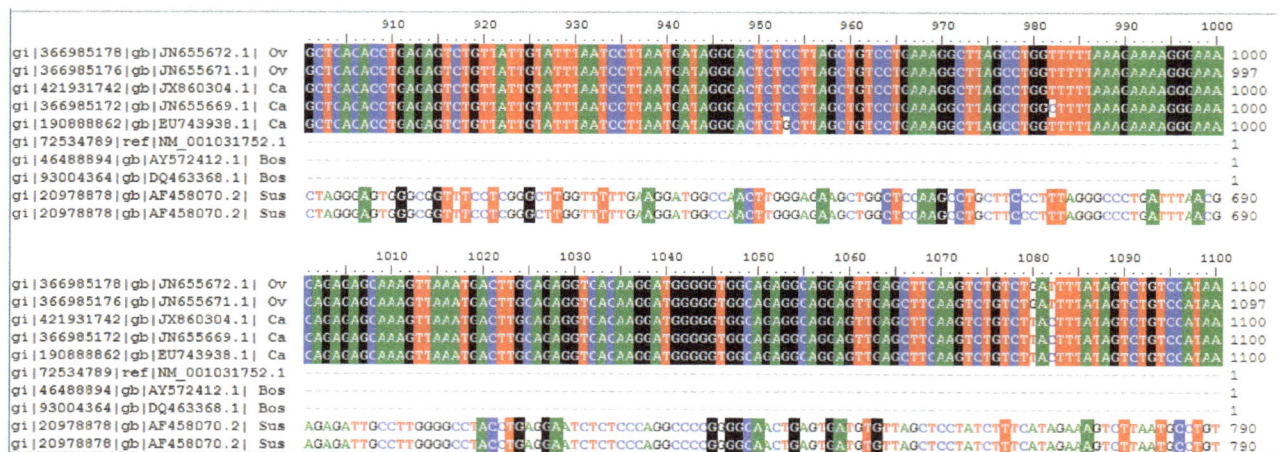

Figure 3 Sequence matching from base 900 up to base 1100 of the *BMP15* gene in some species being studied

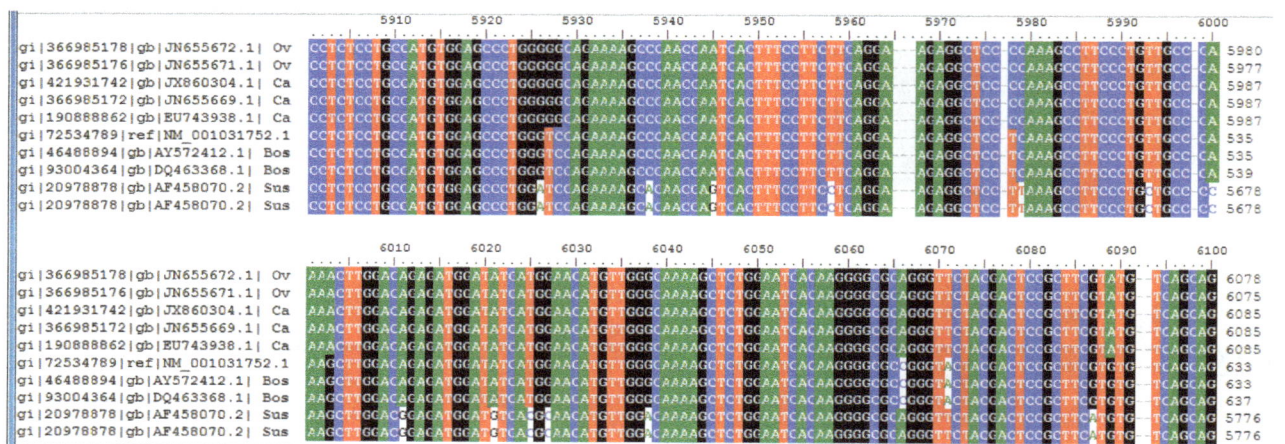

Figure 4 Sequence matching from base 5900 up to base 6100 of the *BMP15* gene in some species being studied

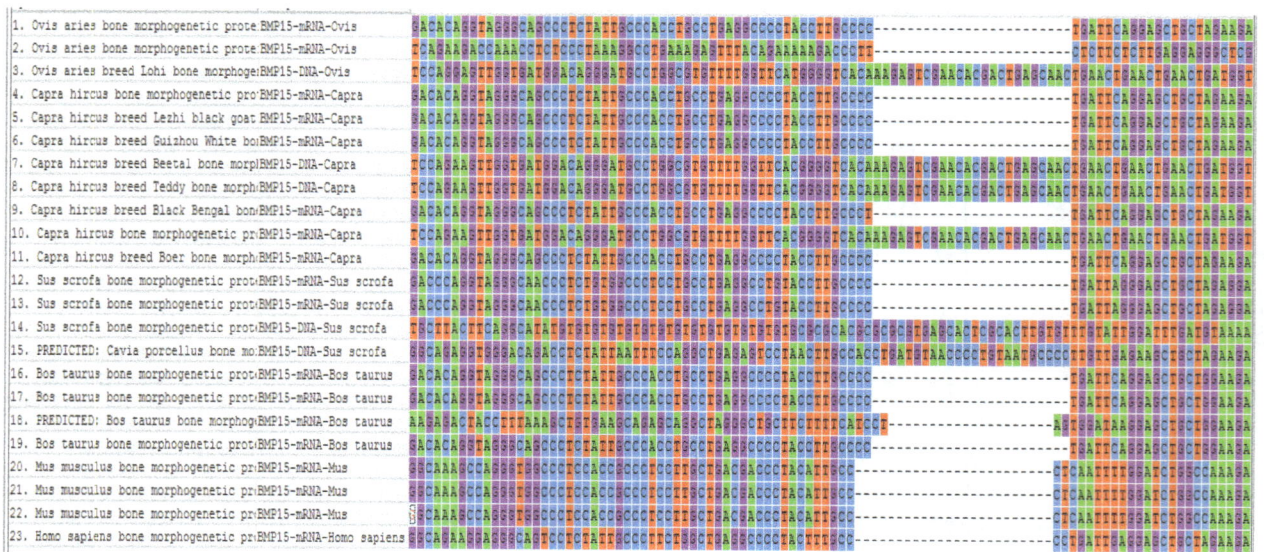

Figure 5 Displays of a part of sequence matching and deletions (introns)

Ovis aries BMP15 gene, partial cds
Sequence ID: gi|560188187|gb|KF644561.1| Length: 222

Range 1: 1 to 222 GenBank Graphics

Score	Expect	Identities	Gaps	Strand
399 bits(216)	6e-108	220/222(99%)	0/222(0%)	Plus/Plus

```
CDS:bone morphogenet  1      Y  C  K  G  V  C  P  R  V  L  H  Y  G  L  T  S  P  N  H  A
Query                 1    CTACTGTAAGGGAGTATGTCCTCGGGTACTACACTATGGTCTCACTTCTCCCAATCATGC   60
                           ||||||||||||||||||||||||||||| ||||||| |||||| |||||||||||||||
Sbjct                 1    CTACTGTAAGGGAGTATGTCCTCGGGTACTATACTATGGTCTCAATTCTCCCAATCATGC   60
CDS:BMP15, partial [  1      Y  C  K  G  V  C  P  R  V  L  Y  Y  G  L  N  S  P  N  H  A

CDS:bone morphogenet  21     I  I  Q  N  L  V  S  E  L  V  D  Q  N  V  P  Q  P  S  C  V
Query                 61   CATCATCCAGAACCTTGTCAGTGAGCTGGTGGATCAGAATGTCCCTCAGCCTTCCTGTGT   120
                           ||||||||||||||||||||||||||||||||||||||||||||||||||||||||||||
Sbjct                 61   CATCATCCAGAACCTTGTCAGTGAGCTGGTGGATCAGAATGTCCCTCAGCCTTCCTGTGT   120
CDS:BMP15, partial [  21     I  I  Q  N  L  V  S  E  L  V  D  Q  N  V  P  Q  P  S  C  V

CDS:bone morphogenet  41     P  Y  K  Y  V  P  I  S  I  L  L  I  E  A  N  G  S  I  L  Y
Query                 121  CCCTTATAAGTATGTTCCCATTAGCATCCTTCTGATTGAGGCAAATGGGAGTATCTTGTA   180
                           ||||||||||||||||||||||||||||||||||||||||||||||||||||||||||||
Sbjct                 121  CCCTTATAAGTATGTTCCCATTAGCATCCTTCTGATTGAGGCAAATGGGAGTATCTTGTA   180
CDS:BMP15, partial [  41     P  Y  K  Y  V  P  I  S  I  L  L  I  E  A  N  G  S  I  L  Y

CDS:bone morphogenet  61     K  E  Y  E  G  M  I  A  Q  S  C  T  C
Query                 181  CAAGGAGTATGAGGGTATGATTGCCCAGTCCTGCACATGCAG   222
                           ||||||||||||||||||||||||||||||||||||||||||
Sbjct                 181  CAAGGAGTATGAGGGTATGATTGCCCAGTCCTGCACATGCAG   222
CDS:BMP15, partial [  61     K  E  Y  E  G  M  I  A  Q  S  C  T  C
```

Figure 6 A comparison between the two sequences of the *BMP* gene of *Ovis* species

Capra hircus breed Black Bengal bone morphogenetic protein 15 (BMP15) mRNA, complete cds
Sequence ID: gi|196122235|gb|EU888137.1| Length: 1182 Number of Matches: 1

Range 1: 957 to 1178 GenBank Graphics   Next Match   Previous Match

Score	Expect	Identities	Gaps	Strand
394 bits(213)	3e-106	219/222(99%)	0/222(0%)	Plus/Plus

```
CDS:bone morphogenet  1      Y  C  K  G  V  C  P  R  V  L  H  Y  G  L  T  S  P  N  H  A
Query                 1    CTACTGTAAGGGAGTATGTCCTCGGGTACTACACTATGGTCTCACTTCTCCCAATCATGC   60
                           |||||||||||||||||||||||||||||  |||||||| ||||||  |||||||||||||
Sbjct                 957  CTACTGTAAGGGAGTATGTCCTCGGGTACTATACTATGGTCTCAATTCTCCCAATCATGC   1016
CDS:bone morphogenet  320    Y  C  K  G  V  C  P  R  V  L  Y  Y  G  L  N  S  P  N  H  A

CDS:bone morphogenet  21     I  I  Q  N  L  V  S  E  L  V  D  Q  N  V  P  Q  P  S  C  V
Query                 61   CATCATCCAGAACCTTGTCAGTGAGCTGGTGGATCAGAATGTCCCTCAGCCTTCCTGTGT   120
                           ||||||||||||||||||| |||||||||||||||||||||||||||||||||||||||||
Sbjct                 1017 CATCATCCAGAACCTTGTCAATGAGCTGGTGGATCAGAATGTCCCTCAGCCTTCCTGTGT   1076
CDS:bone morphogenet  340    I  I  Q  N  L  V  N  E  L  V  D  Q  N  V  P  Q  P  S  C  V

CDS:bone morphogenet  41     P  Y  K  Y  V  P  I  S  I  L  L  I  E  A  N  G  S  I  L  Y
Query                 121  CCCTTATAAGTATGTTCCCATTAGCATCCTTCTGATTGAGGCAAATGGGAGTATCTTGTA   180
                           ||||||||||||||||||||||||||||||||||||||||||||||||||||||||||||
Sbjct                 1077 CCCTTATAAGTATGTTCCCATTAGCATCCTTCTGATTGAGGCAAATGGGAGTATCTTGTA   1136
CDS:bone morphogenet  360    P  Y  K  Y  V  P  I  S  I  L  L  I  E  A  N  G  S  I  L  Y

CDS:bone morphogenet  61     K  E  Y  E  G  M  I  A  Q  S  C  T  C
Query                 181  CAAGGAGTATGAGGGTATGATTGCCCAGTCCTGCACATGCAG   222
                           ||||||||||||||||||||||||||||||||||||||||||
Sbjct                 1137 CAAGGAGTATGAGGGTATGATTGCCCAGTCCTGCACATGCAG   1178
CDS:bone morphogenet  380    K  E  Y  E  G  M  I  A  Q  S  C  T  C  R
```

Figure 7 A comparison between the sequence of the *BMP15* gene in *Ovis* and *Capra* species

These differences in the *BMP15* protein sequences in species being compared are shown in the black line distinct from others.

Also, the important mutations that occurred in this gene in different species are shown in red lines. The *BMP15* protein plays an important role in women's fertility. A number of mutations have occurred in the *BMP15* gene in humans, but none of them is common to those in sheep.

It is interesting that all the mutations that have been reported in humans so far have been associated with diminished ovarian syndrome in heterozygotes. Therefore, heterozygous carriers in humans had an ovarian phenotype similar to that of infertile homozygous FecX ewes. In contrast, heterozygous ewes, FecX[Gr] and FecX[O] carriers, had normal ovaries and a greater number of ovulation and lamb in each delivery (Demars *et al.* 2013).

Bos taurus bone morphogenetic protein 15 mRNA, complete cds
Sequence ID: gi|93004364|gb|DQ463368.1| Length: 1189 Number of Matches: 1

Range 1: 964 to 1185 GenBank Graphics

Score	Expect	Identities	Gaps	Strand
394 bits(213)	3e-106	219/222(99%)	0/222(0%)	Plus/Plus

```
CDS:bone morphogenet  1          Y  C  K  G  V  C  P  R  V  L  H  Y  G  L  T  S  P  N  H  A
Query                 1    CTACTGTAAGGGAGTATGTCCTCGGGTACTACACTATGGTCTCACTTCTCCCAATCATGC   60
                           ||||||||||||||||||||||||||||||||||||||||| |||||||||||||||||||
Sbjct                 964  CTACTGTAAGGGAGTATGTCCTCGGGTACTACACTATGGTCTCAATTCTCCCAATCATGC   1023
CDS:bone morphogenet  321        Y  C  K  G  V  C  P  R  V  L  H  Y  G  L  N  S  P  N  H  A

CDS:bone morphogenet  21         I  I  Q  N  L  V  S  E  L  V  D  Q  N  V  P  Q  P  S  C  V
Query                 61   CATCATCCAGAACCTTGTCAGTGAGCTGGTGGATCAGAATGTCCCTCAGCCTTCCTGTGT   120
                           ||||||||||||||||||||||||||||||||||||||||||||||||||||||||||||
Sbjct                 1024 CATCATCCAGAACCTTGTCAATGAGCTGGTGGATCAGAGTGTCCCTCAGCCTTCCTGTGT   1083
CDS:bone morphogenet  341        I  I  Q  N  L  V  N  E  L  V  D  Q  S  V  P  Q  P  S  C  V

CDS:bone morphogenet  41         P  Y  K  Y  V  P  I  S  I  L  L  I  E  A  N  G  S  I  L  Y
Query                 121  CCCTTATAAGTATGTTCCCATTAGCATCCTTCTGATTGAGGCAAATGGGAGTATCTTGTA   180
                           ||||||||||||||||||||||||||||||||||||||||||||||||||||||||||||
Sbjct                 1084 CCCTTATAAGTATGTTCCCATTAGCATCCTTCTGATTGAGGCAAATGGGAGTATCTTGTA   1143
CDS:bone morphogenet  361        P  Y  K  Y  V  P  I  S  I  L  L  I  E  A  N  G  S  I  L  Y

CDS:bone morphogenet  61         K  E  Y  E  G  M  I  A  Q  S  C  T  C
Query                 181  CAAGGAGTATGAGGGTATGATTGCCCAGTCCTGCACATGCAG   222
                           ||||||||||||||||||||||||||||||||||||||||||
Sbjct                 1144 CAAGGAGTATGAGGGTATGATTGCCCAGTCCTGCACATGCAG   1185
CDS:bone morphogenet  381        K  E  Y  E  G  M  I  A  Q  S  C  T  C  R
```

Figure 8 A comparison between the sequence of the *BMP15* gene in *Ovis aries* and *Bos taurus*

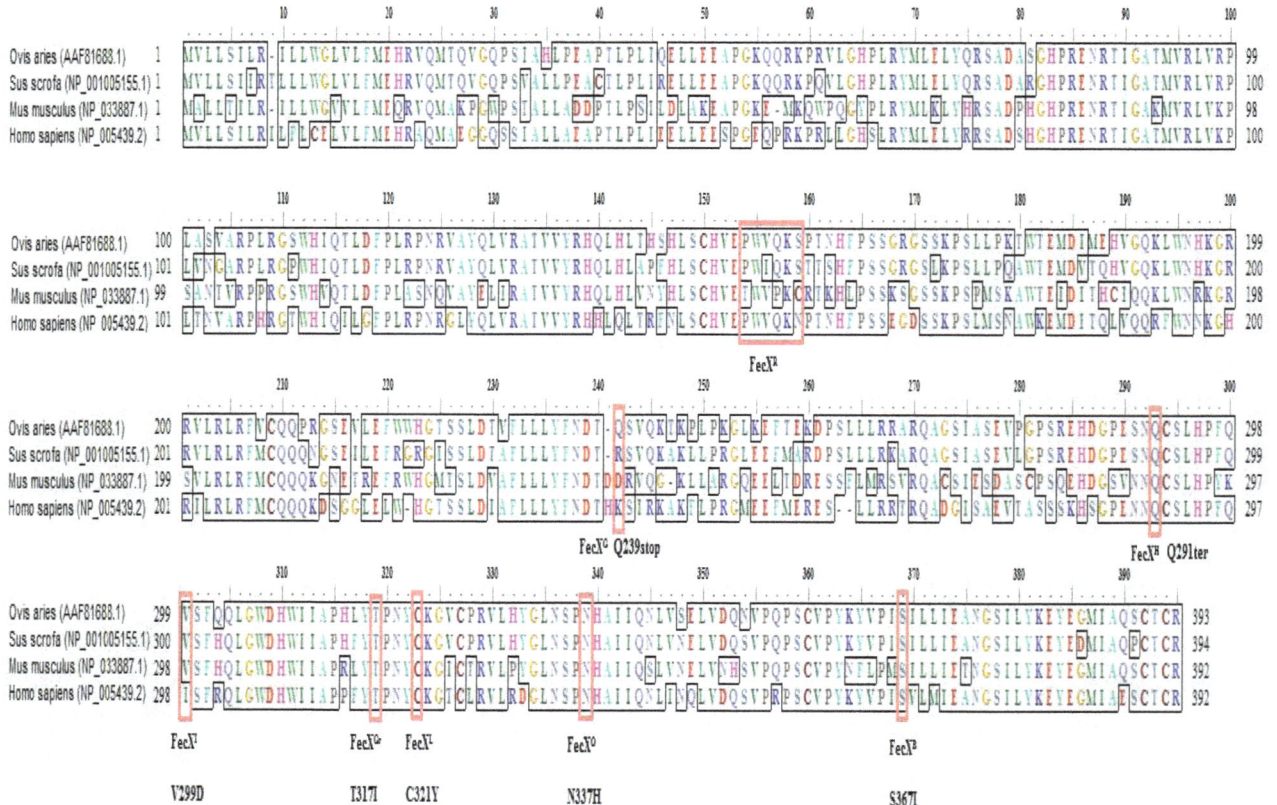

Figure 9 *BMP15* multi-species sequences alignment and position of sheep mutations

According to the information of the *NCBI* gene bank, so far 59, 7, 17 and 6 single nucleotide polymorphisms (SNPs) are identified in the *BMP15* gene of humans, cows, mice and sheep, respectively, and some of them are mentioned in Table 1 and Figure 9.

Recent innovations in DNA sequencing technology have surprisingly enhanced our ability to determine the sequence of a large quantity of DNA. Also, extensive analysis of complementary DNAs-cDNA, the nucleotide sequence of messenger RNA (mRNA), has specified a large quantity of non-coding RNAs in eukaryotic cells and tissues that are involved in the regulation of gene expression. The comprehensive studies of the DNA, which were conducted with the aim of identifying functional parts of the human genome, reported cases of genes overlapping and genes with common exons and different transcription start sites (TSS). This fact and the other results challenged the traditional definition of the gene, a sequence of the DNA that encodes a chain of amino acids (Gojobori *et al.* 2009).

It was previously shown that *BMP15* evolved more quickly than the other members of the TGF-β family, with evidence for positive selection in *BMP15*, especially in *Hominidae* (Auclair *et al.* 2013). The *BMP15* protein has been described in some species mainly human and sheep to play critical roles in female fertility or disorders. A large number of mutations in the *BMP15* gene have been identified in women with premature ovarian failure (Di Pasquale *et al.* 2004; Di Pasquale *et al.* 2006; Dixit *et al.* 2006) and ovarian hyper stimulation syndrome (Hanevik *et al.* 2011; Moron *et al.* 2006), but none are in common between women and sheep.

According to the research by McNatty *et al.* (2005) in mammals with low ovulation rates, the follicular growth and ovulation rate are affected by the *BMP15* released from the oocyte to the somatic cells of the follicle. However, in rodents with high ovulation rates, the follicular cells are relatively insensitive to *BMP15* changes (Yan *et al.* 2001).

This indicates that the mechanisms through which an oocyte controls this process are different in species with a low and high ovulation rates. Although, the role of *BMP15* has not been fully investigated in pigs, it has been shown that there is an active BMP system in their ovaries (Brankin *et al.* 2003; Brankin *et al.* 2004).

In a study on Chinese Hu sheep which have high prolificacy, no polymorphism was reported in the locus of *BMP15* (Guan *et al.* 2006). Also, studies on polymorphism of the locus of *BMP15* in six breeds of Chinese goats have not shown an effective mutation which is associated with prolificacy in the exon parts of this position (He *et al.* 2006).

The results of investigating the mutations in 5 regions of exon 2 of the *BMP15* gene have shown a correlation with fertility in several breeds of sheep (Montgomery *et al.*

2001). Mutations in the *BMP15* fertility gene play an important economic role in sheep and probably in the reproduction of ruminants (Galloway *et al.* 2000; Hanrahan *et al.* 2004; McNatty *et al.* 2005). The function of a main single gene is responsible for the high ovulation rate in sheep Booroola Merino, Inverdale, Belclare, and Cambridge, but there is no reason for the existence of a main gene responsible for prolificacy in other fertile sheep, such as Finish Landrac and Romanov (Gordon *et al.* 2004; Gordon *et al.* 2005). These findings suggest that, at least two mechanisms of genetic control play a role in the high fertility of sheep. The biological effects of mutations varied among mammal species (Yan *et al.* 2001). In addition, Hashimoto *et al.* (2005) have suggested that species-specific differences in the processing of *BMP15* may be correlated with the differences existing in fertility rates among species. In addition to existing variation in DNA and amino acids sequences between *BMP15* in different species this study highlighted specific segments significantly more conserved in *BMP15* from mono-ovulating as compared to super-ovulating species. Thus, further analyses are required in addition to DNA sequencing, for understanding and perceiving the complexities of the genome.

CONCLUSION

There are several mechanisms of genetic control for regulation of fertility in mammalian species in which *BMP15* play an important role. The biological consensuses of mutation in *BMP15* varied among mammal species and are species-specific. Various mRNA processing of *BMP15* in different mammalian species may cause a wide range of performance despite same DNA sequences. In addition to existing variation in DNA and amino acids sequences be-tween *BMP15* in different species this study highlighted specific segments significantly more conserved in *BMP15* from mono-ovulating as compared to super-ovulating species. Thus, further analyses are required in addition to DNA sequencing, for understanding and perceiving the complexities of the genome.

ACKNOWLEDGEMENT

We thanks form all of Ph D student for sharing information about phylogenetic analysis with DNA sequences and computer center of University of Zabol for using bioinformatics packages.

REFERENCES

Auclair S., Rossetti R., Meslin C., Monestier O., Di Pasquale E., Pascal G., Persani L. and Fabre S. (2013). Positive selection in bone morphogenetic protein 15 targets a natural mutation associated with primary ovarian insufficiency inhuman. *PLoS One*. **8**, e78199.

Bibinu B.S., Yakubu A., Ugbo S.B. and Dim N.I. (2016). Computational molecular analysis of the sequences of *BMP15* gene of ruminants and non-ruminants. *Open J. Genet*. **6**, 39-50.

Brankin V., Quinn R.L., McGarr C., Webb R. and Hunter M.G. (2003). BMPs 2, 6 and 15 are regulators of porcine granulosa cell function *in vitro*. *Reprod. Abstr. Ser*. **30**, 94.

Brankin V., Quinn R.L., McGarr C., Webb R. and Hunter M.G. (2004). The function of co-cultured porcine granulosa and theca cells is modulated by BMPs 2, 6 and 15 in conjunction with oocytes in a serum free system. *Reprod. Abstr. Ser*. **31**, 35.

Demars J., Fabre S., Sarry J., Rossetti R., Gilbert H., Luca P., Gwenola T., Mulsant P., Nowak Z., Drobik W., Martyniuk E. and Bodin L. (2013). Genome-wide association studies identify two novel *BMP15* mutations responsible for an atypical hyperprolificacy phenotype in sheep. *PLoS Genet*. **9(4)**, e1003482.

Di Pasquale E., Beck-Peccoz P. and Persani L. (2004). Hypergonadotropic ovarian failure associated with an inherited mutation of human bone morphogenetic protein-15 (*BMP15*) gene. *Am. J. Hum. Genet*. **75**, 106-111.

Di Pasquale E., Rossetti R., Marozzi A., Bodega B., Borgato S., Cavallo L., Einaudi S., Radetti G., Russo G., Sacco M., Wasniewska M., Cole T., Beck-Peccoz P., Nelson L.M. and Persani L. (2006). Identification of new variants of human *BMP15* gene in a large cohort of women with premature ovarian failure. *J. Clin. Endocrinol. Metab*. **91**, 1976-1979.

Dixit H., Rao L.K., Padmalatha V.V., Kanakavalli M., Deenadayal M., Gupta N., Chakrabarty B. and Singh L. (2006). Missense mutations in the *BMP15* gene are associated with ovarian failure. *Hum. Genet*. **119**, 408-415.

Drouilhet L., Mansanet C., Sarry J., Tabet K., Bardou P., Woloszyn F., Harichaux G., Monniaux D., Bodin L., Philippe M. and Fabre S. (2013). The highly prolific phenotype of Lacaune Sheep is associated with an ectopic expression of the *B4GALNT2* gene within the ovary. *PLoS Genet*. **9(9)**, e1003809.

Dube J.L., Wang P., Elvin J., Lyons K.M., Celeste A.J. and Matzuk M.M. (1998). The bone morphogenetic protein 15 gene is x-linked and expressed in oocytes. *Mol. Endocrinol*. **12**, 1809-1817.

Fabre S., Pierre A., Mulsant P., Bodin L., DiPasquale E., Persani L., Monget P. and Monniaux D. (2006). Regulation of ovulation rate in mammals: contribution of sheep genetic models. *Reprod. Biol. Endocrinol*. **4**, 20-25.

Galloway S.M., Gregan S.M., Wilson T., McNatty K.P., Juengel J.L., Ritvos O. and Davis G.H. (2002). *BMP15* mutations and ovarian function. *Mol. Cell. Endocrinol*. **191**, 15-18.

Galloway S.M., McNatty K.P., Cambridge L.M., Laitenen M.P.E., Juengel J.L., Jokiranta T.S., McLaren R.J., Luiro K., Dodds K.G., Montgomery G.W., Beattie A.E., Davis G.H. and Ritvos O. (2000). Mutations in an oocyte-derived growth factor (*BMP15*) cause increased ovulation rate and infertility in a dosage-sensitive manner. *Nat. Genet*. **25**, 279-283.

Gojobori T., Nakagawa S. and Clemente J.C. (2009). DNA Sequence Analysis, Encyclopedia of Life Sciences. John Wiley and Sons, Ltd: Chichester, United Kingdom.

Gordon D.B., Nekludova L., McCallum S. and Fraenkel E. (2005). TAMO: a flexible, object-oriented framework for analyzing transcriptional regulation using DNA-sequence motifs. *Bioinformatics*. **21(14)**, 3164-3175.

Gordon G. and Joiner-Bey H. (2004). The omega-3 Miracle the Icelandic Longevity Secret that Offers Super Protection Against Heart Disease. Freedom Press, California, USA.

Guan F., Liu S.R., Shi G.Q., Ai J.T., Mao D.G. and Yang L.G. (2006). Polymorphism of *FecB* gene in nine sheep breeds or strains and its effects on litter size, lamb growth and development. *Acta Genet. Sinica*. **33**, 117-124.

Hanevik H.I., Hilmarsen H.T., Skjelbred C.F., Tanbo T. and Kahn J.A. (2011). A single nucleotide polymorphism in BMP15 is associated with high response to ovarian stimulation. *Reprod. Biomed. Online*. **23**, 97-104.

Hanrahan J.P., Gregan S.M., Mulsant P., Mullen M., Davis G.H., Powell R. and Galloway S.M. (2004). Mutations in the genes for oocyte-derived growth factors *GDF9* and *BMP15* are associated with both increased ovulation rate and sterility in Cambridge and Belclare sheep (*Ovis aries*). *Biol. Reprod*. **70**, 900-909.

Hashimoto O., Moore R.K. and Shimasaki S. (2005). Posttranslational processing of mouse and human *BMP15* potential implication in the determination of ovulation quota. *Proc. Natl. Acad. Sci*. **102**, 5426-5431.

He Y.Q., Chu M.X., Wang J.Y., Fang L. and Ye S.C. (2006). Polymorphism on *BMP15* as a candidate gene for prolificacy in six goat breeds Chinese. *J. Anhui Agric. Univ*. **33**, 61-64.

Librado P. and Rozas J. (2009). DnaSP v5: software for comprehensive analysis of DNA polymorphism data. *Bioinformatics*. **25**, 1451-1452.

McNatty K.P., Juengel J.L., Reader K.L., Lun S., Myllyma S., Lawrence S.B., Western A., Meerasahib M.F., Mottershead D.G., Groome N.P., Ritvos O. and Laitinen M.P. (2005). Bone morphogenetic protein 15 and growth differentiation factor 9 co-operate to regulate granulosa cell function in ruminants. *Reproduction*. **129**, 481-487.

Monteagudo L.V., Ponz R., Tejedor M.T., Lavina A. and Sierra I. (2009). A 17 bp deletion in the bone morphogenetic protein 15 (*BMP15*) gene is associated to increased prolificacy in the Rasa Aragonesa sheep breed. *Anim. Reprod. Sci*. **110**, 139-146.

Montgomery G.W., Galloway S.M., George H., Davis G.H. and McNatty K.P. (2001). Genes controlling ovulation rate in sheep. *Reproduction*. **121**, 843-852.

Moron F.J., de Castro F., Royo J.L., Montoro L., Mira E., Sáez M.E., Real L.M., González A., Mañes S. and Ruiz A. (2006). Bone morphogenetic protein 15 (*BMP15*) alleles predict over-response to recombinant follicle stimulation hormone and iatrogenic ovarian hyperstimulation syndrome (OHSS). *Pharmacogenet. Genom*. **16**, 485-495.

Otsuka F., Yamamoto S., Erickson G.F. and Shimasaki S. (2001). Bone morphogenetic protein-15 inhibits follicle-stimulating hormone (FSH) action by suppressing FSH receptor expression. *J. Biol. Chem.* **276**, 11387-11392.

Silva J.R.V., van den Hurk R., van Tol H.T.A., Roelen B.A.J. and Figueiredo J.R. (2004). Experssion of growth differentiation factor 9 (GDF-9) and bone morphogenetic protein 15 (*BMP-15*) and BMP receptors in the ovaries of goats. *Mol. Reprod. Dev.* **70**, 11-19.

Souza C.J.H., Campbell B.K., McNeilly A.S. and Baird D.T. (2002). Effect of bone morphogenetic protein 2 (BMP2) on oestradiol and inhibin A production by sheep granulosa cells, and localization of BMP receptors in the ovary by immune histochemistry. *Reproduction.* **123**, 363-369.

Tamura K., Nei M. and Kumar S. (2004). Prospects for inferring very large phylogenies by using the neighbor-joining method. *Proc. Natl. Acad. Sci.* **101**, 11030-11035.

Tamura K., Stecher S., Peterson D., Filipski A., and Kumar S. (2013). MEGA6: molecular evolutionary genetics analysis version 6.0. *Mol. Biol. Evol.* **30(12)**, 2725-2729.

Yan C., Wang P., DeMayo J., DeMayo F.J., Elvin J.A., Carino C., Prasad S.V., Skinner S.S., Dunbar B.S., Dube J.L., Celeste A.J. and Matzuk M.M. (2001). Synergistic roles of bone morphogenetic protein 15 and growth differentiation factor 9 in ovarian function. *Mol. Endocrinol.* **15**, 854-866.

Yang Z. and Kumar S. (1996). Approximate methods for estimating the pattern of nucleotide substitution and the variation of substitution rates among sites. *Mol. Biol. Evol.* **13**, 650-659.

10

The Effects of Various Essential Oils of Medical Plant Seeds and Spices on Digestion Characteristics and Population Changes of Ruminal Anaerobic Fungi in *in vitro* Condition

M. Sadjadian[1], M. Danesh Mesgaran[1*] and A.R. Vakili[1]

[1] Department of Animal Science, Faculty of Agriculture, Ferdowsi University of Mashhad, Mashhad, Iran

*Correspondence E-mail: danesh@um.ac.ir

ABSTRACT

The effect of essential oils (EO) of medical plant seeds and spices on rumen microbial fermentation of alfalfa hay, sugar beet pulp and barley grain (as substrate) were evaluated under *in vitro* conditions. *In vitro* incubations were carried out using the gas production method with glass syringes. Treatments were as follows; a control (no additive), monensin, EO of cinnamon, black pepper seed, cumin seed, fennel seed and garlic oil (200 and 400 µL/g DM). Monensin was used as a positive control in the medium at 5 µmol. Data on gas production were fitted using an exponential equation. Results showed that compared to control treatments, monensin had a significant increase on gas production (P<0.05), and cumin seed EO decreased gas production of the feed samples (200 and 400 µL). The effects of treatments on *in vitro* ruminal fermentation characteristics were tested using an *in vitro* culture inoculated by mixed rumen microbes. The test treatments were as follows; control (no additive), EO of cinnamon, black pepper seed, cumin seed and fennel seed. Evaluations were made for medium pH, ammonia nitrogen concentration and dry matter disappearance after a 48 h incubation period. To evaluate the effect of EO on *in vitro* ruminal fungi populations, a sample was taken from the medium after a 120 h incubation period and fungal population was determination by real-time polymerase chain reaction. Compared to the control treatment, cumin and cinnamon additions resulted in a significant decrease (P<0.05) on disappearance of dry matter in the feed samples. In the present study, additions of all tested EO to alfalfa hay treatment showed a significant increase in the final pH of the culture (P<0.05). However, cinnamon addition resulted in a significant decrease in medium ammonia nitrogen concentration for each of the feed samples (P<0.05). Results of the present study also demonstrate that addition all of the tested EO to alfalfa hay had a significantly decrease on *in vitro* ruminal fungal population (P<0.05).

KEY WORDS alfalfa hay, essential oil, fungal population, real-time PCR.

INTRODUCTION

In ruminants, symbiotic relationship is established with rumen microorganisms, by which the animal provides the nutrients and optimal environmental conditions to allow fermentation of their feed by the rumen microorganisms. Instead, microorganisms degrade fiber and synthesize microbial protein to supply energy and protein needs of the host animal (Van Soest, 1982). However, this symbiotic relationship is somewhat inefficient through loss of energy (methane) and protein (N-NH$_3$) (Van Nevel and Demeyer, 1988). These losses not only reduce production performance, but they also contribute to release of environmental pollutants (Tamminga, 1996). Ruminant nutritionists have long been interested in modulating competition among different microbial populations with the objective of improving the efficiency of energy and protein utilization in the rumen. This has been achieved through optimized diet formulations and use of feed additives that modify fermentation in the rumen and enhance or inhibit specific microbial populations (Calsamiglia *et al.* 2006). Antibiotic ionophores have been very successful in reducing energy and protein losses in the rumen (Van Nevel and Demeyer, 1988). However, the use of antibiotics in animal feed is becoming less socially acceptable because of residue and production of resistant strains of bacteria. The use of antibiotics in animal feed has been banned in the European Union since January 2006 (Directive 1831/2003/CEE, European Commission, 2003). For this reason, scientists have become interested in developing alternative ways to modulate rumen fermentation such as application of yeasts, organic acids, plant extracts, probiotics and antibodies (Calsamiglia *et al.* 2006). Essential oils (EO) are complex mixtures of secondary metabolites and volatile compounds extracted from plants by methods of distillation that seem to have no direct function in their growth and development (Balandrin and Klocke, 1985; Benchaar *et al.* 2008). However, they are responsible odor and color in plants (Guenther, 1948). Application of EO can be a useful strategy to improve efficiency of nutrient utilization by ruminants (Benchaar *et al.* 2008; Tajodini *et al.* 2014). EO have an important role in plants, for example protection against invasive insects, bacteria, viruses and fungi, as well as in attracting pollinators (Bakkali *et al.* 2008). EO have demonstrated antimicrobial activity against a variety of microorganisms, including gram positive and gram-negative bacteria, protozoa, yeast and fungi (Helander *et al.* 1998; Greathead, 2003; Reichling *et al.* 2009). The mechanism by which EO are thought to exert their antimicrobial activity is by disrupting the cell wall structure, affecting electron transport, ion gradients, protein translocation, phosphorylation steps and other enzyme-dependent reactions (Ultee *et al.* 1999; Dorman and Deans, 2000). The antimicrobial activity of plant extract is attributed to a number of secondary plant metabolites, including saponins, terpenoids and phenylpropanoids present in the EO fraction of many plants (Dorman and Deans, 2000). A number of recent *in vitro* studies have evaluated the effects of various EO on ruminal microorganisms and ruminal metabolism and have reported varied results. Many studies have reported the stimulatory and inhibitory effects of EO on a large variety of microorganisms under *in vitro* and *in vivo* conditions (Newbold *et al.* 2004; Benchaar *et al.* 2007; Nanon *et al.* 2014; Khorrami *et al.* 2015).

Fernandez *et al.* (1997) showed that a commercial product of a blended of EO compounds inhibits protein degradation in the rumen, thus has the potential to increase protein supply to the post-ruminal tract.

McIntosh *et al.* (2003), reported that EO can cause a decrease in the ruminal fungal population. The objective of this study was to evaluate the effects of various medicinal plant EO on *in vitro* ruminal fermentation of alfalfa hay, sugar beet pulp and barley grain and *in vitro* ruminal fungal population.

MATERIALS AND METHODS

Substrates and EO preparation
The tested feed samples were as follows; alfalfa hay (AH), barley grain (BG) and sugar beet pulp (BP); these were ground to pass through a 1mm screen (AOAC, 1990). Treatments were feed samples without EO (as controls) or substrates plus 200 or 400 µL/g DM dose of EO of cinnamon, black pepper, cumin, fennel or garlic oil. Monensin was also used as a positive control in the medium at 5 µmol. These values were selected based on our previous experiments. The EO content was obtained for each plant with hydro-distillation of ground samples using Clevenger apparatus (Jahani-Azizabadi *et al.* 2014). The EO samples were stored in a refrigerator (4 °C) until they were used in the experiment.

Gas production technique
Two sheep (45±2 kg, body weight) fitted with rumen cannulae were used as donors for rumen fluid. They were fed 1.5 kg DM alfalfa hay and 0.4 kg DM concentrates (165 g CP/kg DM) per head per day. Rumen content was collected before feeding in the morning. Rumen fluid was strained through 4 layers of cheesecloth and incubated at 39 °C.
The gas production method of Menke and Steingass (1988) was used.

Rumen fluid was immediately strained through four layers of cheesecloth and mixed in a 2:1 with buffer, then, 40 mL of diluted fluid was added to the syringes. Each syringe was gassed with CO_2 then incubated at 38.6 °C. The volume of gas produced was determined at 2, 4, 8, 12, 24, 36, 72 and 96 h after incubation. Gas production data were fitted using an exponential equation of

$$P = b(1-e^{-ct})$$

Where:
b: volume of gas produced.
c: fractional rate constant of gas production (/h).
t: incubation time (h).
P: volume of gas produced at time t.

Ruminal quantification of fungi using batch culture
Feed samples were alfalfa hay (AH), barley grain (BG) and sugar beet pulp (BP). Treatments were substrates without

EO (as control) and substrates plus 200 µL/g DM of EO of cinnamon, black pepper, cumin and fennel.

Samples were incubated in a medium prepared as described in Arroquy et al. (2005). The fermentation medium was pre-reduced anaerobically sterilized ruminal fluid medium consisting 150 mL/L of mineral mixtures I (3 g/L K_2HPO_4), 150 mL/L mineral mixture II (3 g/L KH_2PO_4, 6 g/L $(NH_4)_2SO_4$, 6 g/L NaCl, 0.6 g/L $MgSO_4$_$7H_2O$, 0.6 g/L $CaCl_2$), 0.05 g/L of cellobiose, 400 mL/L of cell-free ruminal fluid, 1 mL/L resazurin, 300 mL/L of distilled water, 4 g/L $NaHCO_3$ and 0.5 g/L of cysteine-HCL. The fermentation medium was autoclaved for about 20 minutes, cooled and then 4 g/L sodium bicarbonate was added. An amount of 45 ml of medium was supplied to a 100 mL bottle that contained 0.45 g of a feed sample (4 replicates per each sample).

Then, each bottle was inoculated under carbon dioxide with 5 mL of mixed rumen microbes. Rumen fluid was obtained from three sheep (49.5±2.5 kg) each fitted with a rumen fistulae before the morning feed, and immediately strained through four layers of cheesecloth.

The animals were fed with 1.5 kg of DM alfalfa hay and 0.4 kg DM concentrates (165 g CP/kg of DM) per head per day. The bottles were incubated for 120 h at 39 °C. Then, each bottle content was filtered through a 42 µm filter, and unfiltered residue was dried using a forced-air oven at 60 °C for 48 h, weighted and analyzed for DM and NDF content.

Evaluations for ammonia nitrogen (N-NH₃) concentration, medium pH and dry matter (DM) disappearance were carried out after 48 h of incubation. At the end of the incubation period, 2 mL of medium content (liquid and solid phase) were taken and stored at -20 °C until the next analysis.

DNA extraction

After thawing, samples were shaken and transferred to 1.5 ml micro tubes containing glass beads and vortexed twice for 2 min with incubation on ice between shakings. Tubes were centrifuged at 200 × g for 5 min at 4 °C for the sedimentation of feeds particles. The supernatants (200 µL) were transferred to a fresh 1.5 mL micro tubes and DNA extraction was performed using a genomic DNA Extraction Kit (AccuPrepTM, Bioneer Corporation) following the manufacturer's instructions.

Real time-PCR

Fungi rDNA concentrations were measured using real time PCR relative to total bacteria amplification (∆∆Ct) and the SYBR Green PCR Master Mix Kit (SYBR Green I qPCR Master Mix, Syntol, Russia). The 16s rRNA gene-targeted primer sets used in the present study are described in Table

1. Templates (1 µL) were added to amplification reactions (25 µL) containing 0.6 µL of primer mixture containing 10 pmol of each primer, 11.5 µL of SYBR Green I qPCR Master Mix (Syntol) and 12 µL of deionized water. SYBR Green I qPCR. Master Mix contained KCl, Tris-HCL (pH 8.8), 6.25 mM $MgCl_2$, dNTP, Taq DNA polymerase, Tween, and SYBR Green I. A no-template (sterile distilled water) negative control was loaded on each plate run to screen for contamination and dimmer formation and to set the background fluorescence for plate normalization. Amplification and detection were performed using an applied biosystems (ABI) 7300 sequence detection system under the following conditions: initial denaturation at 95 °C for 5 min was followed by 40 cycles of denaturation at 95 °C for 15 s, annealing at 61 °C for 15 s, extension at 72 °C for 30 s, and then by the melting curve program (60-95 °C with a heating rate of 0.2 °C per second and a continuous fluorescence measurement). For total bacteria the threshold cycle of each standard dilution was determined during the exponential phase of amplification and regressed against the logarithm of known total bacterial. Total bacteria population size is reported as nano gram (ng) per µL of extracted DNA.

The copy number of total bacteria 16 S ribosomal RNA gene was determined as: Log10 copy number= Ct-(y-intercept/efficiency), where the formula parameters were derived from a standard curve of total bacteria. The population of fungi were expressed relative to the estimated abundance of total bacterial 16 S ribosomal RNA gene.

Chemical analysis

The chemical composition of the diet is reported in Table 2. DM content determined for each substrate by drying samples for 48 h in a 65 °C forced air oven (AOAC, 1990). Dry samples of the feed were maintained overnight at 550 °C in a furnace and organic matter (OM) was subsequently calculated as 100 minus the percentage ash (AOAC, 1990). Total nitrogen of each diet was determined by the kjeldahl method (AOAC, 1990), crud protein was calculated as N × 6.25. Ether extract was determined following the AOAC (1990).

The neutral detergent fiber (NDF) and acid detergent fiber (ADF) of the feed samples were analyzed by the detergent system using the sequential procedure of Van Soest et al. (1991).

For N-NH₃ concentration determination, a 5 mL of each sample was acidified with 5 mL of 0.2 N HCL and was measured by the kjeldahl method. The medium pH was measured at the end of incubation with a pH meter (Metrohm744, Switzerland).

The disappearance of DM was calculated as original weight of the DM minus weight of dry residue (after incubation) divided by weight of the original sample.

Table 1 PCR primers utilized for amplifying the target bacteria

Target species	Forward/reverse	Primer sequence	Reference
Total bacteria	F	GTGSTGCAYGGYTGTCGTCA	Maeda *et al.* (2003)
	R	ACGTCRTCCMCACCTTCCTC	
Fungi	F	GAGGAAGTAAAAGTCGTAACAAGGTTTC	Zhang *et al.* (2008)
	R	CAAATTCACAAAGGGTAGGATGATT	

Table 2 Dry matter and chemical composition of alfalfa hay, sugar beet pulp and barley grain

Composition (% of DM)	Alfalfa hay	Sugar beet pulp	Barley grain
Dry matter (DM)	90.1	85.2	92.3
Crude protein (CP)	14.2	12.4	9.4
Ash	7.2	9.5	3.1
Neutral detergent fiber	45.1	43.1	25.1
Acid detergent fiber	29.9	23.6	7.5
Ether extract	2.1	0.5	2.1

Statistical analyses

Statistical analysis of data from *in vitro* gas production test were made using SAS (1999) with the following model:

$$Y_{ijk} = \mu + A_i + \beta_j + T_k + e_{ijk}$$

Where:

Y: depended variable.

μ: overall mean.

A_i: effect of EO.

B_j: effect of feed sample.

T_k: effect of concentration.

e_{ijk}: residual error, were conducted in a $3 \times 6 \times 3$ factorial design.

Statistical analysis of rumen fermentation characteristics were conducted as a randomized complete design using SAS program with the following model:

$$Y_{ij} = \mu + A_i + e_{ij}$$

Where:

Y: depended variable.

μ: overall mean.

A_i: effect of EO.

e_{ij}: residual error.

Evaluations of significance of difference between means of treatments and controls were determined at (P<0.05) using the Dunnett test.

RESULTS AND DISCUSSION

The effect of treatments on *in vitro* gas production parameters

The effects of EO on *in vitro* gas production parameters are shown in Table 3. Addition EO to AH showed that cinnamon significantly decreased c parameter at 200 μL (P<0.05), but it did not produce gas at 400 μL.

Also, supplementation of cumin significantly decreased gas production parameters and b parameter at 200 and 400 μL (P<0.05), respectively. Moreover, addition of garlic oil and fennel significantly increased c parameter at 200 μL (P<0.05) and black pepper significantly increased c parameter at both 200 and 400 μL (P<0.05) and monensin significantly increased both gas production parameters (P<0.05). In addition EO, BP showed that cinnamon and black pepper significantly decreased c parameter but increased b parameter at 200 μL, while, cinnamon did not produce gas at 400 μL. Moreover, cumin addition significantly decreased gas production parameters and b parameter at 200 and 400 μL (P<0.05), respectively. Also, fennel addition significantly decreased b parameter at both 200 and 400 μL (P<0.05), and garlic oil decreased b parameter at 200 μL (P<0.05), while, monensin significantly increased it. Addition evaluations for EO to BG showed that cinnamon and garlic oil additions significantly increased b parameter at 200 μL, while it did not produce gas at 400 μL. Also, black pepper addition and monensin significantly decreased c parameter but increased b parameter (P<0.05), while fennel decreased b parameter at 200 μL (P<0.05).

The effects of treatments on rumen fermentation characteristics on *in vitro* condition

The effects of treatments on disappearance of concentrations of DM, pH and $N-NH_3$ are shown in Table 4 after 48 h incubation, respectively. Additions of all EO to AH significantly increased pH of culture (P<0.05), but decreased disappearance of DM except in fennel. Black pepper and cinnamon significantly decreased $N-NH_3$ concentration (P<0.05).

In addition, addition of cinnamon and cumin significantly decreased disappearance of DM and $N-NH_3$ concentration of BP (P<0.05), while, cinnamon significantly increased pH of culture (P<0.05). Moreover, all treatments significantly decreased disappearance of DM of BG except fennel. Also, cinnamon and cumin significantly increased pH of culture but decreased $N-NH_3$ concentration (P<0.05).

Table 3 Effect of treatments on gas production of alfalfa hay, sugar beet pulp and barley grain

Treatments	Amount[1]	Parameters	
	Concentration (µL/g DM)	b (mL/0.3 g DM)	c (/h)
Alfalfa hay (AH)	-	49.3	0.07
AH + cinnamon	200	49.1	0.03*
AH + cumin	200	28.7*	0.03*
	400	20.4*	0.06
AH + fennel	200	44.7	0.17*
	400	45.6	0.09
AH + garlic	200	44.7	0.10*
	400	40.6	0.08
AH + black pepper	200	45.8	0.13*
	400	52.2	0.11*
AH + monensin[1]	-	60.9*	0.12*
Sugar beet pulp (SBP)	-	84.0	0.09
SBP + cinnamon	200	98.1*	0.05*
SBP + cumin	200	47.5*	0.03*
	400	44.5*	0.06
SBP + fennel	200	73.8*	0.10
	400	69.7*	0.01
SBP + garlic	200	71.5*	0.08
	400	75.0	0.07
SBP + black pepper	200	93.8*	0.01*
	400	80.9	0.07
SBP + monensin[1]	-	94.7*	0.07
Barley grain (BG)	-	98.7	0.04
BG + cinnamon	200	110.6*	0.04
BG + cumin	200	107.1	0.03
	400	104.8	0.02
BG + fennel	200	87.8*	0.04
	400	106.4	0.06
BG + garlic	200	113.9*	0.04
	400	108.4	0.04
BG + black pepper	200	108.6*	0.07*
	400	100.7	0.04
BG + monensin[1]	-	109.6*	0.07*
SEM	-	1.49	0.004
P-value	-	0.05	0.05

* (P<0.05).
SEM: standard error of the means.
[1] Monensin used as 5 µmol in medium.

The effects of treatments on ruminal anaerobic fungal population

Effects of treatments on ruminal anaerobic fungal population are shown in Figure 1. Results of the present study demonstrate difference between treatments for evaluations of ruminal fungal population. All treatments significantly decreased (P<0.05) both fungal population of AH and fungal population of BG in rumen except cumin.

In general, rumen microbial activity and gas production were affected by the use of EO. These results are consistent with those reported in Fraser et al. (2007) and Macheboeuf et al. (2008) in that cinnamon decreased gas production at incubation periods of 24 h and 16 h. Also, Jahani-Azizabadi et al. (2009) reported that addition of cumin powder (4% of incubated DM) on in vitro rumen microbial fermentation of alfalfa hay decreased production of gas and methane.

On the other hand, results of the present study did not conform to those reported in Hodjatpanah-Montazeri et al. (2015) in which cumin and cinnamon increased gas production 24 h post incubation. Other reports have demonstrated that garlic decreased gas production (Busquet et al. 2005; Kilic et al. 2011).

Furthermore, it has been reported that garlic oil has antimicrobial activity and a high dose of garlic could have a detrimental effect on ruminal fermentation (Feldberg et al. 1988). Cinnamon is a monophenolic compound that binds proteins and disturbs energy metabolism in cells (Wendakoon and Sakaguchi, 1995). Oussalah et al. (2006) also reported that cinnamon had an inhibitory effect on pseudomonas putida. The chemical structure determines the activity and antimicrobial effect of an EO (Dorman and Deans, 2000).

Table 4 Effect of treatments on rumen fermentation characteristics in *in vitro* condition of alfalfa hay, sugar beet pulp and barley grain

Treatments	Parameters[1]		
	disappearance of DM	pH	N-NH$_3$ (mg/100 mL)
Alfalfa hay (AH)	0.73	6.76	47.20
AH + cinnamon	0.41*	7.05*	29.70*
AH + cumin	0.55*	6.95*	48.97
AH + fennel	0.734	6.89*	51.14
AH + black pepper	0.57*	6.88*	26.82*
Beet pulp (BP)	0.88	6.63	30.86
SBP + cinnamon	0.47*	7.27*	22.58*
SBP + cumin	0.67*	6.66	22.87*
SBP + fennel	0.89	6.65	29.05
SBP + black pepper	0.91	6.67	29.86
Barley grain (BG)	0.90	6.56	30.60
BG + cinnamon	0.38*	7.03*	26.28*
BG + cumin	0.78*	6.9*	25.53*
BG + fennel	0.88	6.58	33.55
BG + black pepper	0.81*	6.68	29.19
SEM	0.022	0.02	2.73
P-value	0.05	0.05	0.05

* ($P<0.05$).

SEM: standard error of the means.

Various results have been reported on the effects of EO on ruminal pH. The findings of the present study show that cinnamon and cumin caused a significant ($P<0.05$) increase in pH of culture. Results presented here are consistent with those reported in Fraser *et al.* (2007) in which cinnamon addition increased the fermenter pH in a RUSITEC. Similar results were obtained by Busquet *et al.* (2006) in which addition of cinnamon oil increased fermenter pH an *in vitro* fermentation system. Moreover Hodjatpanah-Montazeri *et al.* (2015) showed the pH value of corn silage increased by cumin in *in vitro* rumen fermentation. Benchaar *et al.* (2007) reported an increase in ruminal pH when dairy cows received EO. In contrast to the effects observed in this study, other reports show that EO had no effect on pH (Yang *et al.* 2007). Whiles Jahani-Azizabadi *et al.* (2014) investigated EO of cumin at its 280 µL L^{-1} reduced pH. Increase in value pH may be related to the higher amount of ruminal NH$_3$ concentration in our experiment. Results of different studies for the impact of EO on ruminal N-NH$_3$ concentration are variable. In some studies, EO showed good potential to alter rumen microbial fermentation and ammonia producing bacteria in the rumen (McIntosh *et al.* 2003; Patra and Yu, 2014). The results of the current study on cinnamon and cumin impact showed a significant ($P<0.05$) decrease in N-NH$_3$ concentration. Several authors have reported a reducing effect on N-NH$_3$ concentration in the incubation medium by cinnamon (Jahani-Azizabadi *et al.* 2011; Jahani-Azizabadi *et al.* 2014). Similar results were obtained by Cardozo *et al.* (2005) when supplementing ocinnamon in an *in vitro* fermentation system fed a high-concentrate finishing diet.

In addition, Busquet *et al.* (2006) demonstrated that cinnamon inhibited N-NH$_3$ concentration at high concentration (3000 mg/L). Fraser *et al.* (2007) reported that cinnamon addition decreased N-NH$_3$ in a RUSITEC. Macheboeuf *et al.* (2008) reported that cinnamon decreased ammonia in an *in vitro* fermentation system. Ferme *et al.* (2004) showed that cinnamon addition to an *in vitro* rumen simulation system resulted in a reduction in *Prevotella* spp. a group of bacteria known to be involved in deamination, providing evidence of a mechanism of activity. These results suggest that cinnamon extract inhibited peptidolysis. The results of this study indicated that black pepper significantly ($P<0.05$) decreased N-NH$_3$ concentration of AH. These results are consistent with results reported in Cardozo *et al.* (2004) in which pepper addition decreased N-NH$_3$ concentration in a continuous culture. Busquet *et al.* (2006) demonstrated that capsicum oil (active component of black pepper) inhibited N-NH$_3$ concentration at high concentration (3000 mg/L). Cardozo *et al.* (2005) reported that capsaicin decreased N-NH$_3$ concentration in an *in vitro* fermentation system fed a diet of straw/concentrate 10:90. Although several studies have demonstrated antimicrobial activity of capsicum oil, the lack of the effect of capsicum oil compared with other EO could be related to its low content of oxygenated hydrocarbons. Some studies have observed that hydrocarbons monoterpenes are in general significantly less active than oxygenated monoterpenes (Cox *et al.* 2001). Decreased N-NH$_3$ concentration with EO suggests that these types of additives decrease deamination activity of bacteria that rely on peptides and AA as a source of N (Taghavi-Nezhad *et al.* 2014).

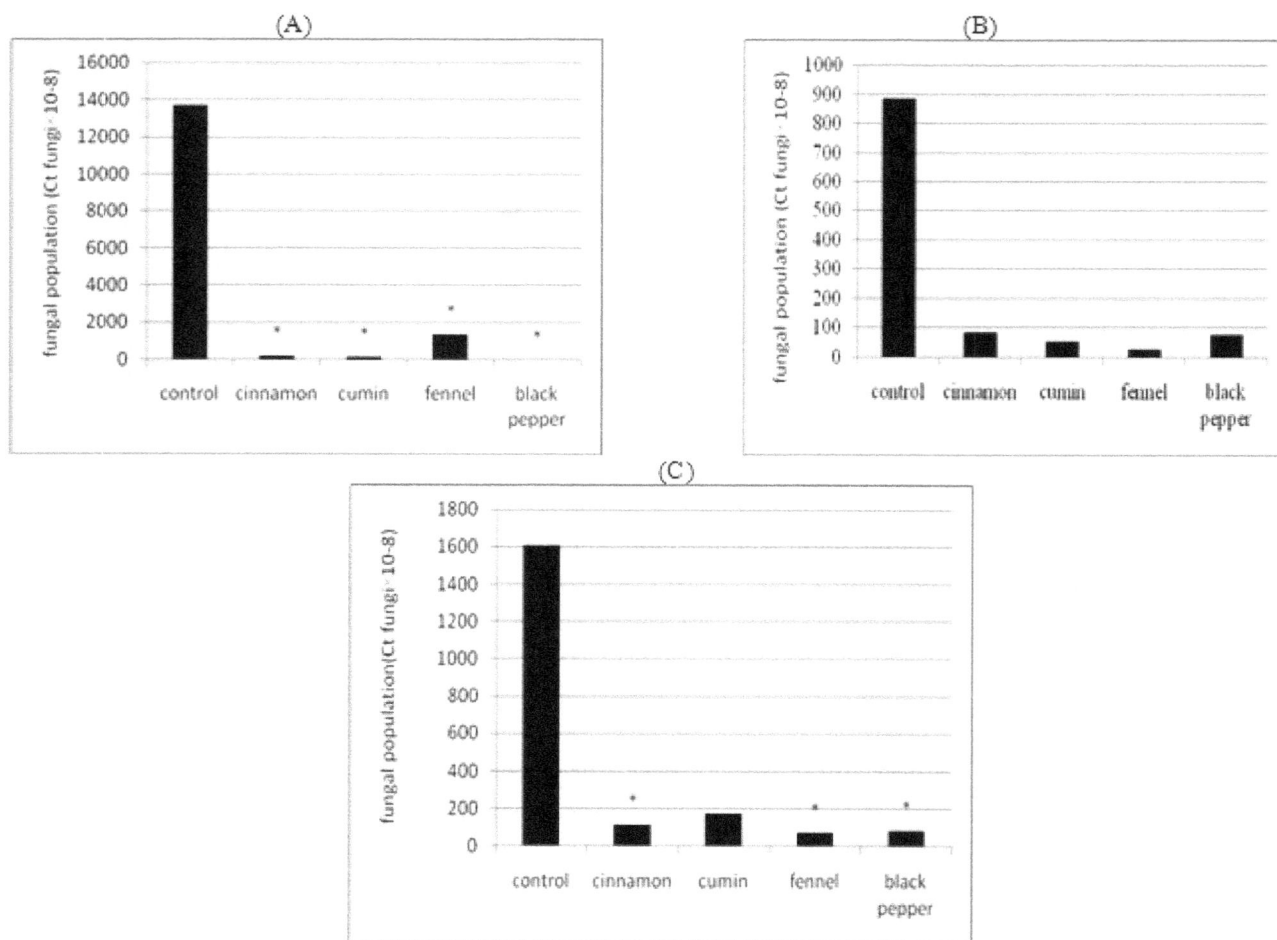

Figure 1 Effect of treatments on quantity of the fungal population existing in the *in vitro* culture relative to total bacteria population (Ct fungi×10-8) of alfalfa hay (A), sugar beet pulp (B), barley grain (C)
Asterisk differ significantly from the control (P<0.05)

Wallace *et al.* (2002) suggested that anti-microbial properties of EO can be exploited to modulate activity of rumen microbial populations by reducing degradation of dietary protein and thereby enhancing rumen N escape. Also, recent studies indicated that application of EO resulted in a significant reduction in protozoal population in the rumen (Patra and Yu, 2014; Khorrami *et al.* 2015; Lin *et al.* 2013). Rumen protozoa are responsible for a considerable amount of rumen ammonia due to predation of rumen bacteria and consumption of feed proteins (Bonhomme, 1990; Firkins *et al.* 2007). Therefore, a portion of observed reduction in ammonia concentration may be due to anti-protozoa effects of the EO (Lorenz *et al.* 2011). Perhaps the different effects of EO on feed samples reflect that there are different species responsible for digestion of the various proteins. Beside these differences among studies, results may have been affected by method of experimentation.

In the present study, cinnamon and cumin decreased disappearance of DM confirming the findings of some previous studies (Jahani-Azizabadi *et al.* 2011; Jahani-Azizabadi *et al.* 2014; Hodjatpanah-Montazeri *et al.* 2015). These findings are similar with those of Fraser *et al.* (2007) who observed a decrease in DM digestibility when cinnamon leaf oil was supplied at 500 mg/L in a RUSITEC. Because fiber disappearance was reduced by cinnamon, this can be lead to the conclusion that fungi were affected by cinnamon addition. However, the magnitude of the reduction in DM disappearance was greater for barley grain than for alfalfa. This implies that microbial populations involved in concentrate digestion were affected more by cinnamon than those involved in forage digestion. Reduction in DM disappearance might be nutritionally unfavorable for an animal but a decrease in concentration of N-NH$_3$ because of an increase in ruminal escape dietary protein can improve efficiency of the nitrogen use by the ruminants (Van Nevel and Demeyer, 1988).

Moreover, this study showed that EO resulted in a decrease in the rumen fungal population.

These results are consistent with those reported in McIntosh *et al.* (2003) showing that addition of a blend of EO inhibited *N. frontalis* fungal activity in the rumen with H_2 production at the concentration of 40 ppm. Similar results were obtained by Talebzadeh *et al.* (2013) in that fungal growth (*Neocalimastix* spp.) were inhibited and activity of fibrolytic fungal enzymes were reduced by adding ajowan EO. Kongmun *et al.* (2010) reported that ruminal anaerobic fungi population was not influenced by coconut oil and garlic powder. Although, garlic oil had fungicidal properties, as reported by Pai and Platt (2008). The antifungal effect of EO may also have contributed to decreased fiber digestibility. However very little information is available on the effect of EO on rumen fungal population. The results observed in the present trial suggest that careful selection of plant extracts may help to improve efficiency of rumen microbial fermentation. However, future research may help to establish the efficacy of EO as rumen microbial modulators.

CONCLUSION

The results obtained in the present study suggest that EO are able to modify rumen fermentation by changing gas production, protein metabolism or by causing disappearance of DM. In terms of protein metabolism, the mechanism of action may be related to inhibition of deamination, although the inhibition of peptidolysis has also been suggested for some EO. However, further research is necessary to determine optimal doses of EO, the potential for adaptation of rumen microflora to the actions of these additives, the fate of these products in an animals and the presence of residue in meat or milk, and effects on animal performance.

ACKNOWLEDGEMENT

The authors gratefully acknowledge the technical and financial support of Ferdowsi University of Mashhad.

REFERENCES

AOAC. (1990). Official Methods of Analysis. Vol. I. 15th Ed. Association of Official Analytical Chemists, Arlington, VA, USA.

Arroquy J.I., Cocran R.C., Nagarja T.G., Titgemeyer E.C. and Johnson D.E. (2005). Effect of type non-fiber carbohydrate on *in vitro* forage fiber digestion of low-quality grass hay. *Anim. Feed Sci. Technol.* **120**, 93-106.

Bakkali F., Averbeck S., Averbeck D. and Idaomar M. (2008). Biological effects of essential oils: a review. *Food Chem. Toxicol.* **46**, 446-475.

Balandrin M.F. and Klocke J.A. (1985). Natural plant chemicals: sources of industrial and medicinal materials. *Science.* **228**, 1154-1160.

Benchaar C., Calsamiglia S., Chaves A.V., Fraser G.R., Colombatto D., McAllister T.A. and Beauchemin K.A. (2008). A review of plant-derived essential oils in ruminant nutrition and production. *Anim. Feed Sci. Technol.* **145**, 209-228.

Benchaar C., Petit H.V., Berthiaume R., Ouellet D.R., Chiquette J. and Chouinard P.Y. (2007). Effects of essential oils on digestion, ruminal fermentation, rumen microbial populations, milk production, and milk composition in dairy cows fed alfalfa silage or corn silage. *J. Dairy Sci.* **90**, 886-897.

Bonhomme A. (1990). Rumen ciliates: their metabolism and relationships with bacteria and their hosts. *Anim. Feed Sci. Technol.* **30**, 203-216.

Busquet M., Calsamiglia S., Ferret A., Cardozo P.W. and Kamel C. (2005). Effects of cinnamaldehyde and garlic oil on rumen microbial fermentation in a dual flow continuous culture. *J. Dairy Sci.* **88**, 2508-2516.

Busquet M., Calsamiglia S., Ferret A. and Kamel C. (2006). Plant extracts affect *in vitro* rumen microbial fermentation. *J. Dairy Sci.* **89**, 761-771.

Calsamiglia S., Castillejos L. and Busquet M. (2006). Alternatives to antimicrobial growth promoters in cattle. Pp. 129-167 in Recent Advances in Animal Nutrition. P.C. Garnsworthy and J. Wiseman, Eds. Nottingham University Press, Nottingham, UK.

Cardozo P.W., Calsamiglia S., Ferret A. and Kamel C. (2004). Effects of natural plant extracts on protein degradation and fermentation profiles in continuous culture. *J. Anim. Sci.* **82**, 3230-3236.

Cardozo P.W., Calsamiglia S., Ferret A. and Kamel C. (2005). Screening for the effects of natural plant extracts at different pH on *in vitro* rumen microbial fermentation of a high-concentrate diet for beef cattle. *J. Anim. Sci.* **83**, 2572-2579.

Cox S.D., Mann C.M. and Markam J.L. (2001). Interaction between components of the essential oil of *Melaleuca alternifolia. J. Appl. Microbiol.* **91**, 492-497.

Dorman H.J.D. and Deans S.G. (2000). Antimicrobial agents from plants: antibacterial activity of plant volatile oils. *J. Appl. Microbiol.* **88**, 308-316.

European Commission. (2003). Regulation (EC) No. 1831/2003 of the European Parliament and of the Council of 22 September 2003 on Additives for Use in Animal Nutrition. Off. J. Eur. Union L L268/29.

Feldberg R.S., Chang S.C., Kotik A.N., Nadler M., Neuwirth Z., Sundstrom D.C. and Thompson N.H. (1988). *In vitro* mechanism of inhibition of bacterial cell growth by allicin. *Antimicrob. Agents Chem.* **32**, 1763-1768.

Ferme D., Banjac M., Calsamiglia S., Busquet M., Kamel C. and Avgustin G. (2004). The effects of plant extracts on microbial community structure in a rumen-simulating continuous-culture system as revealed by molecular profiling. *Folia Microbiol.* **49**, 151-155.

Fernandez M., Serrano E., Frutos P., Gir´aldez F.J., Mantec´on A.R. and Llach J.R. (1997). Efecto del aditivo Crina HC sobre la actividad degradativa ruminal en la especie ovina. *Inf. Tech. Econ. Agraria.* **18**, 160-162.

Firkins J.L., Yu Z. and Morrison M. (2007). Ruminal nitrogen metabolism: perspectives for integration of microbiology and nutrition for dairy. *J. Dairy Sci.* **90**, 1-16.

Fraser G.R., Chaves A.V., Wang Y., McAllister T.A., Beauchemin K.A. and Benchaar C. (2007). Assessment of the effects of cinnamon leaf oil on rumen microbial fermentation using two continuous culture systems. *J. Dairy Sci.* **90**, 2315-2328.

Greathead H. (2003). Plants and plant extracts for improving animal productivity. *Proc. Nutr. Soc.* **62**, 279-290.

Guenther E. (1948). The Essential Oils. Publisher D. Van Nostrand Company Inc., New York.

Helander I.M., Alakomi H.L., Latva-Kala K., Mattila-Sandholm T., Pol L., Smid E.J., Gorris L.G.M. and von Wright A. (1998). Characterization of the action of selected essential oil components on gram negative bacteria. *J. Agric. Food Chem.* **46**, 3590-3595.

Hodjatpanah-Montazeri A., Danesh-Mesgaran M., Vakili A.R. and Tahmasebi A.M. (2015). Effect of essential oils of various plants as microbial modifier to alter corn silage fermentation and *in vitro* methane production. *Iranian J. Appl. Anim. Sci.* **6(2)**, 269-276.

Jahani-Azizabadi H., Danesh-Mesgaran M., Vakili A.R. and Heravi-Moussavi A.R. (2009). Screening the activity of medicinal plants or spices on *in vitro* ruminal methane production. *J. Dairy Sci.* **92**, 277-278.

Jahani-Azizabadi H., Danesh-Mesgaran M., Vakili A.R., Rezayazdi K. and Hashemi M. (2011). Effect of various semiarid native medicinal plant essential oils on rumen fermentation characteristics of a high forage diet using *in vitro* batch culture. *African J. Microbiol. Res.* **27**, 4812-4819.

Jahani-Azizabadi H., Danesh-Mesgaran M., Vakili A.R. and Rezayazdi K. (2014). Effect of some plant essential oils on *in vitro* ruminal methane production and on fermentation characteristics of a mid-forage diet. *J. Agric. Sci. Technol.* **16**, 1543-1554.

Khorrami B., Vakili A.R., Mesgaran M.D. and Klevenhusen F. (2015). Thyme and cinnamon essential oils: potential alternatives for monensin as a rumen modifier in beef production systems. *Anim. Feed Sci. Technol.* **200**, 8-16.

Kilic U., Boga M., Gorgulu M. and Şahan Z. (2011). The effects of different compounds in some essential oils on *in vitro* gas production. *J. Anim. Feed Sci.* **20**, 626-636.

Kongmun P., Wanapat M., Pakdee P. and Navanukraw C. (2010). Effect of coconut oil and garlic powder on *in vitro* fermentation using gas production technique. *Livest. Sci.* **127**, 38-44.

Lin B., Lu Y., Salem A.Z.M., Wang J.H., Lianga Q. and Liu J.X. (2013). Effects of essential oil combinations on sheep ruminal fermentation and digestibility of a diet with fumarate included. *Anim. Feed Sci. Technol.* **184**, 24-32.

Lorenz M.M., Karlsson L., Hetta M. and Uden P. (2011). Recycling of microbial N and estimation of protein degradation by *in vitro* gas production. *Anim. Feed Sci. Technol.* **170**, 111-116.

Macheboeuf D., Morgavi D.P., Papon Y., Mousset J.L. and Arturo-Schaan M. (2008). Dose response effects of essential oils on *in vitro* fermentation activity of the rumen microbial population. *J. Anim. Feed Sci.* **145**, 335-350.

Maeda H., Fujimoto C., Haruki Y., Maeda T., Kokeguchi S., Petelin M., Arai H. and Tanimoto I. (2003). Quantitative real-time PCR using TaqMan and SYBR green for *Actinobacillus actin Omycetemcomitans, Porphyromonas gingivalis, Prevotella intermedia*, tetQ gene and total bacteria. *FEMS Immunol. Med. Microbiol.* **39**, 81-86.

McIntosh F.M., Williams P., Losa R., Wallace R.J., Beever D.A. and Newbold C.J. (2003). Effects of essential oils on ruminal microorganisms and their protein metabolism. *Appl. Environ. Microbiol.* **69**, 5011-5014.

Menke K.H. and Steingass H. (1988). Estimation of the energetic feed value obtained from chemical analysis and *in vitro* gas production using rumen fluid. *Anim. Res. Dev.* **28**, 7-55.

Nanon A., Suksombat W. and Yang W.Z. (2014). Effects of essential oils supplementation on *in vitro* and *in situ* feed digestion in beef cattle. *Anim. Feed Sci. Technol.* **196**, 50-59.

Newbold C.J., McIntosh F.M., Williams P., Losa R. and Wallace R.J. (2004). Effects of a specific blend of essential oil compounds on rumen fermentation. *Anim. Feed Sci. Technol.* **114**, 105-112.

Oussalah M., Caillet S., Saucier L. and Lacroix M. (2006). Antimicrobial effects of selected plant essential oils on the growth of a *Pseudomonas putida* strain isolated from meat. *Meat Sci.* **73**, 236-244.

Pai S.T. and Platt M.W. (2008). Antifungal effects of *Allium sativum* (garlic) extract against the *Aspergillus* species involved in otomycosis. *Appl. Microbiol. Lett.* **20**, 14-18.

Patra A.K. and Yu Z. (2014). Effects of vanillin, quillaja saponin, and essential oils on *in vitro* fermentation and protein-degrading microorganisms of the rumen. *Appl. Microbiol. Biotechnol.* **98**, 897-905.

Reichling J., Schnitzler P., Suschke U. and Saller R. (2009). Essential oils of aromatic plants with antibacterial, antifungal, antiviral, and cytotoxic properties: an overview. *Forsch. Komplementmed.* **16(2)**, 79-90.

SAS Institute. (1999). SAS®/STAT Software, Release 8.2. SAS Institute, Inc., Cary, NC. USA.

Taghavi-Nezhad M., Alipour D., Flythe M.D., Zamani P. and Khodakaramian G. (2014). The effect of essential oils of *Zataria multiflora* and *Mentha spicata* on the *in vitro* rumen fermentation, and growth and deaminative activity of amino acid-fermenting bacteria isolated from Mehraban sheep. *J. Anim. Prod. Sci.* **54(3)**, 299-307.

Tajodini M., Moghbeli P., Saeedi H.R. and Effati M. (2014). The effect of medicinal plants as a feed additive in ruminant nutrition. *Iranian J. Appl. Anim. Sci.* **4(4)**, 681-686.

Talebzadeh R. and Alipour D. (2013). The effect of ajowan (*Carum copticum*) essential oils on eukaryotic ruminal microorganisms of Mehraban sheep. *Iranian J. Microbiol.* **4**, 418-421.

Tamminga S. (1996). A review on environmental impacts of nutritional strategies in ruminants. *J. Anim. Sci.* **74**, 3112-3124.

Ultee A., Kets E.P. and Smid E.J. (1999). Mechanisms of action of carvacrol on the food-borne pathogen *Bacillus cereus*. *Appl. Environ. Microbiol.* **65**, 4606-4610.

Van Nevel C.J. and Demeyer D.I. (1988). Manipulation of rumen fermentation. Pp. 387-443 in The Rumen Microbial Ecosystem. P.N. Hobson, Ed. Elsevier Applied Science, New York.

Van Soest P.J. (1982). Nutritional Ecology of the Ruminant. Comstock, Cornell Univ. Press, New York, NY.

Van Soest P.J., Robertson J.B. and Lewis B.A. (1991). Methods for dietary fiber, neutral detergent fiber and nonstarch polysaccharides in relation to animal nutrition. *J. Dairy Sci.* **74,** 3583-3597.

Wallace R.J., McEwan N.R., McIntosh F.M., Teferedegne B. and Newbold C.J. (2002). Natural products as manipulators of rumen fermentation. *Asian-Australasian J. Anim. Sci.* **15,** 1458-1468.

Wendakoon C.N. and Sakaguchi M. (1995). Inhibition of amino acid decarboxylaseactivity of *Enterobacter aerogenes* by active components in spices. *J. Food Prot.* **58,** 280-283.

Yang W.Z., Benchaar C., Ametaj B.N., Chaves A.V., He M.L. and McAllister T.A. (2007). Effects of garlic and juniper berry essential oils on ruminal fermentation and on the site and extent of digestion in lactating cows. *J. Dairy Sci.* **90,** 5671-5681.

Zhang C., Guo Y., Yuan Z., Wu Y., Wang J., Liu J. and Zhu W. (2008). Effect of octadeca carbon fatty acids on microbial fermentation, methanogenesis and microbial flora *in vitro*. *Anim. Feed Sci. Technol.* **146,** 259-269.

11

Chemical Composition, Physical Characteristics, Rumen Degradability of NDF and NDF Fractionation in Rice Straw as an Effective Fibre in Ruminants

A. Teimouri Yansari[1*]

[1] Department of Animal Science, Faculty of Agricultural Science, Sari University of Agricultural Science and Natural Resources, Sari, Iran

*Correspondence E-mail: astymori@yahoo.com

ABSTRACT

In order to determine of physical characteristics of rice straw as an effective source of fiber in ruminants, alfalfa hay, four varieties of rice straw (Taroum Neda, Taroum Neamat, Taroum Sangi, and Asgari), and four rations that contained four varieties of rice straws were investigated. The chemical (dry matter (DM), organic matter (OM), neutral detergent fiber (NDF), nonfiber carbohydrates (NFC) and crude protein (CP)), and physical characteristics (bulk density, water holding capacity (WHC), and soluble and insoluble DM and ash of samples, kinetics of hydration and change in functional specific gravity (FSG) and feed particle size, physically effective factor (pef)) of forages and total mixed ration (TMR) were determined. Except on ether extract and ash content, the DM, OM, NDF, NFC, and CP content of four rice straw and rations were similar but there was different among alfalfa and rice straws. The rice straws had a bulk density lesser than alfalfa. However, TMR had a similar bulk density, WHC, hydration rate, insoluble DM and ash and greater than alfalfa hay. Alfalfa had lesser WHC than rice straws and there were not different in straws. The soluble DM and FSG of rice straws were similar and lesser than those of alfalfa hay. The TMR had similar physical characteristics. Alfalfa hay had greater FSG than rice straw at all incubation times. Four rice straws and four rations were similar in indegradable NDF (iNDF) and total tract NDF digestibility (TTNDFD). Results showed that regardless the system, rice straws were similar in physically effectiveness and physically more effective than alfalfa because of having greater NDF and iNDF content, geometric mean and pef than alfalfa.

KEY WORDS effective fibre, in degradable NDF, physical characteristic, rice straw.

INTRODUCTION

Rice is the world's second largest cereal crop after wheat, with an annual production of about 750 million metric tons (FAO, 2013). It is the staple food of more than half of the world's population. About 91% of it is grown and consumed in Asia. For every 4 tons of rice grain, about 6 tons of straw are produced, therefore this amounts to about 550 million tons of straw and 110 million tons of husks each year. Rice straw has low nutritive values because of low DM digestibility and low protein content (Van Soest, 2006). Rice straw is lesser in lignin and great in silica compared with the other straws. Until today, a lot of investigations have conducted using a variety of chemical and biological treatments to improve rice straw in ruminant nutrition. These treatments involve sodium hydroxide, ammonia, urea, pressure and heat in combinations with steam, pressure and ammonia, urine, enzymes, acids and fungi. However, the main goal of these treatments was enhancement of digestibility dry and organic matter. Nowadays, some new concepts in ruminant nutrition, such as physically effective fibre (peNDF) are being introduced (Mertens, 1997;

Mertens, 2000) to relate the physical characteristics of fibre (primary particle size) to its effects on chewing activity and the biphasic nature of rumen contents. Although particle size measurement is central to all effective fibre systems, nonetheless, some the physical characteristics such as functional specific gravity (FSG), bulk density, water holding capacity (WHC), insoluble ash, etc. influence effectiveness of fire and rate of passage (Teimouri Yansari et al. 2004; Teimouri Yansari and Pirmohammadi, 2009). In addition, plant breeding has been devoted to maximizing grain yield with less interest in the straw. This has resulted in short varieties in which the proportions of straw and leaf blades are reduced (Capper, 1988; Bainton et al. 1991). However, for more forages the physical characteristics and effectiveness have not been investigated.

Rice straw is important forage for in Northern Iran that was produced 1450 million metric tons rice grain (FAO, 2013). Using rice straw for animal production can save grains and provide additional income to farmers and decrease environmental pollution due to the burning of straw after harvest. Development and application of chemical treatments for upgrading straw have stimulated intense interest, but there are still some blind spots on the mechanism with which the treatments improve the nutritive value of straw. It seems that rice straws are sources of indigestible NDF that may retain in the rumen, make a consistence ruminal mat, stimulate rumination, chewing activity and saliva secretion and ultimately buffer rumen pH and increase the concentration of ruminal acetate and milk fat. On the contrary, they had a great ruminal filling factor, therefore; it is often considered as low-quality forage. Four rice varieties including Taroum Neda and Taroum Neamaat as short varieties and Taroum Sangi and Asgari as tall varieties are abundant. However, until now the quality had not studied and their effectiveness was not compared. Thus, the aim of the current experiment was determination of chemical composition, physical characteristics, ruminally degradability parameters of NDF and NDF fractionation of feeds and TMR that contained four different varieties of rice straw.

MATERIALS AND METHODS

Alfalfa at 15% flessering, four different varieties of rice straw (Taroum Neda and Taroum Neamaat as short varieties and Taroum Sangi and Asgari as tall varieties) were harvested, dried and chopped on the same day, at maturity 14 cm above the ground in August 2014 from the agricultural research center of Agricultural and Natural Resource University (SANRU), Sari, Mazandaran, Iran. Individual small rectangular bales (average weight 10 kg) were chopped with a forage field harvester (Jaguar # 62, Class Company, Germany) for theoretical cut length 19 mm.

Feeds were weighed, sub-sampled, dried at 55 °C, ground through a Wiley mill (1 mm screen) and analyzed for DM, OM, Kjeldahl N, ether extract (AOAC, 2002), NDF (Van Soest et al. 1991; using amylase and inclusive of residual ash), ADF (Van Soest et al. 1991) and ash at 605 °C. Non-fibre carbohydrate in g/kg was calculated as: 1000 - [CP + NDF + Ash + EE].

Bulk density (g/mL), WHC (g/g insoluble DM), and soluble and insoluble DM and ash (g/kg) of alfalfa and rice straws were measured as described by Giger-Reverdin (2000). Kinetics of hydration and change in FSG of forages were measured with 100 mL pycnometer at 39.0 ± 0.5 °C (Wattiaux, 1990; Teimouri Yansari et al. 2004). The mixed rumen fluids from two sheep fed only alfalfa were collected before to feeding and rinsed with eight layers of cheese cloth, centrifuged at 3000 × g, for 10 min and the supernatant (with density 1.0068±0.0005 g/mL) were used as hydration solution. Sodium azide (0.50 g/L) and penicillin G (25000 units/L) were added to the hydration solution to prevent microbial growth. About 1.5 g of each sample, in 5 replicates were weighed in pycnometers. The pycnometers were half-filled to allow vigorous shaking after initial soaking of samples and for removal of gas bubbles. The first reading of the total weight of pycnometers was taken after 6 min (0.1 h) of initial soaking, which was the shortest interval necessary to eliminate all gas bubbles. After completely filling the pycnometers, they were again put on the stirring plate for gentle and continual stirring. Pycnometers were refilled and weights were recorded at 0.5, 1.0, 1.5, 2, 4, 6, 12, 24, 36, 48 and 72 h.

During measurements of hydration kinetics, very small gas bubbles accumulated near the junction between adapter and flask of pycnometers, connecting a vacuum pump to pycnometer for 2 min dislodged gas bubbles from the junction. Data were used to estimate the rate of hydration and water uptake or WHC using NLIN procedures of SAS® (SAS, 1998; Wattiaux, 1990). A biexponential model as was described by the function below was used to estimate hydration parameters:

$$Y_t = Ae^{-k_a t} + Be^{-k_b t}$$

Where:

Y_t: water uptake over time (g/g of insoluble DM).

A and B: represent pool sizes of hydration.

k_a and k_b: represent respective fractional rates of hydration (min⁻¹).

Total WHC (g/g of insoluble DM) was calculated as the sum of total solution uptake (sum of A+B) and initial moisture content of samples. A mean for hydration rate that was weighted for pool sizes from biexponential models was calculated: $[(A \times k_a) + (B \times k_b)] / (A+B)$. As mentioned above, in this study, the WHC were measured using filtration method (Giger-Reverdin, 2000; Table 1) and nonlinear curve fitting method (Wattiaux, 1990).

Feed particle size and distribution were determined by dry sieving in four replicates, using the Penn State particle separator. The physical effective factor (pef) of TMR were determined as the sum of retained particle on two 19 and 8mm sieves (pef$_{>8}$; Lammers et al. 1996), and three 19, 8, and 1.18mm sieves (pef$_{>1.18}$; Kononoff, 2002). The NDF of all materials retained on each sieve were measured (Van Soest et al. 1991). The peNDF$_{>8}$ and peNDF$_{>1.18}$ were calculated by multiplying NDF content of each portion on each sieve on pef$_{>8}$ and pef$_{>1.18}$, respectively (Table 2). The geometric mean and its standard deviation were calculated (American Society of Agricultural Engineers, 2002).

Using two ruminally fistulated Zel ewes (BW=30.5±1.8 kg); 5 g sample in 4 replications was weighed in sealed nylon bags (7 cm×8 cm, polyamide, with 15±2 μ pore size) and incubated in the rumen for 240 h (Huhtanen et al. 1994). Sheep housed on front shed, fed a total mixed ration (TMR) containing 50% chopped alfalfa hay, 25% rice straw, 25% barely grain, and mineral/vitamin supplement according to their requirements. On removal, bags were washed using cold water, dried at 55 °C for 48 h, residues for the periods were homogenized and analyzed for Kjeldahl N, NDF, and acid detergent lignin (ADL; Van Soest et al. 1991; Table 1), and multiplied by a fixed factor of 2.4 calculated as ADL × 2.4 (iNDF$_{2.4}$). The pdNDF calculated using the following equation: pdNDF= NDF − iNDF (Cotanch et al. 2014; Raffrenato and Van Amburgh, 2010).

Experimental data were analyzed using the PROC MIXED of SAS (1998) as a completely randomized design with 5 replications by the following model:

$$Y_{ij} = \mu + T_i + e_{ij}$$

Where:
Y_{ij}: dependant variable.
μ: overall mean.
T_i: random effect of treatment.
e_{ij}: experimental error.

The data of particle size was analyzed as a completely randomised design with model effects of forage and two methods of particle size measurement using the REML variance component and PROC MIXED of SAS (1998).

The data of particle size was analyzed as a completely randomized design with model effects of forage and two methods of particle size measurement using the REML variance component and PROC MIXED procedure of SAS

(1998) (Table 1). Mean separation was determined using the PDIFF procedure, and significance was declared at ($P<0.05$).

RESULTS AND DISCUSSION

Dry matter, OM, NDF, NFC and CP content of the four rice straw varieties were similar, except EE and ash content, but there was significant difference among alfalfa and rice straws (Table 1).

Rice straw had greater NDF and ash and lesser NFC and CP than alfalfa. The TMR that contained four different varieties of rice straw were also similar on DM, OM, NDF, NFC and CP content, however, their EE and ash content was significantly different. Previous researches have evaluated rice varieties for their composition and nutritive value and found that there is considerable variation among varieties relative to straw quality (Singh and Singh, 1995; Vadivelloo, 1995; Vadivelloo, 2000; Vadivelloo and Phang, 1996). In addition, short and tall varieties are different in chemical composition and digestibility relative to leaf, sheath and stem proportions.

Leafiness is associated with height among varieties in contrast to other grasses (Vadivelloo, 1995). Leaves tend to be less digestible than stems (Vadivelloo, 1995; Vadivelloo, 2000; Vadivelloo and Phang, 1996). The total mixed rations that contained Taroum Neda had lesser EE and ash content than others (Table 1). However, the quality of rice straw varieties is highly dependent on soil type and any genetic study will require control of the soil type (Van Soest, 1994). In the current study, since all varieties cultivated at the similar condition, the similarity in chemical composition was expected. Bulk density or packing density is the ratio of the mass of a collection of discrete pieces of solid material to a sum of the volume of the solid in each piece, the voids within the pieces, and the voids among the pieces of the particular collection (D3766, D32, ASTM Committee EO2 On Terminology, 2000). The rice straws had a bulk density lesser than alfalfa. The values of bulk density for rice straws and alfalfa were lesser than 1 and confirmed that as other forages these materials easily bounced over ruminal particulate post feeding (Table 2). There is a negative correlation among NDF and bulk density. Singh and Narang (1991) and, Giger-Reverdin (2000) reported that feedstuffs with high NDF content had low bulk density, and might have more effect on rumen fill than feedstuffs with high bulk density. Hence, forages that occupy larger volumes per unit of DM weight should have a greater effect on fill than another feeds (Wattiaux, 1990). Wattiaux (1990); Van Soest, (1994) and Van Soest, (2006) reported that bulk density influences dry matter intake (DMI), passage rate, and ruminal mean retention time.

Table 1 Chemical composition (% of DM) of feeds and total mixed rations that contained four different varieties of rice straws

Item	DM	OM	NDF	NFC	CP	EE	Ash
Feeds							
Alfalfa hay	89.20[a]	91.90[a]	48.63[b]	27.30[a]	14.16[a]	1.57[b]	8.33[c]
Taroum Neda	90.50[b]	86.77[b]	76.80[a]	3.21[b]	5.37[b]	1.41[a]	13.23[ab]
Taroum Neamaat	90.53[b]	87.23[b]	77.13[a]	2.33[b]	5.50[b]	1.53[ab]	13.50[a]
Taroum Sangi	90.43[b]	87.50[b]	77.20[a]	3.02[b]	5.10[b]	1.63[a]	12.63[b]
Taroum Asgari	90.70[b]	86.83[b]	78.63[a]	2.76[b]	5.37[b]	1.60[a]	12.63[b]
SEM	0.333	0.306	0.388	0.509	0.126	0.010	0.124
P-values	0.0208	< 0.0001	< 0.0001	< 0.0001	< 0.0001	0.0058	< 0.0001
Total mixed rations that contained four different varieties of rice straw							
Taroum Neda	91.83	92.23	53.13	22.60	11.20	1.40[b]	11.67[c]
Taroum Neamaat	92.93	92.13	52.80	22.90	10.70	1.53[ab]	12.07[bc]
Taroum Sangi	92.67	92.87	52.93	21.57	11.10	1.63[a]	12.50[ab]
Taroum Asgari	92.70	92.67	53.20	21.47	10.90	1.60[a]	12.83[a]
SEM	1.580	0.688	0.637	0.947	0.111	0.009	0.127
P-values	0.4366	0.6738	0.9210	0.2518	0.3279	0.0301	0.0185

DM: dry matter; OM: organic matter; NDF: neutral detergent fiber; NFC: nonfiber carbohydrates; CP: crude protein and EE: ether extracts.
The means within the same row with at least one common letter, do not have significant difference (P>0.05).
SEM: standard error of the means.

Table 2 Physical characteristics of feeds and total mixed rations that contained four different varieties of rice straws

Item	Bulk density (g/mL)	Water holding capacity (g/g insoluble DM)	Hydration rate (g/g insoluble DM /min)[1]	Hydration rate (g/g insoluble DM /min)[2]	Functional specific gravity	Soluble DM (g/g DM)	Insoluble DM (g/g DM)	Insoluble ash (% of ash)	Insoluble ash (g/g DM)
Feeds									
Alfalfa hay	0.838[a]	3.09[b]	0.058[b]	0.068[b]	1.143[a]	0.288[a]	0.712[b]	91.30[a]	8.70[b]
Taroum Neda	0.645[b]	6.52[a]	0.069[a]	0.075[a]	1.005[b]	0.128[b]	0.872[a]	88.87[b]	11.13[a]
Taroum Neamaat	0.655[b]	6.62[a]	0.066[a]	0.076[a]	1.004[b]	0.130[b]	0.870[a]	89.50[b]	10.50[a]
Taroum Sangi	0.644[b]	6.65[a]	0.067[a]	0.073[a]	1.003[b]	0.129[b]	0.871[a]	89.16[b]	10.13[a]
Taroum Asgari	0.670[b]	6.66[a]	0.069[a]	0.075[a]	1.006[b]	0.137[b]	0.866[a]	88.87[b]	10.84[a]
SEM	0.032	0.055	0.004	0.003	0.015	0.002	0.002	0.002	0.002
P-values	< 0.0001	0.0455	< 0.0001	< 0.0001	0.0036	< 0.0001	< 0.0001	0.0041	0.0041
Total mixed rations that contained four different varieties of rice straw									
Taroum Neda	0.621	5.97	0.038[b]	0.042[b]	1.174	0.210	0.790	79.00	8.97[b]
Taroum Neamaat	0.621	5.89	0.035[a]	0.043[a]	1.186	0.200	0.800	80.30	9.77[ab]
Taroum Sangi	0.622	5.93	0.032[a]	0.042[a]	1.172	0.213	0.787	78.67	11.40[a]
Taroum Asgari	0.618	6.10	0.034[a]	0.041[a]	1.175	0.214	0.786	78.77	10.17[ab]
SEM	0.046	0.038	0.001	0.002	0.235	0.022	0.033	0.042	0.022
P-values	0.056	0.0654	< 0.0001	< 0.0001	0.0765	0.0632	0.0578	0.0672	0.0001

[1] Water holding capacity that measured using filtration methods (Giger-Reverdin, 2000).
[2] Water holding capacity that calculated using curve fitting methods (Wattiaux, 1990).
The means within the same row with at least one common letter, do not have significant difference (P>0.05).
SEM: standard error of the means.

As presented in Table 1, since rice straws had a high NDF content than alfalfa and their bulk densities were lesser than alfalfa, Taroum Asgari had relatively greater bulk density compared to others because of greater ash content. However, TMR that contained four different varieties of rice straw had similar bulk density. The WHC, hydration rate, insoluble DM, and ash contents of rice straws significantly greater than alfalfa hay, however, there were no difference among straws. In the current experiment, hydration rate is measured using two methods.

The values obtained using filtration methods (Giger-Reverdin, 2000) were lesser than using curve fitting methods (Wattiaux, 1990).

Nonetheless, in both the methods, alfalfa had significantly lesser WHC than rice straws and there were no significant different amongst straws. It seems that greater values for WHC in straws varieties were the result of a high NDF content and lesser bulk density. On the contrary, soluble DM and FSG of rice straws were significantly lesser than alfalfa hay (Table 2).

In addition, except insoluble ash content in ration, TMR that contained four different varieties of rice straw were similar for others physical characteristics. The FSG of alfalfa and rice straws over incubation time in pycnometer is presented in Table 3. Alfalfa hay had significantly greater FSG than four varieties of rice straw at all incubation times. Using the original two sieves Penn State particle separator, the distribution of the particle for rice straws and TMR on different sieves was significantly different. In this system, the geometric means of the particle were significantly different (Table 4). Using three sieves of Penn State particle separator, the distribution of particle for rice straws and TMR on different sieves also, were significantly different. However, in this system, the geometric means of particle for rice straws and TMR were similar (Table 4). In addition, the values for $pef_{>8}$ and $peNDF_{>8}$ were significantly greater for rice straws than alfalfa; however these values were similar for all TMR that contained one variety of rice straws. Also, the values of $pef_{>1.18}$ and $peNDF_{>1.18}$ had a similar trend. Comparison of $pef_{>8}$ and $pef_{>1.18}$ showed that $pef_{>1.18}$ were significantly greater than $pef_{>8}$ for alfalfa, rice straws and TMR that confirmed with pervious researchers (Teimouri et al. 2004). The distribution of particle size showed that regardless the system, rice straws were more physically effective than alfalfa because they had greater geometric mean and $pef_{>8}$, $pef_{>1.18}$, $peNDF_{>8}$, and $peNDF_{>1.18}$ than alfalfa (Table 4). These characteristics confirmed that different varieties of rice straws had no significant difference on $pef_{>8}$, $pef_{>1.18}$, $peNDF_{>8}$, and $peNDF_{>1.18}$. Therefore, their physical properties especially physically effectiveness were similar, measured using the original version of Penn State particle separator (Lammers et al. 1996) and the new version of Penn State particle separator (Kononoff, 2002). Rice straws had greater lignin and silica and were limiting factor to rice straw quality. As a viewpoint, rice straws are good source of indigestible NDF of effective NDF that may retain in the rumen, made a consistence ruminal mat, stimulate rumination, chewing activity, and saliva secretion, and ultimately buffer rumen pH and increase concentration of ruminal acetate and milk fat. Contrarily, they had a great ruminal filling factor; therefore, it is often considered as low-quality forage.

Alfalfa had greater soluble, slowly degradable, potential degradable fraction, and rate of degradability for NDF in the rumen, pdNDF and total-tract NDF digestibility (TTDNDF) than four rice straws. Also, the slowly degradable, the potential degradable fraction, and rate of degradability for NDF in the rumen, the content of NDF, ADL, $iNDF_{288}$, pdNDF and TTDNDF four rice straws were similar (Table 5). The TMR that contained four different varieties of rice straw were similar in rate of degradability, NDF, ADF, ADL, $iNDF_{288}$, $iNDF_{2.4}$, and TTDNDF but the ration

that contained Taroum Neda had lesser soluble, slowly degradable, and potential degradable fraction than other rations. Although $iNDF_{288}$ (% of DM) of four rations had not significantly different but the $iNDF_{288}$ as proportion of NDF were significantly different. Taroum Asgari and Taroum Neda had the greatest and lowest the iNDF288 as proportion of NDF, respectively. Fiber digestion occurs primarily in the rumen and is the result of a dynamic process that is affected by the chemical nature of the plant fiber that controls the digestion and passage of fiber within the animal's digestive tract. Rate of fiber digestion (K_d) and the proportion of NDF that is pdNDF vary considerably between and within forage types (Van Soest, 1994). Rate of passage of fiber is primarily affected by level of intake of the animal, and, consequently, fiber digestibility increases with longer retention time of feed in the rumen. Recently, a model was developed to use an in vitro NDF fermentation assay to measure the proportion pdNDF and rate of digestion of NDF to predict TTNDFD (Cotanch et al. 2014). The digestibility of forage and the capacity of ruminants to consume it are largely influenced by its content of NDF that is directly related to pdNDF as the NDF fraction which disappears after a long incubation period and leaving the iNDF which is unavailable for microbial digestion. According to some studies, the determination of iNDF should be included in all basic feedstuff analysis because it is an ideal fraction which has zero digestibility, uses for the estimation of pdNDF, and recommended that there should be a defined proportion of iNDF in the diet (Cotanch et al. 2014; Zali et al. 2015). In addition, Lippke (1986) suggested that maximum iNDF consumption is about 20 g/kg $BW^{0.75}$ per day, however, more research is required to resolve if this value is relevant for different production systems and different forages.

The forages can have the same NDF content but differ vastly in iNDF. In the current experiment, without significant differences, four varieties of rice straw had high NDF and iND_{F288} content. Nutritional models predict dietary iNDF to rumen digesta load and feed intake because there are strongly negative relationships between iNDF and feed intake when iNDF content exceeds 15% of TMR (Raffrenato and Van Amburgh, 2010), and the iNDF as a predictor of OM digestibility in forage-based diets (Cotanch et al. 2014; Zali et al. 2015). The relationship between DM intake and NDF is greater than just NDF content in the diet but also dependent on the pdNDF (Lippke, 1986). The pdNDF fraction is the difference between the NDF and iNDF. The iNDF component is the rate-limiting constituent of forages at greater NDF level. The iNDF is unavailable to microbial digestion in ruminants even if the total tract residence time of fibre is extended to effectively an infinite time.

Table 3 Functional specific gravity of alfalfa and rice straws according to incubation time (h) in pycnometer

Feeds	Incubation time in pycnometer (h)										
	0.1	0.5	1	2	4	6	12	24	36	48	72
Alfalfa hay	1.116[a]	1.237[a]	1.258[a]	1.275[a]	1.299[a]	1.318[a]	1.367[a]	1.443[a]	1.475[a]	1.443[a]	1.455[a]
Taroum Neda	1.005[b]	1.009[b]	1.011[b]	1.021[b]	1.020[b]	1.024[b]	1.026[b]	1.029[b]	1.031[b]	1.047[b]	1.046[b]
Taroum Neamaat	1.004[b]	1.013[b]	1.019[b]	1.026[b]	1.021[b]	1.021[b]	1.023[b]	1.025[b]	1.024[b]	1.034[b]	1.045[b]
Taroum Sangi	1.003[b]	1.012	1.018[b]	1.028[b]	1.034[b]	1.035[b]	1.038[b]	1.040[b]	1.042[b]	1.044[b]	1.046[b]
Taroum Asgari	1.006[b]	1.009[b]	1.017[b]	1.024[b]	1.033[b]	1.039[b]	1.043[b]	1.047[b]	1.049[b]	1.052[b]	1.057[b]
SEM	0.023	0.033	0.032	0.021	0.016	0.023	0.041	0.054	0.045	0.044	0.034
P-values	0.005	0.001	0.032	0.023	0.022	0.001	0.001	0.002	0.003	0.001	0.003

The means within the same row with at least one common letter, do not have significant difference (P>0.05).
SEM: standard error of the means.

Table 4 Determination of particle size distribution using Penn State particle separators

Separator sieves	Feeds							Total mixed rations that contained different rice straw					
	Alfalfa hay	Taroum straw				SEM	P-value	Neda	Neamaat	Sangi	Asgari	SEM	P-value
		Neda	Neamaat	Sangi	Asgari								
Using original Penn State particle separator													
19 mm	15.0[c]	16.0[b]	15.0[c]	17.0[a]	15.0[c]	0.30	0.0001	13.0[b]	14.0[a]	13.0[b]	14.0[a]	0.27	0.0001
8 mm	35.0[c]	56.0[b]	57.0[a]	54.0[d]	55.0[c]	0.21	0.0001	35.0[b]	36.0[a]	34.0[c]	35.0[b]	0.19	0.0001
pan	50.0[a]	28.0[d]	28.0[d]	29.0[c]	30.0[b]	0.12	0.0001	52.0[b]	50.0[c]	53.0[a]	51.0[bc]	0.55	0.0012
GM (mm)	7.80[c]	8.21[c]	8.13[d]	8.45[a]	8.21[b]	0.003	0.0001	6.34[b]	6.21[b]	7.35[a]	7.38[a]	0.03	0.0001
SDGM (mm)	3.33	3.33	3.51	3.61	3.54	-	-	3.32	3.22	3.40	3.14	-	-
pef$_{>8}$[2]	0.50[b]	0.72[a]	0.72[a]	0.71[a]	0.70[a]	0.240	0.0001	0.48	0.50	0.47	0.49	0.28	0.0342
peNDF$_{>8}$[1]	34.04[b]	54.53[a]	55.53[a]	55.58[a]	55.04[a]	0.082	0.0001	25.54	26.47	24.82	26.03	1.95	0.3441
Using new Penn State particle separator													
19 mm	15.0[b]	15.0[b]	14.0[c]	13.0[d]	17.0[a]	0.28	0.0001	14.0[a]	12.0[c]	13.0[b]	14.0[a]	0.22	0.0003
8 mm	35.0[c]	55.0[b]	57.0[a]	58.0[a]	54.0[b]	0.57	0.0001	34.0[a]	33.0[b]	33.0[b]	31.0[c]	0.23	0.0011
1.18 mm	35.0[a]	28.0[b]	25.0[b]	24.0[b]	25.0[b]	1.97	0.0001	37.0[b]	38.0[a]	36.0[c]	37.0[b]	0.26	0.0034
pan	15.0[a]	2.0[b]	4.0[b]	5.0[b]	4.0[b]	1.76	0.0001	15.0	17.0	18.0	18.0	0.43	0.0003
GM (mm)	7.68	8.85	9.03	9.04	9.28	0.986	0.0653	7.40	7.07	7.30	7.27	0.321	0.1239
SDGM (mm)	2.56	2.39	2.36	2.34	2.42	-	-	2.55	2.51	2.54	2.57	-	-
pef$_{>1.18}$[2]	0.85[b]	0.95[b]	0.96[b]	0.95[b]	0.96[b]	0.51	0.0001	0.85	0.87	0.82	0.82	0.25	0.0733
peNDF$_{>1.18}$[3]	41.34[b]	72.96[a]	74.04[a]	74.11[a]	74.70[a]	0.987	0.0054	45.22	46.05	43.30	43.57	1.654	0.0603

pef$_{>8}$= physically effective factor determined as the proportion of DM retained on sieves of the original version of Penn State particle separator (Lammers et al. 1996).
pef$_{>1.18}$= physically effective factor determined as the proportion of DM retained on sieves of the new version of Penn State particle separator (Kononoff, 2002).
The peNDF was calculated by multiplying NDF content of each portion on each sieve on each pef.
GM: geometric mean and SDGM: standard deviation of geometric mean.
The means within the same row with at least one common letter, do not have significant difference (P>0.05).
SEM: standard error of the means.

Table 5 Ruminally degradability parameters of NDF and NDF fractionation of feeds and total mixed rations that contained four different varieties of rice straws

Item	K_d[1]	a[2] (% of DM)	b[2] (% of DM)	a+b[2] (% of DM)	NDF (% of DM)	ADF (% of DM)	ADL (% of DM)	iNDF$_{288}$[3] (% of DM)	iNDF$_{288}$ (% of NDF)	iNDF$_{2.4}$[4] (% of DM)	pdNDF[4] (% NDF)	TTDNDF[5] (% of NDF)
Alfalfa hay	0.076[a]	21.34[a]	43.6[a]	64.94[a]	48.63[b]	44.92[a]	6.76[a]	23.34[b]	48[b]	16.22[a]	52[a]	46.54[a]
Taroum Neda	0.032[b]	14.61[b]	34.55[b]	49.16[b]	76.8[a]	40.33[c]	5.23[b]	39.94[a]	52[a]	12.55[b]	48[b]	12.30[b]
Taroum Neamaat	0.033[b]	14.21[b]	35.21[b]	49.42[b]	77.13[a]	41.32b[c]	4.98[b]	41.65[a]	54[a]	11.95[b]	46[b]	12.31[b]
Taroum Sangi	0.031[b]	14.33[b]	33.22[b]	47.55[b]	77.2[a]	42.43[b]	5.43[b]	40.92[a]	53[a]	13.03[b]	47[b]	11.49[b]
Taroum Asgari	0.033[b]	13.44[b]	35.12[b]	48.56[b]	78.63[a]	42.87[b]	5.13[b]	40.89[a]	52[a]	12.31[b]	48[b]	12.85[b]
SEM	0.008	1.542	2.543	2.432	3.388	0.934	0.51	0.333	0.621	1.306	0.388	0.126
P-values	< 0.0001	< 0.0001	0.212	< 0.0001	< 0.0001	< 0.0001	0.51	0.0208	< 0.0001	< 0.0001	< 0.0001	< 0.0001
Total mixed rations that contained four different varieties of rice straw												
Taroum Neda	0.038	17.45[b]	45.34[b]	62.79[b]	53.13	34.23	5.65	34.53	65[c]	13.56	35[a]	11.53
Taroum Neamaat	0.042	19.55[a]	46.32[a]	65.87[a]	52.8	34.54	5.76	35.38	67[b]	13.824	33[b]	12.63
Taroum Sangi	0.0391	18.56[a]	47.43[a]	65.99[a]	52.93	35.12	5.61	34.93	66[bc]	13.464	34[ab]	11.68
Taroum Asgari	0.041	19.03[a]	46.31[a]	65.34[a]	53.2	36.32	5.69	36.71	69[a]	13.656	31[c]	11.44
SEM	0.008	0.581	0.731	1.271	0.637	2.643	0.947	1.584	0.527	0.688	0.637	0.111
P-values	0.0761	0.0371	0.0471	0.0423	0.9213	0.3291	0.2518	0.4366	< 0.0001	0.6738	0.0021	0.3279

[1] pdNDF K_d= potential digestible NDF fraction digestion rate calculate from TTNDFD model.
[2] a, b, and a + b are the soluble, slowly degradable, and potential degradable fraction of NDF in rumen.
[3] The indigestible NDF that determined after 288 h ruminal incubation of samples and the iNDF$_{24}$ was calculated as 2.4 × ADL.
[4] The estimation of potentially digestible NDF (pdNDF=NDF–iNDF; Raffrenato and Van Amburgh, 2010).
[5] TTNDFD= predicted total-tract NDF digestibility using in vitro TTNDFD model.
NDF: neutral detergent fiber; ADF: acid detergent fiber; ADL: acid detergent lignin and DM: dry matter.
The means within the same row with at least one common letter, do not have significant difference (P>0.05).
SEM: standard error of the means.

The lack of digestibility in the iNDF fraction of forage is attributable to the cross-linking between cell wall lignin and hemicellulose. Also, a greater iNDF intake limits a ruminant's ability to consume sufficient forage to meet nutrient requirements (Cotanch et al. 2014; Lippke, 1986). The intake of forage-based diets by ruminants is often controlled by rumen fill and the rate of disappearance. The rate of disappearance is largely influenced by the inherent rate of digestion and passage rate. The indigestible portion is removed from the rumen by passage only and will accumulate in the rumen relative to the potentially digestible portion, therefore having a longer rumen retention time (Cotanch et al. 2014; Van Soest, 1994; Zali et al. 2015). A longer retention time in the rumen results in a lesser intake.

CONCLUSION

Chemical composition of four different rice straw had not significant differences, except on EE and ash content, but there was significant difference among alfalfa and rice straws. Rice straw had greater NDF and ash, and lesser NFC and CP than alfalfa. Bulk density and WHC of rice straws were similar and lesser than alfalfa. Alfalfa hay had significantly greater FSG than four varieties of rice straw at all incubation times. Rice straw varieties were more physically effective than alfalfa because they had greater NDF, geometric mean, $pef_{>8}$, $pef_{>1.18}$, $peNDF_{>8}$ and $peNDF_{>1.18}$ than alfalfa. However, the physical properties especially pef were similar. Alfalfa had greater soluble, slowly degradable, potential degradable fraction, and rate of degradability for NDF, pdNDF and TTDNDF in the rumen than the four rice straws. Also, the slowly degradable, the potential degradable fraction, and rate of degradability for NDF in the rumen, content of NDF, ADL, $iNDF_{288}$, pdNDF and TTDNDF of four rice straws were similar. The total mixed rations that contained four different varieties of rice straw were similar in rate of degradability for NDF in the rumen, NDF, ADF, ADL, $iNDF_{288}$, $iNDF_{2.4}$ and TTDNDF but the ration that contained Taroum Neda had lesser soluble, slowly degradable, and potential degradable fraction than other rations. In conclusion, for high yielding ruminant the ratio of forage to concentrate is decreased to enhancement of energy and nutrients content, and for physical effectiveness, fibre also increased. Under the circumstances, inclusion of rice straw even at low level may be useful to balance high yielding dairy rations and meeting physical effectiveness.

ACKNOWLEDGEMENT

This work was supported by the research grant of Sari Agricultural and Natural Resources University (SANRU), Mazandaran, Iran.

REFERENCES

American Society of Agricultural Engineers. (2002). Method of Determining and Expressing Particle Size of Chopped Forage (S424.1). American Society of Agricultural and Biological Engineers, St. Joseph, Michigan.

AOAC. (1990). Official Methods of Analysis. Vol. 17th Ed. Association of Official Analytical Chemists, Arlington, VA, USA.

ASTM Committee EO2 On Terminology. (2000). ASTM Dictionary of Engineering Science and Technology. West Conshohocken, Pennsylvania.

Bainton S.J., Plumb V.E., Juliano B.O., Perez C.M., Roxas D.B., Khush G.S., de Jesus J.C. and Gomez K.A. (1991). Variation in the nutritional value of rice straw. Anim. Feed Sci. Technol. 34, 261-277.

Capper B.C. (1988). Genetic variation in the feeding value of cereal straw. Anim. Feed Sci. Technol. 21, 127-140.

Cotanch K., Grant R., Van Amburgh M., Zontini A., Fustini M., Palmonari A. and Formigoni A. (2014). Applications of uNDF in ration modeling and formulation. Pp. 114-131 in Proc. Cornell Nutr. Conf. Dept. Anim. Sci. Cornell Univ. Ithaca, New York.

FAO. (2013). Food Wastage Footprint: Impacts on Natural Resources. Summary report. Natural Resources Management and Environment Department, Food and Agriculture Organization (FAO). Rome, Italy.

Giger-Reverdin S. (2000). Characterization of feedstuffs for ruminants using some physical parameters. Anim. Feed Sci. Technol. 86, 53-69.

Kononoff P.J. (2002). The effect of ration particle size on dairy cows in early lactation. Ph D. Thesis. The Pennsylvania State Univ., Pensilvania. USA.

Lammers B.P., Buckmaster D.R. and Heinrichs A.J. (1996). A simple method for the analysis of particle sizes of forages and total mixed rations. J. Dairy Sci. 79, 922-928.

Lippke H. (1986). Regulation of voluntary intake of ryegrass and sorghum forages in cattle by indigestible neutral detergent fiber. J. Anim. Sci. 63, 1459-1468.

Mertens D.R. (2000). Physically effective NDF and its use in dairy rations explored. Feedstuffs. 10, 11-14.

Mertens D.R. (1997). Creating a system for meeting the fiber requirements of dairy cows. J. Dairy Sci. 80, 1463-1481.

Raffrenato E. and Van Amburgh M. (2010). Development of a mathematical model to predict sizes and rates of digestion of a fast and slow degrading pool and the indigestible NDF fraction. Pp. 52-65 in Proc. Cornell Nutr. Conf. Syracuse, New York.

SAS Institute. (1998). SAS®/STAT Software, Release 8. SAS Institute, Inc., Cary, NC. USA.

Singh B. and Narang M.P. (1991). Some physico-chemical characteristics of forages and their relationship to digestibility. Indian J. Anim. Nutr. 8, 179-186.

Singh M. and Singh H.P. (1995). Genetic variation in chemical composition and digestibility of nutrients in rice straw. Int. Rice Res. Notes. 20, 4-8.

Teimouri Yansari A. and Pirmohammadi R. (2009). Effect of particle size of alfalfa hay and reconstitution with water on intake, digestion and milk production in Holstein dairy cows. *Animal*. **3**, 218-227.

Teimouri Yansari A., Valizadeh R., Naserian A., Christensen D.A., Yu P. and Eftekhari Shahroodi F. (2004). Effects of alfalfa particle size and specific gravity chewing activity, digestibility and performance of Holstein dairy cows. *J. Dairy Sci.* **87**, 3912-3924.

Vadivelloo J. (1995). Factors contributing to varietal differences in the nutritive value of rice straw. *Anim. Feed Sci. Technol.* **54**, 45-53.

Vadivelloo J. (2000). Nutritional properties of the leaf and stem of rice straw. *Anim. Feed Sci. Technol.* **83**, 57-65.

Vadivelloo J. and Phang O.C. (1996). Differences in the nutritive value of two rice straw varieties as influenced by season and location. *Anim. Feed Sci. Technol.* **61**, 347-352.

Van Soest P.J. (2006). Rice straw, the role of silica and treatments to improve quality. *Anim. Feed Sci. Technol.* **130**, 137-171.

Van Soest P.J. (1994). Nutritional Ecology of the Ruminant. Durham and Downey Inc., Portland, Oregon, USA.

Van Soest P.J., Robertson J.B. and Lewis B.A. (1991). Methods for dietary fiber, neutral detergent fiber and non-starch polysaccharide in relation to animal nutrition. *J. Dairy Sci.* **74**, 3583-3597.

Wattiaux M.A. (1990). A mechanism influencing passage of forage particles through the reticulo-rumen: change in specific gravity during hydration and digestion. Ph D. Thesis. University of Wisconsin, Madison, Wisconsin, USA.

Zali S., Teimouri Yansari A. and Jafari Sayyadi A. (2015). Effect of particle size and fragility of corn silage and alfalfa hay on intake, digestibility, performance, and chewing activity of fattening male lambs. *Res. Rev: J. Vet. Sci.* **1**, 1-11.

12

Use of an *in vitro* Rumen Gas Production Technique to Evaluate the Nutritive Value of Five Forage to Concentrate Ratios

S. Arbabi[1], T. Ghoorchi[1*] and S. Ramzanpour[1]

[1] Department of Animal and Poultry Nutrition, Faculty of Animal Science, Gorgan University of Agricultural Science and Natural Resources, Gorgan, Iran

*Correspondence E-mail: ghoorchi@gau.ac.ir

ABSTRACT

This study presents the chemical composition and *in vitro* fermentation of five diets with different forage (alfalfa) to concentrate (faba bean) (F:C) ratios, F0:C100, F25:C75, F50:C50, F75:C25, F100:C0 on a dry matter (DM) basis. Results indicate that the chemical composition in term of organic matter (OM), ether extract (EE), crude protein (CP), acid detergent fiber (ADF), neutral detergent fiber (NDF), neutral detergent insoluble protein (NDIP), acid detergent insoluble protein (ADIP) and total phenols (TPh), tannin (TT) and condensed tannin (TCT) were varies among five diets (P<0.05). After an initial gas test to evaluate 96 h gas production profiles of diets, the time to half maximal gas production was calculated and a second incubation was conducted with fermentation stopped at substrate specific half-time ($t_{1/2}$) and 24 h for each substrate. *In vitro* true DM degradability (*iv*TDDM), OM degradability (*iv*OMD) were increased (P<0.01) by addition proportions of concentrate in diets. Microbial mass (g/kg DM), metabolizable energy (ME) (MJ/kg DM), were greater in diets which those had been higher degradability. The efficiency of microbial production (PF) (mg/mL) were calculated for both substrate specific $t_{1/2}$ and 24 h and was not shown differences between experimental diets at 24 h but F100:C0 was lesser (1.17 mg/mL) at substrate specific $t_{1/2}$ (P<0.01). Gas produced from fermentable fraction (B) and the rate of gas production (c) were (P<0.01) greater in diets with grater concentrate ratio. Increasing the F:C ratio increased ruminal pH and N ammonia and affected concentrations of short-chain fatty acid (SCFA) (P<0.01). Amount of CH_4 emission from 13 to 17.16 g/kg DM and the great value was related to F100:C0.

KEY WORDS faba bean, *in vitro* gas production, methane, tannin.

INTRODUCTION

Researchers have been looking for alternative protein sources to replace soybean meal in animal nutrition (Volpelli *et al*. 2010). One of the possible protein sources successfully used in the feeding of ruminants and non-ruminants is faba beans, such as *Vicia faba* (Moschini *et al*. 2005; Volpelli *et al*. 2010). Despite the fact that faba bean is not rich in protein as soybean, it can be considered as a "dual purpose" feed, since it has greater starch contents, similar to barley (Masoero *et al*. 2006). Furthermore, the use of high levels of raw faba bean has no negative effects on palatability and digestibility in sheep (Liponi *et al*. 2009). On the other hand, faba bean is planted on 35000 hectares s in Iran where over 35% of those is located in Golestan Province. This makes Golestan the largest producer of faba bean in Iran (Sabaghpour, 1995). Due to the negative effects of ruminant methane production on the global warming, extensive efforts have been made to measure or/and reduce it (Steinfeld *et al*. 2006). Several of enteric methane production experiments are done measuring emission rates from animals in open circuit respirometers and in strictly controlled environments (Murray *et al.* 1999). In contrast, *in vitro* techniques can be used, which are less expensive and time-consuming than *in vivo* trials. These techniques also allow one to preserve experimental conditions more precisely (Menke *et al.* 1979). The use of an *in vitro* gas technique in evaluating feedstuffs (through the measurement of variables like methane, microbial mass and short-chain fatty acids (SCFA) is a very effective and robust way of estimating energy loss from diets, microbial and feed nitrogen supply to ruminants (Anele *et al.* 2010). In the gas technique gases were produced directly as a result of fermentation containing CO_2 and CH_4 and the indirect gas produced from the buffering of SCFA (Blümmel and Ørskov, 1993).

The objective of this study was to estimate differences in the nutritive value of five forage (alfalfa) to concentrate (faba bean) ratios based and to determine how diets differing in forage to content based on faba bean, would affect *in vitro* true degraded dry matter (DM), degraded organic matter, methane production, microbial mass, partitioning factor, short chain fatty acids, pH and N ammonia.

MATERIALS AND METHODS

Sample preparations and chemical analyses

Five diets with different forage (alfalfa) to concentrate (faba bean) (F:C) ratios, F0C100, F25C75, F50C50, F75C25, F100C0 on a dry matter (DM) basis were ground in mills with a 1-mm screen before being analyzed. The DM, OM, crude fat (ether extract), crude protein (CP, Kjeldahl N×6.25) and ash were determined according to the AOAC (2000). Neutral detergent fibre (NDF) and acid detergent fibre (ADF) were determined (Van Soest *et al.* 1991), and were corrected for residual acid insoluble ash; Neutral detergent fibre was assayed without sodium sulfite but with a heat-stable α-amylase due to the high levels of starch. Both aNDFom and ADFom were expressed without residual ash. Non-fibre carbohydrates (NFC) were calculated as: NFC=100 - % CP - % Ash - % EE - % NDF, All results were expressed on DM basis. Neutral detergent insoluble protein (NDIP) and acid detergent insoluble protein (ADIP) were measured as explained by Licitra *et al.* (1996). Total phenols and total tannins were determined using colorimetric assay with Folin-Ciocalteu phenol reagent according to Makkar (2003). Condensed tannins were consecutive extracted using the butanol-HCl method (Porter *et al.* 1986).

In vitro gas production measurement

The *in vitro* gas production measurement was done as described by Menke and Steingass (1988).

Rumen fluids were collected prior to feeding from three fistulated Dalagh sheep ($\approx$50 kg body weight) fed a standard diet (500 g grass hay/500 g concentrate). Rumen fluid was poured into a prewarmed insulated container, and strained through sterile gauze. Then it was mixed well and continuous flow of CO_2 flushed rumen fluid. The buffered mineral solution was added to rumen fluid. The mixture was kept stirred under CO_2 in a water bath at 39 °C. Samples were weighed and 200 mg (basis DM) was added to each syringe. Buffered rumen fluid (30 mL) was dispensed into the syringes and those were immediately placed in a shaking water bath at 39 °C. Cumulative gas volume measurements of samples were read manually from the three replicates each at 0, 4, 6, 8, 12, 24, 48, 72 and 96 h of incubation.

Fermentation syringes without samples (blanks) were included to allow correction for gas produced directly from rumen fluid. After subtraction of gas production from blanks, data were fitted to exponential model Ørskov and McDonald (1979) equation:

$$y= B(1-exp-c \times [t - lag])$$

Where:
y: cumulative volume of gas produced at time 't' (h).
B: asymptotic gas volume.
c: rate constant.
lag: time(h) between inoculation and commencement of gas production.

Halftime ($t_{1/2}$) production of the asymptotic gas volume (B; mL) was calculated as:

$$t_{1/2}= (ln2/c) + lag$$

Using the method of Menke *et al.* (1979), metabolizable energy:

$$ME \ in \ MJ= 2.20 + 0.136 \ GP + 0.057 \ CP$$

And organic matter digestibility:

$$ivOMD \% = 14.88 + 0.889 \ GP + 0.45 \ CP + 0.0651 \ XA$$

Where:
GP: total gas volume.
CP: crude protein.
XA: ash.

The SCFA production (mmol) was calculated using equation described by Getachew *et al.* (2000):

SCFA= 0.0239 GP - 0.0601

After ending of incubation at $t_{1/2}$, 24 h and 96 h, fermentation parameters such as pH and ammonia N of the fluid were determined (Hu *et al.* 2005).

In vitro truly degraded dry matter (*iv*TDDM)

In vitro truly degraded dry matter (*iv*TDDM) was determined using methods described by Anele *et al.* (2010). After the initial 96 h gas run, substrate specific $t_{1/2}$ was calculated and a second incubation of the samples was conducted to obtain degradability measures at substrate-specific $t_{1/2}$ and 24 h. The incubations were cut off at $t_{1/2}$ and 24 h and the volume of gas was recorded. In order to measure true substrate degradability of diets at $t_{1/2}$ and 24 h, refluxing the incubation residue with ND solution for 1 h with subsequent recovery of the truly undergirded substrate in dacron bags (3 cm×10 cm, 45 μm pore size) (Van Soest *et al.* 1991). *iv*TDDM= feed (DM) incubated − residue (DM) recovered in the crucibles/feed (DM) incubated.

Partitioning factor (PF, a measure of fermentation efficiency) was calculated as:

PF= *iv*TDDM (mg) / GP$_{24}$ (mL)

Using the method of Blümmel *et al.* (1997) the *in vitro* microbial mass production can be calculated as:

Microbial mass (mg)= *iv*TDDM (mg) - (GP$_{24}$ (mL)×2.2)

Where:
2.2: stoichiometrical factor according to the amounts (mg) of C, H and O required for production of 1 mmol of SCFA and associated 1 mL gas.

Methane measurement

For the measurement of methane, at the end of incubation ($t_{1/2}$ and 24 h), the clip was carefully unscrewed and 4 mL of 10 molar NaOH was introduced through the silicon tube into syringes.

So, CO_2 absorbed by the NaOH and piston of syringe dipped. The gas volume remaining in the syringe considered to be CH_4 (Demeyer *et al.* 1988).

Statistical analysis

Chemical composition and *in vitro* fermentation data were analyzed using the general linear models (GLM) procedure of SAS (1999). Data were analyzed within a completely randomized experimental design. Treatment means were separated using Duncan test at 0.05 and 0.01 probability levels.

RESULTS AND DISCUSSION

Chemical composition

As would be expected, there was different in chemical composition between five diets (Table 1). Organic matter content of faba bean seed (F0C100) were (P<0.05) greater than those of other diets. Diet with 75:25 ratio of concentrate to forage (F25C75) (P<0.05) greater in crude protein content except F50C50. The CP content of F0C100 (26.05) was lesser than of F25C75 (30.07). Ether extract varies (P<0.05) among diets. There was (P<0.05) differences in ADFom content among F0C100 and F100C0. The values of NDFom also differ (P<0.05), F0C100 and F25C75 seems to have greater values than others. The aNDFom content of diets ranged from 32.65% DM (F0C100) to 18.68% DM (F100C0), whereas that of ADFom ranged from 16.48% DM (F100C0) to 12.43% DM (F0C100). Results indicated a t variation in terms of protein fractionation and concentration addition linearly decreased the protein fractions NDIP and ADIP (P<0.05). Results showed that F0C100 and F25C75 have great levels of total phenol (TPh) compounds (P<0.05), 2.92 and 2.76% DM, respectively. After measuring the non-tannin phenols and subtract from total phenols, the total tannin (TT) was estimated which the treatments F0C100 and F25C75 had the great value and F100C0 had the less (P<0.05). Alfalfa hay (F100C0) had no condensed tannin (CT) and faba bean in levels 100% and 75% of diets had 0.34 and 0.31% DM.

In vitro gas production measurement

Gas production profiles of the experimental diets are shown in Figure 1, and the parameters of the gas production are presented in Table 3. Substrate specific $t_{1/2}$ was 9.56 h (F0C100), 9.06 h (F25C75), 8.76 h (F50C50), 8.52 h (F75C25) and 10.86 h (F100C0). The figure was plotted to show the trend of the incubation. A comparison of the diets showed that first diet (100% faba bean) produced the most gas (65 mL/200 mg DM) after 96 h and fifth diet (100% alfalfa hay) the least (26.91 mL/200 mg DM).

The maximum gas volume from fermentable fraction (B) was great for F0C100 and F25C75 and greater for F50C50 (P<0.01) than for F75C725 or F100C0. The maximum rate of gas production (c) was great for F50C50 and F75C25 (P<0.01), followed by F25C75 (P<0.01), and less for F0C100 and F100C0 (P<0.01). Hourly rumen ammonia-N concentration and pH are shown in Table 2. Reduction trend of the rumen ammonia-N levels and rumen pH on all the diets continued after $t_{1/2}$ h to 96 h. Increasing the concentrate component in the diet reduced rumen pH by around 0.3 pH units and caused a small but significant (P<0.01) increasing in SCFA molar.

Table 1 Ingredients and nutrient composition (% DM) of diets

Item	Treatment[1]					
	F0C100	F25C75	F50C50	F75C25	F100C0	SEM
Chemical composition, %						
Dry matter (DM)	92.85[a]	92.72[ab]	92.29[b]	92.23[b]	92.56[ab]	0.06
Organic matter (OM)	97.06[a]	94.31[b]	92.54[cb]	90.76[cd]	88.92[d]	0.28
Crude protein (CP)	26.05[b]	30.07[a]	28.35[ab]	19.05[c]	15.62[d]	0.40
Neutral detergent fibre (aNDFom)	32.65[a]	26.88[ab]	23.44[bc]	18.93[c]	18.68[c]	0.39
Acid detergent fibre (ADFom)	12.43[c]	13.68[bc]	14.54[bc]	16.20[a]	16.48[a]	0.48
Ether extract (EE)	1.53[b]	1.48[b]	1.72[ab]	1.79[ab]	2.2[a]	0.12
Non-fibre carbohydrates (NFC)	13.51[b]	9.19[b]	10.57[b]	20.75[a]	23.65[a]	0.15
Neutral detergent insoluble protein (NDIP)	1.48[b]	1.67[b]	1.95[b]	2.13[b]	3.26[a]	0.63
Acid detergent insoluble protein (ADIP)	0.43[c]	0.80[b]	1.73[a]	1.79[a]	1.66[a]	0.25
Total phenols (TPh)	2.92[a]	2.76[ab]	2.27[b]	1.69[c]	0.87[d]	0.18
Total tannins (TT)	1.97[a]	1.80[a]	1.48[b]	0.66[c]	0.18[d]	0.30
Condensed tannins (CT)	0.34[a]	0.31[ab]	0.22[b]	0.14[cd]	0.00[d]	0.08

[1] Diets with different forage to concentrate ratio.
The means within the same row with at least one common letter, do not have significant difference (P>0.05).
SEM: standard error of the means.

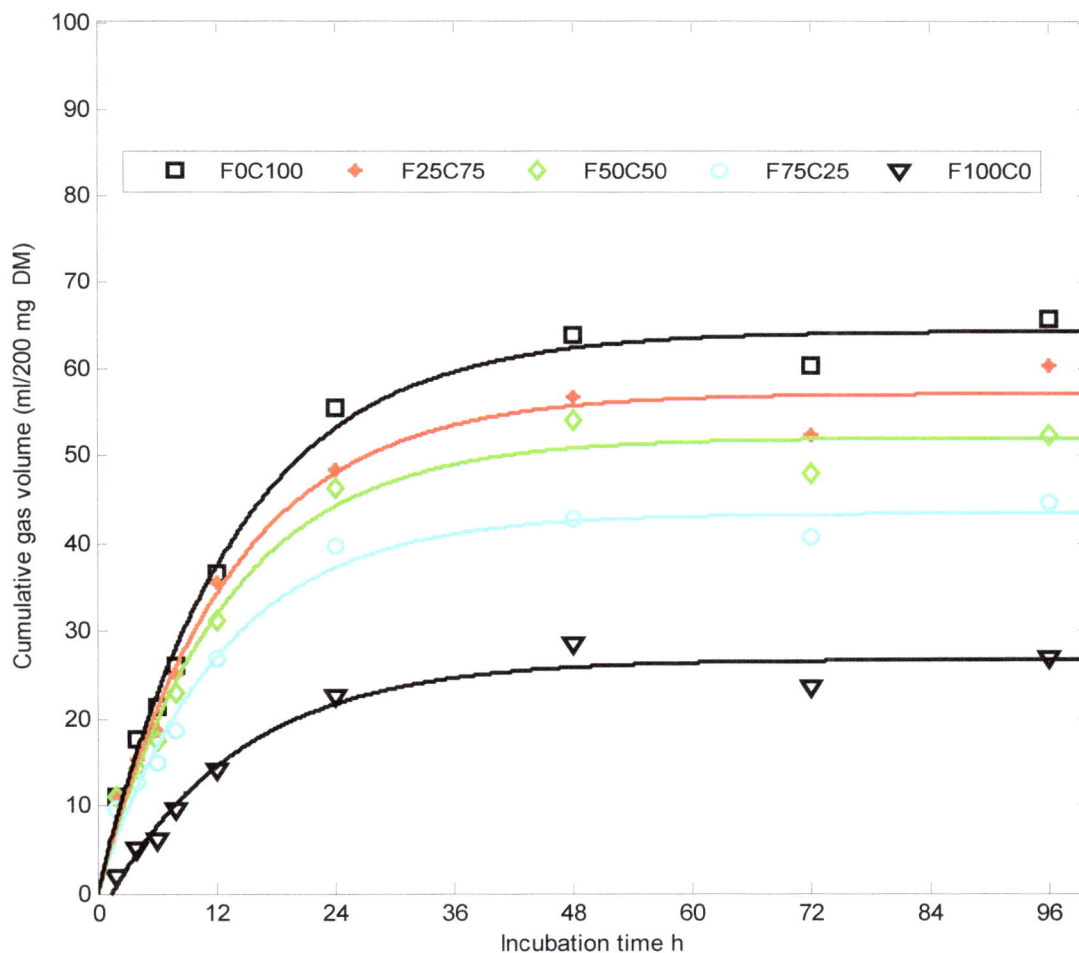

Figure 1 *In vitro* gas production profiles of the experimental diets, (F0C100 (100% concentrate); F25C75 (75% concentrate, 25% forage); F50C50 (50% concentrate, 50% forage); F75C25 (25% concentrate, 75% forage) and F100C0 (100% forage))

In this experiment, ammonia concentration in the fermenter fluid was increased (P<0.01) by supplementation with concentrate.

In vitro truly degraded dry matter (*iv*TDDM)

After 24 h incubation, *iv*TDDM, *iv*DOM, microbial mass, ME, SCFA, PF of the experimental diets were observed in Table 3. Increasing concentrate ratio in the experimental diets led to greater *iv*TDDM and *iv*DOM coefficient. However, only F100C0 had significant difference with other treatments (P<0.01). Both *iv*TDDM and *iv*DOM coefficients at substrate specific $t_{1/2}$ followed similar trend observed at 24 h and were lesser in F100C0 (P<0.01). The PF values of the diets were calculated for both substrate specific $t_{1/2}$ and 24 h and were not shown significant differences between experimental diets at 24 h but F100C0 was grater (2.17 mg/ml) at substrate specific $t_{1/2}$ (P<0.01).

Methane measurement

The methane emission factors found in this study are showed in Figure 2. The CH_4 production from F25C75 (13 g/kg DM) was lesser (P<0.01). It was observed that the decreasing proportion of concentrate in diets generally produced more methane. However, F100C0 and F75C25 had not significant effect on methane production; also the same procedure was observed in three other diets means: F0C100, F25C75 and F50C50.

Chemical composition

In this study, the forage-to-concentrate ratios were F0C100, F25C75, F50C50, F75C25 and F100C0, for the alfalfa hay–faba bean. For F75:C25 and F100C0, the ADFom fraction is a large proportion of the NDFom, which indicates high content of cellulose, lignin and lesser levels of hemicellulose (Abdulrazak *et al.* 2000). Cell wall content of feeds is related to their organic matter digestibility and their nutritive value (Giger-Reverdin, 1995). Indeed, fibre content seems to have reduced fermentability of diets, whose fermentation parameters were consistently the less. Alfalfa (F100C0) had the great fibre content (NDFom, 18.68% DM and ADFom, 16.48% DM) among the substrates investigated. Utilization of alfalfa could not be limited by its fibre content. Tannins intervene in the definition of fibres by forming complex with proteins, and thus those are insoluble in the detergent solution (Makkar *et al.* 1995). Results demonstrated *iv*DOM and *iv*TDDM of diets with a greater ratio of forage were lesser. As can be seen in Figure 1, the gas production has been minus in these diets. There is a close relationship between the extent of potential NDF digestibility and the total volume of gas produced (Huhtanen *et al.* 2008). Archimede *et al.* (1995) indicated the use of concentrate in the ruminant diets causes the maximum utilization of energy, protein and other contents of feeds. The grain supplementation may reduce digestibility of forage of diet. The amount of this effect depends on the ratio of concentrate and its nature. However, a lack of effect of grain supplementation on NDF digestibility (Cerrillo *et al.* 1999) or even and increase on fibre digestibility (Molina Alcaide *et al.* 2000) have also been reported. The CP content of faba bean was 26.05%; this result is similar to results obtained by Volpelli *et al.* (2010) and Cazzato *et al.* (2012) who found 25.3 and 26.52%, respectively. The CP levels in the current study are within the range of 15.62-30.07% DM, which is adequate for maintenance and growth requirements of ruminants (NRC, 1996). However, optimum utilization of CP in diets is dependent on soluble phenolic compounds, such as tannins (Woodward and Reed, 1989). This fact has been reported that faba beans in the diet of non-ruminant animals depress the apparent digestibility of nitrogen compounds (Reddy *et al.* 1985). This harmful effect has been largely ascribed to the presence of condensed tannins in the majority of varieties of faba beans. The condensed tannins appear to be one of the most important antinutritional factors in faba beans, which it can interact and precipitate proteins. Also, the condensed tannins depressing the utilization of dietary protein either by interaction with proteins to form indigestible complexes and by formation of tannin-enzyme complexes resulting in an inhibition of digestive enzyme activity by similar processes (Ortiz *et al.* 1993). On the other hand, Makkar *et al.* (1997) suggested that the presence 2-3 percent of condensed tannins in the diet of ruminants can have beneficial effects. These useful results may be related to concentration, type and structure of tannins. Tannins can reduce proteolysis and loss proteins in the rumen. These functions resulted to increase the by-pass protein, efficiency of microbial protein synthesis and turn-over of urea.

In vitro gas production measurement

Gas produced from fermentable fraction (B) was (P<0.05) greater in F0C100 may be due to their high content of CP and NDF (Khazaal *et al.* 1993). Therefore the greater values obtained for the B fraction in F0C100 and F25C75 will indicate a better nutrient availability for rumen microorganisms. As it was calculated F0C100 to F75C25 had the greater metabolizable energy (ME). The *in vitro* gas production technique has the potential to reflect the *in vivo* digestibility of feeds for ruminants. However, Prasad *et al.* (1994) indicated that the great predictive value for the *in vivo* digestibility of diets was obtained after 45 to 52 h of *in vitro* fermentation. Ruminal micro-organisms fermented Non-structural carbohydrates rapidly, both ruminal pH decline and also ruminal volatile fatty acids (VFA) patterns changed (Kalscheur *et al.* 1997).

Table 2 Effect of different concentration to forage ratio on parameters of gas production (defined by the equation: y= B(1−exp−c×[t − lag])), cumulative gas volume (mL/200 mg DM), ammonia N (mg/L) and pH

Treatment[1]	Parameters of gas production		Cumulative gas volume (mL/200 mg DM)			Ammonia N (mg/L)			pH		
	B[2]	c[3]	$t_{1/2}$	24 h	96 h	$t_{1/2}$	24 h	96 h	$t_{1/2}$	24 h	96 h
F0C100	64.27^a	0.072^c	30.34^a	55.33^a	65.58^a	101.4^c	97.0^c	85.9^c	6.41^c	6.38^b	6.32^b
F25C75	57.01^{ab}	0.076^b	27.67^a	48.33^a	60.25^{ab}	111.5^c	108.4^c	102.7^b	6.52^b	6.50^{ab}	6.48^{ab}
F50C50	51.94^b	0.079^{ab}	24.29^a	46.33^a	52.25^{bc}	153·6^b	147·1^b	138·5^{ab}	6.53^b	6.50^{ab}	6.50^a
F75C25	43.38^c	0.081^a	20.26^a	39.67^{ab}	44.58^c	164·7^{ab}	161·5^a	149·4^{ab}	6.65^a	6.60^a	6.56^a
F100C0	26.73^d	0.073^c	12.77^b	22.67^b	26.91^d	177·3^a	173·8^a	165·3^a	6.70^a	6.66^a	6.62^a
SEM	2.26	0.0012	2.17	2.80	1.78	4.71	4.16	3.63	0.28	0.46	0.34

[1] Diets with different forage to concentrate ratio.
[2] Gas produced from fermentable fraction (mL).
[3] Rate constant of gas production during incubation (mL h⁻¹).
The means within the same column with at least one common letter, do not have significant difference (P>0.05).
SEM: standard error of the means.

Table 3 In vitro truly degraded dry matter (ivTDDM); degraded organic matter (ivDOM) coefficient, microbial protein; efficiency of microbial production (PF) of the experimental diets

Item	Treatment[1]					SEM
	F0C100	F25C75	F50C50	F75C25	F100C0	
Twenty-four hour (24 h)						
ivTDDM	0.683^a	0.566^{ab}	0.511^{ab}	0.466^{ab}	0.328^b	0.02
ivDOM	75.99^a	69.97^a	71.09^a	59.32^{ab}	42.78^b	0.40
Microbial mass (g/kg DM)	561.59^a	464.73^{ab}	405.33^{ab}	379.39^{ab}	278.46^b	0.39
Metabolizable energy (ME) (MJ/kg DM)	11.21^a	10.39^a	10.21^a	8.68^{ab}	6.17^b	0.37
Short-chain fatty acids (SCFA) (μM)	1.22^a	1.02^{ab}	1.06^{ab}	0.87^{ab}	0.49^b	0.06
Efficiency of microbial production (PF) (mg/mL)	1.26^a	1.43^a	1.07^a	1.20^a	1.47^a	0.13
Half life ($t_{1/2}$)						
ivTDDM	0.558^a	0.461^{ab}	0.399^{abc}	0.371^{bc}	0.277^c	0.08
ivDOM	53.77^a	53.38^a	49.71^{ab}	42.06^b	33.98^c	0.33
Microbial mass (g/kg DM)	491.56^a	400.79^{ab}	345.72^b	327.09^{bc}	249.22^c	0.41
Metabolizable energy (ME) (MJ/kg DM)	7.81^a	7.67^a	7.11^{ab}	6.04^{ab}	4.82^b	0.26
Short-chain fatty acids (SCFA) (μM)	0.66^a	0.61^a	0.53^b	0.44^b	0.27^c	0.11
Efficiency of microbial production (PF) (mg/mL)	1.82^b	1.66^b	1.64^b	1.83^b	2.17^a	0.20

[1] Diets with different forage to concentrate ratio.
The means within the same row with at least one common letter, do not have significant difference (P>0.05).
SEM: standard error of the means.

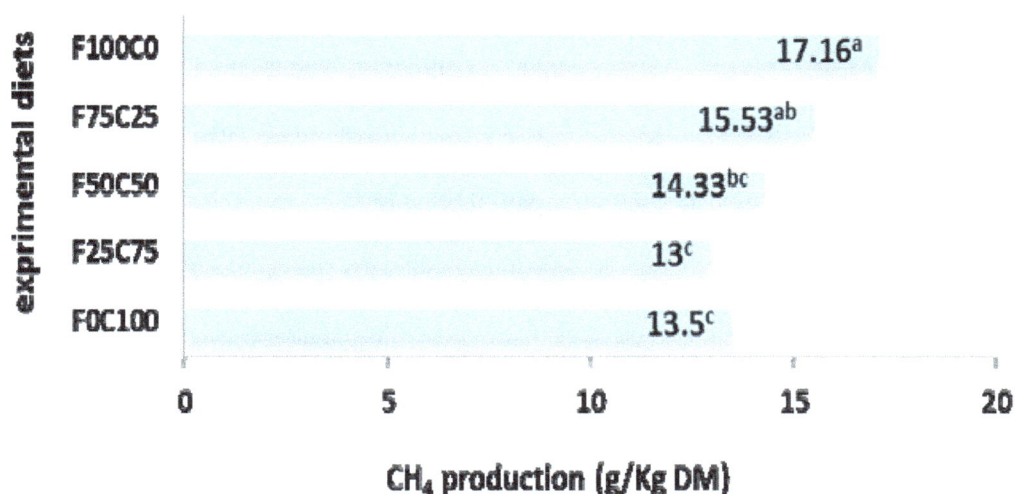

Figure 2 In vitro methane production profiles of the experimental diets (with different forage to concentrate ratio)
[a, c and c]: Means with no common superscript differ significantly (P<0.05)

When the concentrate in the diet increases, the size of the rumen pH decline is highly variable. In the present study, pH value decreased by increasing percentage of concentrate. However, this reduction was small and pH never went below 6.0. In some of the recent research works, it was reported that the reduction of rumen pH below 6·0 occurs when feeding a high proportion (70%) of concentrate in the diet (Carro *et al.* 2000). In contrast, Khorasani *et al.* (2001) and Ueda *et al.* (2003) indicated that the feeding of concentrate resulted in small drops of between 0 and 0·2 pH units. But in the own study used faba bean as its major dietary source, which its rapid fermentation pattern was similar to barley (Masoero *et al.* 2006). In some studies (Loor *et al.* 2003; Sackmann *et al.* 2003) where barley contributed as a major dietary supply of starch, it was shown to result in a greater decline in rumen pH. This small change in ruminal pH in the present study may be due to differences in the proportion of concentrate in the diet. Other factors which may explain this anomaly may include buffering capacity of the feeds, hence buffering capacity of the rumen, ruminal fluid sample (Lee *et al.* 2006). On the other, this small change in ruminal pH probably led to a lesser ruminal cellulolytic activity and as a result a small shift occurred in SCFA molar proportions (Ueda *et al.* 2003). In this study, incremental increase in concentrate led to a reduction in rumen ammonia concentration. Lee *et al.* (2006) demonstrated the feeding of readily available energy sources has decreased rumen ammonia concentrations. This reduction could be explained by a greater utilization of ammonia by the micro-organisms as a readily available energy source (Lee *et al.* 2003); or by reductions in the use of amino acids as an energy source by micro-organisms (Nocek and Russell, 1988). Optimization of microbial yield in the rumen depends largely on the availability of carbohydrates and nitrogen (N) in the rumen (Shabi *et al.* 1998).

In vitro truly degraded dry matter (*iv*TDDM)

Decreasing forage to concentrate ratio in the experimental diets led to greater *iv*TDDM and *iv*DOM coefficient but the greater *iv*TDDM and *iv*DOM were caused to greater microbial mass. The PF value for diets with more proportion of concentrate were lesser than F100C0 and this reason was supported the microbial mass production of those diets (Anele *et al.* 2010). In the ruminant nutrition to improve microbial efficiency, feed conversion into microbial biomass should be maximized because high microbial efficiency improves microbial protein supply to the small intestine and, proportionally, reduces fermentative gaseous carbon losses (Beever, 1993). The tannin content of faba bean in the concentrate section of experimental diets corroborated these results. Observations were showed that F0C100 and F25C75 had greater levels of TPh, TT and CT, but according to results of Makkar *et al.* (1997) amount of CT was low and seems it cannot influence on *iv*DOM and *iv*TDDM, but can have beneficial effects including tannins bind proteins to reduce loss proteins in the rumen. As previously mentioned this function can increase the by-pass protein. On the other hand, the gas volumes at all times of incubation increased with increasing proportions of concentrate to forage ratio in the mixtures and as Hess *et al.* (2006) had expressed the degradation of OM, NDF, ADF increased with legume containing tannin.

Methane measurement

Previous studies indicated, the optimum pH for methane production is 7.0-7.2, but the gas production can occur in the pH range of 6.6-7.6. However, beyond this range, the activity of fiber degraders reduces (Arglyle and Baldwin 1988; Dijkstra *et al.* 1992). Forage: concentrate ratio also influences the acetate: propionate ratio and methane emission decreases drastically from 6-12% (forage-based diet) to 2-3% when diet with the 90% concentrate predominates (Johnson and Johnson 1995).

Whitelaw *et al.* (1984) obtained same results and suggested that the feeding of a high concentrate: low roughage diet produce less methane as compared to low concentrate: high roughage diet. Benchaar *et al.* (2001) reported a decrease in the methane emission with inclusion legumes in the diet of ruminants. This lesser methane emission can be attributed to lesser portion of carbohydrates, faster passage rate, hence, shifting the fermentation pattern towards greater propionate production (Johnson and Johnson, 1995). Also, there are some indications that tannins in the diet reduce ruminal methane production (Waghorn *et al.* 2002). Tannins decrease the degradation of nutrients in the rumen which then may be degraded in the hindgut. This could have contributed to a lesser methane emission because hindgut fermentation resulted in a lesser methane production per unit of fermented nutrients.

CONCLUSION

Although the faba bean contains tannins, the results of the chemical composition and *in vitro* degradability study showed the rumen microbes can be effectively utilized to it and will supply the requirements of energy and amino acids to the ruminants. It was concluded that there were significant variations in chemical composition and gas production characteristics of five F:C ratios in this study. The results indicated that the addition of concentrate in diets increased nutritive value and limited the methane emission. If total condensed tannin was lesser than 2-3% DM, it cannot affect cumulative gas production and estimated parameters of feed.

ACKNOWLEDGEMENT

The authors are grateful to Morteza Ghasemi and Abdolhakim Toghdori for their skilled technical assistance. The project was funded by the Ministry of Science, Research, and Technology of Iran and Gorgan University of Agricultural Science and Natural Resources, Iran.

REFERENCES

Abdulrazak S.A., Fujihara T., Ondiek J.K. and Ørskov E.R. (2000). Nutritive evaluation of some Acacia tree leaves from Kenya. *Anim. Feed Sci. Technol.* **85,** 89-98.

Anele U.Y., Südekum K.H., Hummel J., Arigbede O.M., Oni A.O., Olanite J.A., Bottgera C., Ojo V.O. and Jolaosho A.O. (2010). Chemical characterization, *in vitro* dry matter and ruminal crude protein degradability and microbial protein synthesis of some cowpea (*Vigna unguiculata*) haulm varieties. *J. Feed Sci. Technol.* **163,** 161-169.

AOAC. (2000). Official Methods of Analysis. Vol. I. 17th Ed. Association of Official Analytical Chemists, Arlington, VA, USA.

Archimede H., Sauvant D., Hervieu J., Poncet C. and Dorleans M. (1995). Digestive interactions in the ruminant relationships between whole tract and stomach evaluation. *Anim. Feed Sci. Technol.* **54,** 327-340.

Arglyle J.L. and Baldwin R.L. (1988). Modeling of rumen water kinetics and effects of rumen pH changes. *J. Dairy Sci.* **71,** 1178-1188.

Beever D.E. (1993). Ruminant animal production from forages: present position and future opportunities. Pp. 158-164 in Grasslands for Our World. M.J. Baker, Ed. SIR Publishing, Wellington, New Zealand.

Benchaar C., Pomer C. and Chiquette J. (2001). Evaluation of dietary strategies to reduce methane production in ruminants: a modeling approach. *Canadian J. Anim. Sci.* **81,** 563-574.

Blümmel M., Makkar H.P.S. and Becker K. (1997). *In vitro* gas production-a technique revisited. *J. Anim. Physiol. Anim. Nutr.* **64,** 24-34.

Carro M.D., Valdes C., Ranilla M.J. and Gonzalez J.S. (2000). Effect of forage to concentrate ratio in the diet on ruminal fermentation and digesta flow kinetics in sheep offered food at a fixed and restricted level of intake. *Anim. Sci.* **70,** 127-134.

Cazzato E., Tufarelli V., Ceci E., Stellacci A.M. and Laudadio V. (2012). Quality, yield and nitrogen fixation of faba bean seeds as affected by sulphur fertilization. *Soil. Plant. Sci.* **2,** 1-7.

Cerrillo M.A., Russell J.R. and Crump M.H. (1999). The effects of hay maturity and forage to concentrate ratio on digestion kinetics in goats. *Small Rum. Res.* **32,** 51-60.

Demeyer D., De Meulemeester M., De Graeve K. and Gupta B.W. (1988). Effect of fungal treatment on nutritive value of straw. *Med. Fac. Landbouww. Rijksuniv. Gent.* **53,** 1811-1819.

Dijkstra J., Neal H.D. and Beever S.C. (1992). Stimulation of nutrient digestion, absorption and outflow in the rumen: model description. *J. Nutr.* **122,** 2239-2256.

Getachew G., Makkar H.P.S. and Becker K. (2000). Stoichiometric relationship between short chain fatty acid and *in vitro* gas production in presence and absence of polyethylene glycol for tannin containing browses. Pp. 65-64 in Proc. EAAP Satellite Symp. Gas Prod. Wageningen, The Netherlands.

Giger-Reverdim S. (1995). Review of the main methods of cell wall estimation: interest and limits for ruminants. *Anim. Feed Sci. Technol.* **55,** 295-334.

Hess H.D., Tiemann T.T., Noto F., Carulla J.E. and Kreuzer M. (2006). Strategic use of tannins as means to limit methane emission from ruminant livestock. *Int. Congr. Ser.* **1293,** 164-167.

Hu W.L., Liu J.X., Ye J.A., Wu Y.M. and Guo Y.Q. (2005). Effect of tea saponin on rumen fermentation *in vitro*. *Anim. Feed Sci. Technol.* **120,** 333-339.

Huhtanen P., Seppälä A., Ahvenjärvi S. and Rinne M. (2008). Prediction of *in vivo* neutral detergent fiber digestibility and digestion rate of potentially digestible neutral detergent fiber: comparison of models. *J. Anim. Sci.* **86,** 2657-2669.

Johnson K.A. and Johnson D.E. (1995). Methane emissions from cattle. *J. Anim. Sci.* **73,** 2483-2492.

Kalscheur K.F., Teter B.B., Piperova L.S. and Erdman R.A. (1997). Effect of dietary forage concentration and buffer addition on duodenal flow of trans-C18:1 fatty acids and milk fat production in dairy cows. *J. Dairy Sci.* **80,** 2104-2114.

Khazaal K.A., Dentinho M.T., Ribeiro R. and Ørskov E.R. (1993). A comparison of gas production during incubation with rumen contents *in vitro* and nylon bag degradability as predictors of the apparent digestibility *in vivo* and voluntary intake of hays. *Anim. Prod.* **57,** 105-112.

Khorasani G.R., Okine E.K., Corbett R.R. and Kennelly J.J. (2001). Nutritive value of peas for lactating dairy cattle. *Canadian J. Anim. Sci.* **81,** 541-551.

Lee M.R.F., Merry R.J., Davies D.R., Moorby J.M., Humphreys M.O., Theodorou M.K., MacRae J.C. and Scollan N.D. (2003). Effect of increasing availability of water-soluble carbohydrates on *in vitro* rumen fermentation. *Anim. Feed Sci. Technol.* **104,** 59-70.

Lee M.R.F., Tweed J.K.S., Dewhurst R.J. and Scollan N.D. (2006). Effect of forage: concentrate ratio on ruminal metabolism and duodenal flow of fatty acids in beef steers. *J. Anim. Sci.* **82,** 31-40.

Licitra G., Hernandez T.M. and Van Soest P.J. (1996). Standardization of procedures for nitrogen fractionation of ruminant feeds. *Anim. Feed Sci. Technol.* **57,** 347-358.

Liponi G.B., Casini L., De Vincenzi S. and Gatta D. (2009). Di gestibility and nitrogen balance of diets based on faba bean, pea seeds and soybean meal in sheep. *Italian J. Anim. Sci.* **8(2),** 353-360.

Loor J.J., Hoover W.H., Miller-Webster T.K., Herbein J.H. and Polan C.E. (2003). Biohydrogenation of unsaturated fatty acids in continuous culture fermenters during digestion of orchardgrass or red clover with three levels of ground corn supplementation. *J. Anim. Sci.* **81,** 1611-1627.

Makkar H.P.S. (2003). Quantification of Tannins in Tree and Shrub Foliage. Kluwer Academic Publishers, Dordrecht, The Netherlands.

Makkar H.P.S., Blümmel M. and Becker K. (1995). Formation of complexes between polyvinyl pyrrolidones or polyethylene glycols and tannins, and their implication in gas production and true digestibility in *in vitro* techniques. *Br. J. Nutr.* **73**, 897-913.

Makkar H.P.S., Blümmel M. and Becker K. (1997). *In vitro* rumen apparent and true digestibilities of tannin-rich forages. *Anim. Feed Sci. Technol.* **67**, 245-251.

Masoero F., Moschini M., Fusconi G. and Piva G. (2006). Raw, extruded and expanded pea (*Pisum sativum*) in dairy cows diets. *Italian J. Anim. Sci.* **5**, 237-247.

Menke K.H., Raab L., Salewski A., Steingass H., Fritz D. and Schneider W. (1979). The estimation of the digestibility and metabolizable energy content of ruminant feeding stuffs from the gas production when they are incubated with rumen liquor *in vitro*. *J. Agric. Sci.* **92**, 217-222.

Menke K.H. and Steingass H. (1988). Estimation of the energetic feed value obtained from chemical analysis and gas production using rumen fluid. *Anim. Res. Dev.* **28**, 7-55.

Molina Alcaide E., Martín García A.I. and Aguilera J.F. (2000). A comparative study of nutrient digestibility, kinetics of degradation and passage and rumen fermentation pattern in goats and sheep offered good quality diets. *Livest. Prod. Sci.* **64**, 215-223.

Moschini M., Masoero F., Prandini A., Fusconi G., Morlacchini M. and Piva G. (2005). Raw pea (*Pisum sativum*), raw faba bean (*Vicia faba var. minor*) and raw Lupin (*Lupinus albus var. multitalia*) as alternative protein sources in broiler diets. *Italian J. Anim. Sci.* **4**, 59-69.

Murray P.J., Moss A., Lockyer D.R. and Jarvis S.C. (1999). A comparison of systems for measuring methane emissions from sheep. *J. Agric. Sci.* **133**, 439-444.

Nocek J.E. and Russell J.B. (1988). Protein and energy as an integrated system. Relationship of ruminal protein and carbohydrate availability to microbial synthesis and milk production. *J. Dairy Sci.* **71**, 2070-2107.

NRC. (1996). Nutrient Requirements for Beef Cattle. 7th Ed. National Academy Press, Washington, DC, USA.

Ørskov E.R. and McDonald I. (1979). The estimation of protein degradability in the rumen from incubation measurements weighted according to rate of passage. *J. Agric. Sci. Cambr.* **92**, 499-503.

Ortiz L.T., Centeno C. and Trevifio J. (1993). Tannins in faba bean seeds: effects on the digestion of protein and amino acids in growing chicks. *Anim. Feed Sci. Technol.* **41**, 271-278.

Porter L.J., Hrstich L.N. and Chan B.G. (1986). The conversion of procyanidins and prodelphinidins to cyanidin and delphinidin. *Phytochemistry.* **25**, 223-230.

Prasad C.S., Wood C.D. and Sampath K.T. (1994). Use of *in vitro* gas production to evaluate rumen fermentation of untreated and urea-treated finger millet straw (*Eleusine coracana*) sup-

plemented with different levels of concentrate. *J. Sci. Food Agric.* **65**, 457-464.

Reddy N.R., Pierson M.D., Sathe S.K. and Salunkhe D.K. (1985). Dry bean: a review of nutritional implications. *J. Am. Oil. Chem. Soc.* **62**, 541-549.

Sabaghpour S.H. (1995). Effect of plant density on the yield of faba bean. *Seed. Plant. Improv. J.* **11**, 9-13.

Sackmann J.R., Duckett S.K., Gillis M.H., Realini C.E., Parks A.H. and Eggelston R.B. (2003). Effects of forage and sunflesser oil levels on ruminal biohydrogenation of fatty acids and conjugated linoleic acid formation in beef steers fed finishing diets. *J. Anim. Sci.* **81**, 3174-3181.

SAS Institute. (1999). SAS®/STAT Software, Release 6.03. SAS Institute, Inc., Cary, NC. USA.

Shabi Z., Arieli A., Bruckental I., Aharoni Y., Zamwel S., Bor A. and Tagari H. (1998). Effect of the synchronization of the degradation of dietary crude protein and organic matter and feeding frequency on ruminal fermentation and flow of digesta in the abomasum of dairy cows. *J. Dairy Sci.* **81**, 1991-2000.

Steinfeld H., Gerber P., Wassenaar T., Castel V., Rosales M. and de Haan C. (2006). Livestock's Long Shadow-Environmental Issues and Options. Animal Production and Health Division, Livestock, Environment and Development Initiative (LEAD), Rome, Italy.

Ueda K., Ferlay A., Chabrot J., Loor J.J., Chilliard Y. and Doreau M. (2003). Effect of linseed oil supplementation on ruminal digestion in dairy cows fed diets with different forage: concentrate ratios. *J. Dairy Sci.* **86**, 3999-4007.

Van Soest P.J., Robertson J.B. and Lewis B.A. (1991). Methods for dietary fiber, neutral detergent fiber, and non-starch polysaccharides in relation to animal nutrition. *J. Dairy Sci.* **74**, 3583-3597.

Volpelli L.A., Comellini M., Masoero F., Moschini M., Lo Fiego D.P. and Scipioni R. (2010). Faba beans (*Vicia faba*) in dairy cow diet: effect on milk production and quality. *Italian J. Anim. Sci.* **9**, 138-144.

Waghorn G.C. and Tavendale M.H. (2002). Wood field methanogenesis from forages fed to sheep. *Proc. New Zealand Grassl. Assoc.* **64**, 167-171.

Whitelaw F.G., Eadie J.M. and Bruce L.A. (1984). Methane formation in faunated and ciliate free cattle and its relationship with rumen volatile fatty acid proportions. *Br. J. Nutr.* **52**, 261-275.

Woodward A. and Reed J.D. (1989). The influence of polyphenolics on the nutritive value of browse: a summary of research conducted at ILCA. Pp. 2-11 in Proc. 35th Int. Livest. Centre Africa, Addis Ababa, Ethiopia.

13

Growth Performance of Blackhead Pleven Lambs during the Suckling Period

M. Simeonov[1*] **and D. Pamukova**[2]

[1] Agricultural Institute, Stara Zagora, Bulgaria
[2] Trakia University, Faculty of Agriculture, Stara Zagora, Bulgaria

*Correspondence E-mail: msimeonov78@abv.bg

ABSTRACT

The aim of the present study was to analyze the live weight, body length, and chest perimeter in Blackhead Pleven lambs at an early age with a view to their future use as prognostic indicators of growth performance. The study was conducted with 46 lambs. It found out that at birth and weaning the singletons had higher live weight and higher average daily weight gain in comparison with the twins (P<0.01). Body length and chest perimeter at birth were higher in the singletons lambs (P<0.01). The type of birth influenced significantly (P<0.01) on the live weight ($\eta2$=26.85%), live weight ($\eta2$=21.39%) and the chest perimeter ($\eta2$=27.08%) of the lambs at birth. At 26 days of age i.e. at weaning, only the live weight was influenced by the birth type ($\eta2$=27.53%, P<0.01). The chest perimeter of lambs at weaning was not influenced by the body length at birth and at weaning. There was a moderate to strong phenotypic correlation among the other studied parameters.

KEY WORDS chest perimeter, lambs, live weight.

INTRODUCTION

The growth of animals is directly related to morphological and physiological changes of the organism. Some parameters as live weight, body length, and chest perimeter are used to evaluate the intensity of growth in young animals. Kadlečík and Kasarda (2007) outlined that growth is a process accompanied by changes in the weight and body size of animals, hence these are two processes occurring in the organism which determine the concept of development. Brown *et al.* (1973) noted that body measurements served to determine the growth rate and feed conversion by animals. Growth usually defined as the increase in size or live weight at a given age is one of the important selection criteria for improvement of lamb production (Afolayan *et al.* 2006). Linear body measurements taken on live animals have been widely used in research work as a simple means of recording certain aspect of animal growth and shape (Salako, 2006; Alphonsus *et al.* 2010). Linear measurement can be used in assessing growth rate, weight, feed utilization and carcass characteristic, for tracing the relationship between production performance, visual appraisal and body measurements (Fourie *et al.* 2002).

The live weight is a parameter of growth performance of lambs from all production types. In Blackhead Pleven lambs, Savov (1948) reported that females are born with higher live weight than males. The study of Ivanova and Raicheva (2009) shows that over the years the team on this indicator has not stopped to water and male lambs are born with a higher live weight than females.

The gender and birth type have a substantial effect on live body weight. According to Abbas *et al.* (2010), the gender is the cause for males being heavier than females at all ages in the course of skeletal development. Higher live body weight of male lambs is outlined by El-Toum (2005) and Kumar *et al.* (2008), and in the view of Macit *et al.* (2002) could be attributed to the higher average daily weight gain.

The type of birth has a significantly influence on the live weight compared to the sex of animals (Idris *et al.* 2010). According to Klewies *et al.* (2002); Idris *et al.* (2010), regardless of the breed, lambs born as singletons are heavier than twins both at birth and after that, by reason of the lack of competition during the suckling period (Klewies *et al.* 2002; Idris *et al.* 2010).

Vuchkov and Dimov (2008) outlined that the gender, the birth type, production year and the herd are factors influencing statistically significantly the weight at birth and at 30 and 60 days of age.

Exterior measurements (body length and chest perimeter) are directly related to body weight and growth intensity. Philips *et al.* (2002) reported that male lambs weaned at 60 days of age not only attain higher liver weight at the end of an 86-day experimental period but also exhibited statistically significantly higher body length, chest perimeter (P<0.05) and body height (P<0.01) than females. Koritiaki *et al.* (2013) did not observe considerable differences at birth and at weaning (70 days of age) in body height and length between the genders, but with respect to birth type, differences were substantially smaller in male lambs.

The aim of the present study was to analyze the live weight, body length and chest perimeter in Blackhead Pleven lambs at an early age with a view to their future use as prognostic indicators of growth performance.

MATERIALS AND METHODS

To fulfill the study's aim, the growth performance of 46 early weaned lambs from the Blackhead Pleven breed was monitored at the Institute of Forage Crops, Pleven in 2010. The lambs were born during January with an age difference of 5 days.

The lambs were weaned at 26 days of age at average live weight of 11.080 kg. After the 6th day of age, lambs were allowed creep feeding (Alcock, 2006), i.e. free access to pelleted protein concentrate, pea hay and water.

For monitoring of growth intensity from birth to weaning the each 7 days to weaning was determined body weight, body length and chest. The body weight was determined with electronic balance, and body length and chest perimeter with a measuring tape.

The body length was measured from the anterior point of the scapulohumeral joint to the root of the tail, and chest perimeter was measured just behind the shoulder blade. The correlation among the traits was calculated by means of Pearson's analysis and the power of factors (η^2) on studied traits was determined by analysis of variance using Statistica for windows software (Statistica, 2006). The significance of between-group differences was calculated by the t-test at (P<0.05) and (P<0.01) levels.

RESULTS AND DISCUSSION

During the study period, male lambs had a higher liver weight than females although the gender differences in weights at birth and at weaning were insignificant (Table 1). Idris *et al.* (2010) reported that in Sudan, female lambs were born with lower live weight as compared to males (P<0.05), but after the first week of life females were superior to males with regard to this parameter (P<0.05). This disagrees with the studies of Vuchkov and Dimov (2008) in White Maritsa lambs. The obtained coefficients of variation indicated that the possibility for large-scale selection for live weight at birth would have a positive impact on the herd (Table 1). At birth and at weaning (26 days of age), male lambs had a body length which was insignificantly higher than that of females by 1.06 cm and 3.93 cm respectively (P>0.05, Table 1). During the study period, the body length in male lambs increased by 27.9% from birth to weaning, while that of females only by 21.1%.

At weaning, chest perimeter of male lambs increased by 15.78 cm (37.9%) as compared to the perimeter at birth. For the same period, the chest perimeter of female lambs increased by 13.35 cm (33.1%). As gender was concerned, the chest perimeter of males was by 1.4 cm higher on the average than that of females. At weaning, the average difference was already 3.83 cm, but differences were not statistically significant (P>0.05, Table 1).

The measurement of exterior traits (body length and chest perimeter) demonstrated high coefficient of variation at weaning, which was characteristic for the growth and development of lambs during the suckling period (Table 1). At birth, male lambs had insignificantly higher compactness index; this advantage at weaning was in favour of females, although not statistically significant (P>0.05, Table 1).

Singletons are born heavier than twins by 19.4% (P<0.05, Table 2). This benefit was preserved also at weaning when the difference of 25.3% was statistically significant at (P<0.05). These results are not compatible with data reported by Sava *et al.* (2011) in Tigaia lambs, in which the insignificantly higher body weight by 28 and 90 days of age was in favour of twins. Rajab *et al.* (1992) indicated that twin lambs were born with lower live weight than singletons due to the limited space in the uterus during the pregnancy and the delivery of fewer nutrients from the dam to the fetuses during the embryonic life. At weaning, singletons exhibited higher average daily weight gain by 39.6% than twins (P<0.05, Table 2), due to the lack of competition during the suckling period (Rajab *et al.* 1992).

Table 1 Influence of gender on the growth

Indicators	Gender of the lamb					P-value
	Male (n=23)		Female (n=23)			
	(Mean±SE)	C	(Mean±SE)	C		
Live weight, kg						
At birth	5.017±0.176	16.84	4.594±0.208	21.71		NS
At weaning	11.035±0.675	29.33	10.499±0.575	26.27		NS
Gain	0.231±0.021	43.36	0.227±0.013	27.92		NS
Body length, cm						
At birth	39.01±0.723	8.88	37.95±1.044	13.19		NS
At weaning	49.88±2.468	23.73	45.95±2.080	21.71		NS
Chest perimeter, cm						
At birth	41.67±0.609	7.01	40.27±0.634	7.55		NS
At weaning	57.45±2.841	23.72	53.62±2.369	21.19		NS
Compactness index						
At birth	107.01±1.104	4.94	106.91±1.595	7.15		NS
At weaning	115.27±0.959	3.99	116.84±0.713	2.93		NS

* (P<0.05).
n: number of animals.
NS: non significant.
SE: standard error.
C: coefficient of variation.

Table 2 Influence of birth type on the growth

Indicators	Birth type					P-value
	Single (n=21)		Twins (n=25)			
	(Mean±SE)	C	(Mean±SE)	C		
Live weight, kg						
At birth	5.394±0.224	19.22	4.349±0.109	12.53		*
At weaning	12.485±0.703	25.80	9.324±0.364	19.53		*
Gain	0.273±0.019	31.96	0.165±0.015	30.87		*
Body length, cm						
At birth	40.65±1.004	11.31	36.66±0.610	8.32		*
At weaning	51.15±2.297	26.23	45.20±1.554	17.19		*
Chest perimeter, cm						
At birth	42.73±0.672	7.21	39.48±0.415	5.25		*
At weaning	58.21±3.385	26.65	53.29±1.828	17.15		*
Compactness index						
At birth	105.73±1.528	6.62	108.07±1.198	5.54		NS
At weaning	113.75±0.760	3.06	117.99±0.711	3.01		*

* (P<0.05).
n: number of animals.
NS: non significant.
SE: standard error.
C: coefficient of variation.

The coefficients of variation at birth and weaning were lower in twins while higher for singletons. The amount and quality of dam's milk could also influence the weight gain of lambs. This is supported by the study of Hrouz and Šubrt (2007) proving that during the suckling period, the growth intensity depended on the amount and quality of milk protein. At birth, the body length was statistically significantly higher in singletons *vs.* twins by 3.99 cm (P<0.05, Table 2). At weaning, singletons preserve their superiority with respect to the analyzed trait and the difference attained 5.95 cm (P<0.05). During the experimental period, body length of singletons increased by 25.8%, *vs.* 23.3% in twins (Table 2).

At birth, Singleton lambs had a chest perimeter greater by 3.25 cm than twins at the background of a relatively low coefficient of variation (P<0.05, Table 2).

At weaning, the difference in chest perimeters between both birth types attained 4.92 cm in favour of Singletons (P<0.05). It was established that during the period of the study, the chest perimeter of singleton lambs increased by 15.48 cm (36.2%) while of twins by 13.81 cm (35.0%, Table 2).

Comparing the growth performance of singleton and twin lambs, there was a strong correlation at weaning indicative for the amount and quality of suckled milk during the pre-weaning period (Table 2).

The performed analysis of variance (Table 3) showed that the greatest share of the total variance was that of the birth type, which influenced significantly (P<0.01) the live weight (η^2=26.85%), the body length (η^2=21.39%) and the chest perimeter (η^2=27.08%) at birth.

The gender of lambs, and gender × birth type were factors with relatively low shares of the total variance with statistically insignificant differences (Table 3). At 26 days of age i.e. at weaning, only the live weight was influenced by the birth type (η^2=27.53%, P<0.01), while none of the other studied traits had a substantial effect on the growth performance of animals (Table 3). The analysis of the repeatability and phenotypic correlations between the studied traits showed that the chest perimeter at weaning was not influenced by the body length at birth and at weaning. There was a moderate to strong relationship among the other studied parameters (Table 4).

Behzadi et al. (2007), stated that phenotypic correlations between various stages of body development in Kermani sheep were positive.

Table 3 Analysis of variance and the extent of influence of factors

Indicators	Source of variance					
	Sex of lamb		Type of birth		Sex × type	
	MS	η^2	MS	η^2	MS	η^2
Live weight						
at birth	0.955[ns]	2.48	10.353**	26.85	0.050[ns]	0.13
at weaning	0.531[ns]	0.13	109.219**	27.53	11.78[ns]	2.97
Body length						
at birth	1.88[ns]	0.23	175.50**	21.39	27.35[ns]	3.33
at weaning	138.0[ns]	2.56	332.2[ns]	6.15	123.1[ns]	2.28
Chest perimeter						
at birth	11.59[ns]	2.88	108.90**	27.08	0.80[ns]	0.20
at weaning	143.40[ns]	2.03	217.70[ns]	3.08	150.4[ns]	2.13

** (P<0.01).
MS: variance.
NS: non significant.
η^2: power of factor's impact, percent.

Table 4 Repeatability and phenotypic correlations between the studied traits

Parameters	LWB	LWW	BLB	BLW	CPB
Live weight at birth (LWB)	-	-	-	-	-
Live weight at weaning (LWW)	0.866**	-	-	-	-
Body length at birth (BLB)	0.702**	0.655**	-	-	-
Body length at weaning (BLW)	0.694**	0.683**	0.839**	-	-
Chest perimeter at birth (CPB)	0.565**	0.635**	0.495**	0.441**	-
Chest perimeter at weaning (CPW)	0.325*	0.467**	0.286	0.151	0.884**

* (P<0.05) and ** (P<0.01).

CONCLUSION

At birth and at weaning, singleton lambs had a higher body weight and higher average daily weight gain as compared to twins. The body length and the chest perimeter at birth were greater in singletons and statistically significantly lower for twin lambs. The birth type was a factor with statistically significant effect on live body weight (η^2=26.85%), body length (η^2=21.39%) and chest perimeter (η^2=27.08%) of lambs at birth. The chest perimeter of lambs at weaning was not influenced by the body length at birth and at weaning. There was a moderate to strong phenotypic correlation among the other studied parameters.

REFERENCES

Abbas S.F., Abd Allah F., Allam A.A. and Abon E. (2010). Growth performance of Rahmani and Chios lambs weaned at different ages. *Australian J. Basic Appl. Sci.* **4(7),** 1583-1589.

Afolayan R.A., Adeyinka I.A. and Lakpini C. (2006). The estimation of live weight from body measurements in Yankasa sheep. *Czech J. Anim. Sci.* **51,** 343-348.

Alcock D. (2006). Creep feeding lambs. *Primef. Profit. Sustain. Prim. Indust.* **224,** 1-4.

Alphonsus C., Finangwai H., Yashim S., Agubosi O. and Sam I. (2010). Effect of dam parity on measures of growth in Red Sokoto goats 1, 3, 6 and month of age. *Continental J. Anim. Vet. Res.* **2,** 9-13.

Behzadi M., Shahroudi F. and van Vleck L. (2007). Estimates of genetic parameters for growth traits in Kermani sheep. *J. Anim. Breed. Gen.* **124,** 296-301.

Brown J.E., Brown C.J. and Butts W.T. (1973). Evaluating relationships among immature measures of size, shape and performance on beef bulls I: principal component as measures of size and shape in young Hereford and Angus bulls. *J. Anim. Sci.* **36,** 1010-1020.

El-Toum A. (2005). Effect of pre-partum supplementary feeding on desert ewe productivity under rangelands in north Kordofan, Sudan. MS Thesis. University of Khartonm, Khartonm, Sudan.

Fourie P., Neser F., Olivier J. and der Westhuizen C. (2002). Relationship between production performance, visual appraisal and body measurements of young Dorper rams. *South African J. Anim. Sci.* **32,** 256-262.

Hrouz J. and Šubrt J. (2007). General Zootechnics. Mendel University of agriculture and forestry in Brno, Brno, Czech.

Idris A.O., Kijora C., El-Hag F.M. and Salih A.M. (2010). Effects of supplementation on late pregnancy and early lactation of body weight of desert ewes and their lambs. Livest. Res. Dev. Available at:
http://www.lrrd.org/lrrd22/10/idri22193.htm.

Ivanova T. and Raycheva E. (2009). Weight performance of Blackhead Pleven lambs during the suckling period. *Agric. Sci.* **1,** 11-16.

Kadlečík O. and Kasarda R. (2007). General Zootechnics. Slovak Agricultural University in Nitra, Nitra, Slovak.

Klewies J., Martyniuk E., Gabryszuk M. and Baranowski A. (2002). Growth rate in Booroola × Olkuska crossbred lambs as related to the crossing scheme. *Anim. Sci. Pap. Rep.* **20(2),** 93-101.

Koritiaki N., Ribeiro E., Mizubuti I., da Silva L., Bardosa M., Scerbo D., Muniz C. and Junior F. (2013). Effect of environmental factors on performance of purebred and crossbred Santa Ines lambs from birth to 154 days of age. *Rev. Bras. Zootec.* **42(2),** 87-94.

Kumar S., Mishra A.K., Kolte A.P., Arora A.L., Singh D. and Singh V. (2008). Effects of the Booroola (FecB) genotypes on growth performance, ewe's productivity efficiency and litter size in Garole × Malpura sheep. *Anim. Rep. Sci.* **105(3),** 319-331.

Macit M., Esenbuga N. and Karaoglu M. (2002). Growth performance and carcass characteristics of Awassi, Morkaraman and Tushin lambs grazed on pasture and supported with concentrates. *Small Rumin. Res.* **44,** 241-246.

Philips W., Reuter R., Brown M., Fich J., Rao S. and Mayeux H. (2002). Growth and performance of lambs fed a finishing diet containing either Alfalfa or Kenaf as the roughage source. *Small Rumin. Res.* **46,** 75-79.

Rajab M.H., Cartwright I.C., Dahm P.F. and Figueireda E.A.P. (1992). Performance of three tropical hair sheep breeds. *J. Anim. Sci.* **70,** 3351-3359.

Salako AE. (2006). Application of morphological indices in the assessment of type and Function in sheep. *Int. J. Morphol.* **24,** 13-18.

Sava C., Pascal C., Zaxaria N., Zaxaria R. and Atanasin T. (2011). Mother's age lambing type as influential factors on body growth and development of youth sheep. *Lucrări Ştiinţifice Seria Zootehnie.* **55,** 131-135.

Savov T. (1948). Investigation on the production performance of some sheep breeds and races bred at the Georgi Dimitrov state livestock enterprise near Pleven (129-148).

Statistica. (2006). Statistica for Windows, StatSoft Inc., Tulsa, Oklahoma, USA.

Vuchkov Y. and Dimov D. (2008). Study on live body weight and growth performance of White Maritsa lambs. *Anim. Sci. Sofia.* **4,** 41-45.

14

Characteristics Determination of Rheb Gene and Protein in Raini Cashmere Goat

M.R. Mohammadabadi[1*] **and F. Tohidinejad**[1]

[1] Department of Animal Science, Faculty of Agriculture, Shahid Bahonar University of Kerman, Kerman, Iran

*Correspondence E-mail: mrm@uk.ac.ir

ABSTRACT

The aim of the present study was to determine characteristics of Rheb gene and protein in Raini Cashmere goat. Comparative analyses of the nucleotide sequences were performed. Open reading frames (ORFs), theoretical molecular weights of deduced polypeptides, the protein isoelectric point, protein characteristics and three-dimensional structures was predicted using online standard softwares. The full cDNA nucleotide sequence shares 99%, 99%, 99% and 94% identity with Inner Mongolia Cashmere goat, cattle, horse and human, respectively. The deduced Rheb protein of Raini Cashmere goat consist of 184 amino acid residues and its predicted molecular weight was 20478 g/mol for the unmodified protein and the estimated isoelectric point (pI) was 5.59. Results showed that Rheb has a RAS domain starting at the amino acid 4 and ending at the amino acid 170. In conclusion, our data shows the Rheb cDNA is 555 bp in length, including a complete ORF corresponding to a polypeptide of 184 amino acids. Our results in Raini Cashmere goat showed that there are 1 N-glycosylation sites, 3 protein kinase C phosphorylation sites, 2 casein kinase II phosphorylation sites, 4 microbodies C-terminal targeting signals, 2 ATP/GTP binding sites motif A (P-loop) and a prenyl group binding site for predicted Psites of Rheb. Raini Cashmere goat Rheb protein also had an interaction with other predicted proteins. Hence, can suggest that Rheb has probably role in goat cells and must detect in future investigations.

KEY WORDS base sequence, goat, interaction, isoelectric point.

INTRODUCTION

Goat production is one of the key elements contributing to the economy of farmers living in the arid and semi-arid regions including most areas of Iran (Mohammad Abadi *et al.* 2009). Raini goat is one of the most important Iranian native goats that spread in the southeast of Iran where these animals are kept for both meat and cashmere production. It has long white, black, or yellow hair and the mean weight of the fleece produced is 400 g ranging from 130 to 1,100 g per head (Moghbeli *et al.* 2013). Mean body weight at 12 months for male and female is 18.7 ± 0.04 and 15.5 ± 0.04 respectively (Hasani *et al.* 2010).

However, these economic characters are quantitative in nature, being under the control of several genes plus the environment (Askari *et al.* 2011). Rheb (Ras homolog enriched in brain) gene belongs to Ras family that encodes a carboxyl terminal CAAX box indicating that the protein may undergo post-translational farnesylation. Ras homolog enriched in brain (Rheb) is a key regulator of the mammalian target of rapamycin (mTOR) complex 1 (mTORC1) signaling pathway (Dunlop *et al.* 2009). Rheb is a member of the Ras superfamily GTPases and shares the highest homology with Ras and Rap and originally identified as immediate early gene (IEG) in 1994, encoding 184 amino acids with a deduced molecular mass of 20497

Da in hippocampus (Yamagata *et al.* 1994) The Rheb gene is highly conserved in eukaryotes from yeast to mammals (Patel *et al.* 2003). Genetic studies of fly and fission yeast indicate that Rheb plays an important role in the stimulation of cell growth and regulation of G0/G1 cell cycle progression (Yang *et al.* 2001).

The growth arrest phenotype caused by Rheb mutation in *Schizosaccharomyces pombe* can be complemented by human Rheb (Yang *et al.* 2001), suggesting the conservation of Rheb function from yeast to human. The precise physiological functions of Rheb were unknown in high eukaryotes until recently. Both genetic studies in *Drosophila melanogaster* and biochemical studies in mammalian cells have shown that Rheb is involved in signal transduction pathways that regulate cell growth (Tee *et al.* 2003). Homozygous inactivation of Rheb is lethal, while mosaic analyses of Rheb mutant cells in *Drosophila* show that the inactivation of Rheb decreases cell size. In contrast, overexpression of Rheb increases cell size (Yang *et al.* 2001). Genetic epistatic analysis demonstrates that Rheb functions between TSC1-TSC2 and TOR.

The evidences show that Rheb regulates mTOR through direct binding of Rheb-GTP to mTOR to promote activation of mTOR (Long *et al.* 2005). Other study demonstrated that FKBP38 can bind to mTOR to inhibit its activity and Rheb-GTP interacts directly with FKBP38 to prevent FKBP38 associating with mTOR (Ma *et al.* 2008).

However, comprehensive biochemical characterization of the Rheb/FKBP38 interaction using three different *in vitro* assays has not detected an interaction between Rheb and FKBP38. Therefore, the mechanism of the interaction between Rheb and FKBP38 is still under debate and needs to be further characterized. Although the Rheb gene has been identified in mice (NM_053075) and humans (NM_005614), its physiological function has not been fully investigated and remains to be clearly defined. In farm animals, (Zheng *et al.* 2011) studied Rheb cDNA in Inner Mongolia Cashmere goat and showed that it has 555 bp in length, including a complete ORF corresponding to a polypeptide of 184 amino acids. Rheb gene was expressed in all the tested tissues; the highest level of mRNA accumulation is detected in brain tissue.

Furthermore, it has demonstrated that molecular studies are important in farm animals breeding (Javanmard *et al.* 2008; Mohammadabadi *et al.* 2010; Mohammadi *et al.* 2009; Ruzina *et al.* 2010).

So far, no study concerning the characterization of Rheb gene in Iranian farm animals, especially in Raini Cashmere goat has been published. Hence the aim of the present study was to determine for the first time characteristics of Rheb gene and protein in Raini Cashmere goat and compare with other species.

MATERIALS AND METHODS

Tissues including brain, heart, lung, pancreatic, spleen, kidney, liver and testis were collected from the Raini Cashmir goat after slaughter (3 repeats from any tissue). Extracted RNA were immediately stored at -80 °C. Quality and quantity of RNA were evaluated and cDNA was synthesized and PCR was performed. PCR Products were electrophoresed on 1.5% agarose gel and were evaluated different levels of expression in studied different tissues using SPSS (2011) software at (P<0.01) (Tohidi nezhad *et al.* 2015). The PCR products of Rheb gene from Raini Cashmere goat were sequenced. Comparative analyses of the nucleotide sequences were performed online at National Center for Biotechnology Information (NCBI). Predictions of theoretical molecular weights of deduced polypeptides were made by the protein property calculator (Kyte and Doolittle, 1982). Input data for this software was peptide sequences and outputs were predicted molecular weight, extinction coefficient and approximate volume. The protein isoelectric point was predicted by the calculation of protein isoelectric point (Kozlowski, 2016), in which was added amino acids sequence and was received estimated isoelectric point (pI). The domain of Raini Cashmere goat Rheb protein predicted with SMART software (Schultz *et al.* 1998; Letunic *et al.* 2015) and the switch I region and the switch II region were predicted with NCBI CDD program (Marchler-Bauer *et al.* 2011).

Input data for these softwares was protein sequence and outputs were basic amino composition, total number of negatively and positively charged residues, its domain starting and ending at what amino acid and indicate switch regions. Prediction of protein characteristics and three-dimensional structures provided by Molecular Bioinformatics Center of National Chiao Tung University (Chen *et al.* 2006; Chen *et al.* 2009) in which was added the query sequence in FASTA format and was received three-dimensional structure of Rheb for Raini Cashmere goat and its sites. The STRING program used for representing predicted protein interactions (Szklarczyk *et al.* 2015). Input data for this software was protein name and organism and outputs were interaction with other predicted proteins and description of predicted functional partners.

RESULTS AND DISCUSSION

The deduced Rheb protein of Raini Cashmere goat consist of 184 amino acid residues and its predicted molecular weight was 20478 g/mol for the unmodified protein and the estimated isoelectric point (pI) was 5.59. Extinction coefficient and approximate volumes were 16050 cm^{-1}m^{-1} and 24778 A^3 respectively.

The basic amino composition is given in Table 1 and the total number of negatively charged residues (Asp+Glu) and the total number of positively charged residues (Arg+Lys) were 22 and 19 respectively. Rheb has a RAS domain starting at the amino acid 4 and ending at the amino acid 170, including a switch I region from amino acid 36 to 43 and a switch II region from amino acid 62 to amino acid 80 (Figure 1). Protein characteristics and three-dimensional structures of Rheb for Raini Cashmere goat are shown in Figure 2. Our results in Raini Cashmere goat (Figure 2) showed that there are 1 N-glycosylation sites, 3 protein kinase C phosphorylation sites, 2 casein kinase II phosphorylation sites (SAKE and TAVD), 4 microbodies C-terminal targeting signals, 2 ATP/GTP binding sites motif A (P-loop) and a prenyl group binding site (CAAX box) for predicted Psites of Rheb. Raini Cashmere goat Rheb interaction with other predicted proteins and description of predicted functional partners using the STRING program is given in Figure 3. The most interaction exists functionally between Rheb and tuberous sclerosis complex 2 (TSC2), mammalian target of rapamycin (mTOR) and tuberous sclerosis complex 1 (TSC1) respectively and the least was seen between Rheb and protein kinase B1 substrate 1 (AKT1S1) and elongation initiation factor 4E (EIF4E). Zheng *et al.* (2011) for Inner Mongolia Cashmere goat also achieved that protein consist of 184 amino acid residues and its predicted molecular weight is 20358 Da for the unmodified protein and the estimated isoelectric point (pI) is 6.27. The basic amino acids comprise 10.3% Ser, 9.2% Ile, 9.2% Val, 7.6% Lys and 7.6% Leu that confirmed results in this study. Protein characteristics and three-dimensional structures of Rheb for Raini Cashmere goat are same as results of Zheng

et al. (2011) for Inner Mongolia Cashmere goat. Three-dimensional structures of Rheb for Raini Cashmere goat (Figure 3) is very similar to the structure of Mus musculus Rheb presented by Mazhab-Jafari *et al.* (2012).

Mazhab-Jafari *et al.* (2012) proposed that whereas providing an Asn thumb as a means of accelerating catalysis, TSC2GAP may also stimulate the GTPase activity of Rheb by relieving autoinhibition and aligning Rheb's catalytic machinery.

Interaction of TSC2 with Rheb switch I may disrupt the electrostatic contact between Tyr35 and the g-phosphate, reducing the autoinhibitory effect of this residue on GTP hydrolysis, explaining the functional and thermodynamic similarities between WT Rheb in the presence of the TSC2GAP and the Rheb Y35A mutant alone (Mazhab-Jafari *et al.* 2012) that confirmed results in this study. TSC1 and TSC2 form a dimeric complex that has tumor suppressor activity and TSC2 is a GTPase activating protein (GAP) for Rheb. The TSC1/TSC2 complex inhibits the activation of TOR kinase through Rheb. Rheb has also been shown to induce the formation of large cytoplasmic vacuoles in a process that is dependent on the GTPase cycle of Rheb, but independent of the target of rapamycin (TOR) kinase, suggesting Rheb plays a role in endocytic trafficking that leads to cell growth and cell-cycle progression. Most Ras proteins contain a lipid modification site at the C-terminus, with a typical sequence motif CaaX, where a, indicates an aliphatic amino acid and X, indicates any amino acid. Lipid binding is essential for membrane attachment, a key feature of most Ras proteins. The 15 amino acid in Ras protein is glycine instead of arginine in goat Rheb amino acid sequence.

Table 1 Amino acid composition of Rheb gene in Raini Cashmere goat

Amino acid full name	Amino acid name with 3 letters	Amino acid name with 1 letter	Number	%
Alanine	Ala	A	11	5.9
Arginine	Arg	R	5	2.7
Asparagine	Asn	N	6	3.2
Aspartic acid	Asp	D	9	4.9
Cysteine	Cys	C	1	0.5
Glutamine	Gln	Q	10	5.4
Glutamic acid	Glu	E	13	7.1
Glycine	Gly	G	12	6.5
Histidine	His	H	3	1.6
Isoleucine	Ile	I	18	9.7
Leucine	Leu	L	13	7.1
Lysine	Lys	K	14	7.6
Methionine	Met	M	5	2.7
Phenylalanine	Phe	F	7	3.8
Proline	Pro	P	4	2.1
Serine	Ser	S	19	10.3
Threonine	Thr	T	9	4.9
Tryptophan	Trp	W	1	0.5
Tyrosine	Tyr	Y	8	4.3
Valine	Val	V	16	8.6

4----36===== 43--61====80-------170

switch region I   switch region II

N terminal

C terminal

RAS DOMAIN (4-170)

Raini cashmere goat Rheb Protein with 184 aa

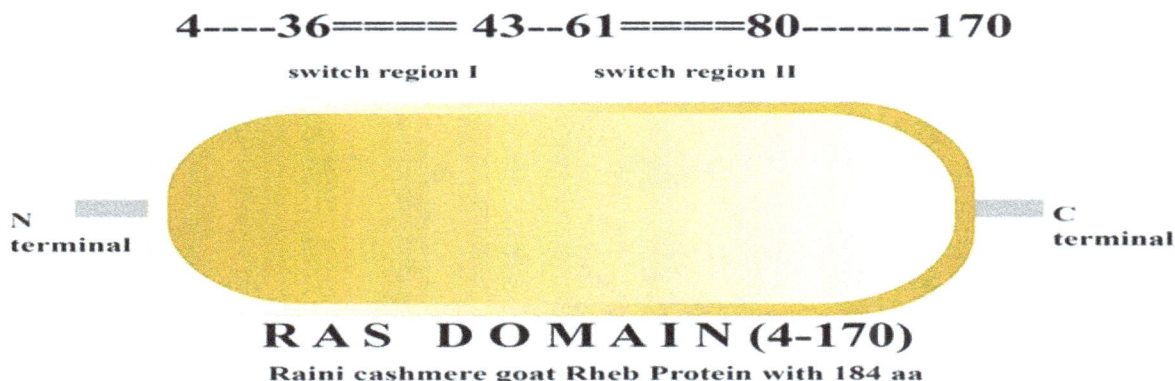

Figure 1 The predicted domain of Raini Cashmere goat Rheb protein (the RAS domain ranges from amino acid 4 to 170, including a switch region I starting from the amino acid 36 to the amino acid 43 and a switch region II from amino acid 61 to 80)

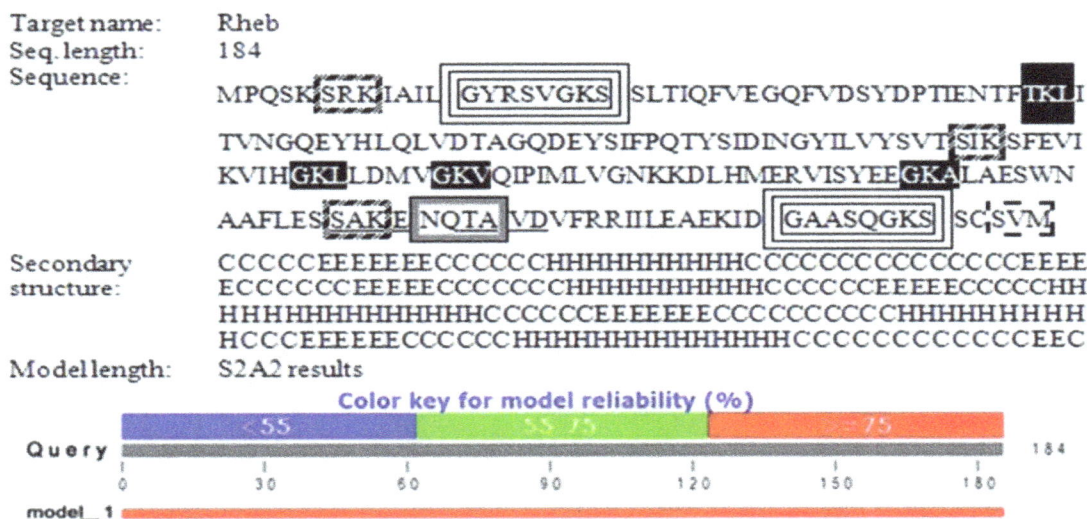

Target name: Rheb
Seq. length: 184
Sequence:

MPQSKSRKIAIL GYRSVGKS SLTIQFVEGQFVDSYDPTIENTFTKLI
TVNGQEYHLQLVDTAGQDEYSIFPQTYSIDINGYILVYSVTSIKSFEVI
KVIHGKLLDMVGKVQIPIMLVGNKKDLHMERVISYEEGKALAESWN
AAFLESSAKIENQTAVDVFRRIILEAEKID GAASQGKS SCSVM

Secondary structure:

CCCCCEEEEEEECCCCCCCHHHHHHHHHHCCCCCCCCCCCCCCCCEEEE
ECCCCCCEEEEEECCCCCCCHHHHHHHHHHHCCCCCEEEEEECCCCCHH
HHHHHHHHHHHHCCCCCCEEEEEECCCCCCCCCCCHHHHHHHHH
HCCCEEEEEECCCCCCCHHHHHHHHHHHHHCCCCCCCCCCCCEEC

Model length: S2A2 results

Color key for model reliability (%)

<55   55-75   >=75

Query   184

0   30   60   90   120   150   180

model_1

PDB Model	Template	Seq-len	Aligned (%)	Identity (%)	Bit-score	E-value
model_1	1xtqA	169	91.85	98.82	337.2	1.2e-12

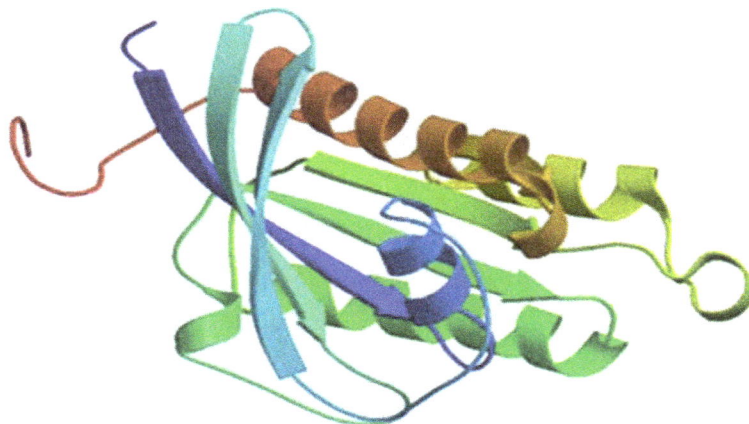

Figure 2 Protein characteristics and three-dimensional structures of Rheb for Raini Cashmere goat

NA indicates N-glycosylation sites; SK indicates protein kinase C phosphorylation sites; underline indicates casein kinase II phosphorylation sites; blacked shadows indicate microbodies C-terminal targeting signals; GS indicates ATP/GTP binding sites motif A (P-loop) and SVM indicates a prenyl group binding site (CAAX box) for predicted Psites

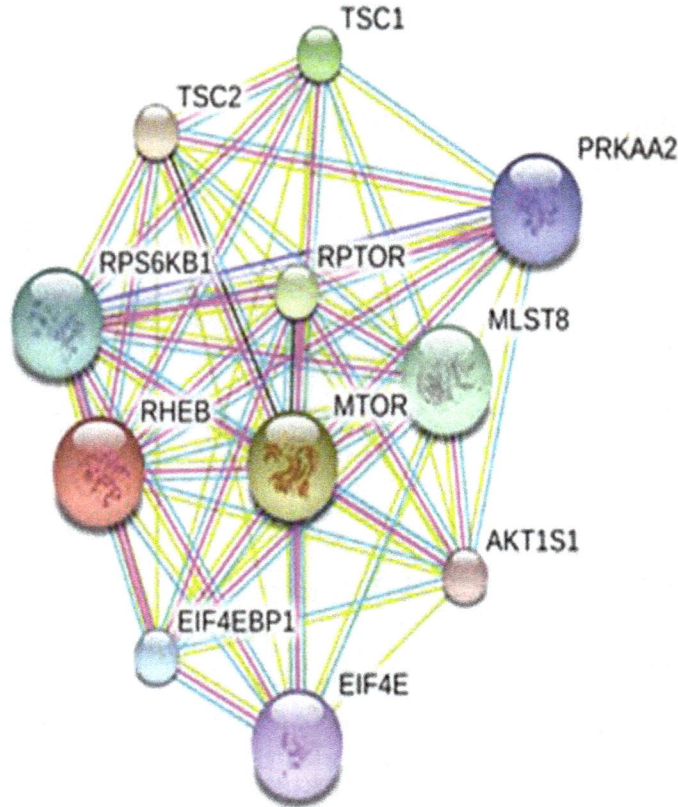

Figure 3 Raini Cashmere goat Rheb interaction with other predicted proteins and description of predicted functional partners using the STRING program
Different line colors represent the types of evidence for the association
Line thickness relates to combined score
Colors describe the type of evidence

The replacement leads to lower basal GTPase activity in Rheb than that of Ras and result in a higher GTP level, which is required for S6K1 phosphorylation via mTOR because the Rheb is an upstream regulator of S6K1 and mTOR. Meanwhile, the goat Rheb protein has a prenyl group binding site (CAAX box) where X is metionine at the C-terminal. The CAAX box in Rheb is farnesylated and the membrane localization of Rheb through farnesylation is important for upstream regulation of S6K1 activity in mTOR signaling pathway (Tee *et al.* 2003), that this replacement was seen in results of this study and confirmed results of other researchers. Besides, the deduced goat Rheb protein has G1 box to G5 box and two switch regions in its Ras family conservative domain (Figures 2 and 3). All indicated features are very similar to reported results about Rheb in Inner Mongolia Cashmere goat and other species, indicating that Rheb gene has been correctly studied from Raini Cashmere goat.

The Rheb switch II is also critical for signaling to the target of mTOR although the mechanism is unclear. As shown above, goat Rheb has a Ras family conservative domain containing switch I region from amino acid 36 to 43. The switch I region and FKBP-C domain may have interaction and regulate the activity of mTOR signaling path way in goat cells, but need to be further verification.

CONCLUSION

In conclusion, Rheb cDNA is 555 bp in length, including a complete ORF corresponding to a polypeptide of 184 amino acids. Rheb employs an autoinhibitory mechanism maybe maintain a high activation state in cells essential for the proper maintenance of mTORC1 signaling and cellular growth. This is the first proposed bioinformatics study of Rheb in Raini Cashmere goat within the Ras subfamily, which may be relevant to some other Ras superfamily and provides a view into the molecular and bioinformatic mechanism of Rheb.

ACKNOWLEDGEMENT

The authors thank all individuals who willingly participated in the study.

REFERENCES

Askari N., Mohammadabadi M.R. and Baghizadeh A. (2011). ISSR markers for assessing DNA polymorphism and genetic characterization of cattle, goat and sheep populations. *Iranian J. Biotechnol.* **9,** 222-229.

Chen C.C., Hwang J.K. and Yang J.M. (2006). (PS)² : protein structure prediction server. *Nucleic Acids Res.* **34,** 152-157.

Chen C.C., Hwang J.K. and Yang J.M. (2009). (PS)²-v2: template-based protein structure prediction server. *BMC Bioinformatics.* **10,** 366-371.

Dunlop E.A., Dodd K.M., Seymour L.A. and Tee A.R. (2009). Mammalian target of rapamycin complex 1-mediated phosphorylation of eukaryotic initiation factor 4E-binding protein 1 requires multiple protein–protein interactions for substrate recognition. *Cell. Signal.* **21,** 1073-1084.

Hasani M.N., Asadi Fozi M., Esmailizadeh A.K. and Mohammadabadi M.R. (2010). A genetic analysis of growth traits in Raieni Cashmere goat using multivariate animal model. *Iranian J. Anim. Sci.* **41,** 323-329.

Javanmard A., Mohammadabadi M.R., Zarrigabayi G.E., Gharahedaghi A.A., Nassiry M.R., Javadmansh A. and Asadzadeh N. (2008). Polymorphism within the intron region of the bovine leptin gene in Iranian Sarabi cattle (*Iranian Bos taurus*). *Russian J. Genet.* **44,** 495-497.

Kozlowski L.P. (2016). IPC-Isoelectric point calculator. *Biol. Direct.* **11,** 55.

Kyte J. and Doolittle R.F. (1982). A simple method for displaying the hydropathic character of a protein. *J. Mol. Biol.* **157,** 105-132.

Letunic I., Doerks T. and Bork P. (2015). SMART: recent updates, new developments and status in 2015. *Nucleic Acids Res.* **43,** 257-260.

Long X., Lin Y., Ortiz-Vega S., Yonezawa K. and Avruch J. (2005). Rheb binds and regulates the mTOR kinase. *Curr. Biol.* **15,** 702-713.

Ma D., Bai X., Guo S. and Jiang Y. (2008). The switch I region of Rheb is critical for its interaction with FKBP38. *J. Biol. Chem.* **283,** 25963-25970.

Marchler-Bauer A., Lu S., Anderson J.B., Chitsaz F., Derbyshire M.K., DeWeese-Scott C., Fong J.H., Geer L.Y., Geer R.C., Gonzales N.R., Gwadz M., Hurwitz D.I., Jackson J.D., Ke Z., Lanczycki C.J., Lu F., Marchler G.H., Li J., Miyamoto K. and McClane B.A. (2011). CDD: a conserved domain database for the functional annotation of proteins. *Nucleic Acids Res.* **39,** 225-229.

Mazhab-Jafari M.T., Marshall C.B., Ishiyama N., Ho J., Di Palma V., Stambolic V. and Ikura M. (2012). An autoinhibited non-canonical mechanism of GTP hydrolysis by Rheb maintains mTORC1 homeostasis. *Structure.* **20,** 1528-1539.

Moghbeli S.M., Barazandeh A., Vatankhah M. and Mohammadabadi M.R. (2013). Genetics and non-genetics parameters of body weight for post-weaning traits in Raini Cashmere goats. *Trop. Anim. Health Prod.* **45,** 1519-1524.

Mohammad Abadi M.R., Askari N., Baghizadeh A. and Esmailizadeh A.K. (2009). A directed search around caprine candidate loci provided evidence for microsatellites linkage to growth and cashmere yield in Rayini goats. *Small Rumin. Res.* **81,** 146-151.

Mohammadabadi M.R., Nikbakhti M., Mirzaee H.R., Shandi A., Saghi D.A., Romanov M.N. and Moiseyeva I.G. (2010). Genetic variability in three native Iranian chicken populations of the Khorasan province based on microsatellite markers. *Russian J. Genet.* **46,** 505-509.

Mohammadi A., Nassiry M.R., Mosafer J., Mohammadabadi M.R. and Sulimova G.E. (2009). Distribution of BoLA-DRB3 allelic frequencies and identification of a new allele in the Iranian cattle breed Sistani (*Bos indicus*). *Russian J. Genet.* **45**, 198-202.

Patel P.H., Thapar N., Guo L., Martinez M., Maris J., Gau C.L., Lengyel J.A. and Tamanoi F. (2003). Drosophila Rheb GTPase is required for cell cycle progression and cell growth. *J. Cell. Sci.* **116**, 3601-3610.

Ruzina M.N., Shtyfurko T.A., Mohammadabadi M.R., Gendzhieva O.B., Tsedev T. and Sulimova G.E. (2010). Polymorphism of the *BoLA-DRB3* gene in the Mongolian, Kalmyk, and Yakut cattle breeds. *Russian J. Genet.* **46**, 456-463.

Schultz J., Milpetz F., Bork P. and Ponting C.P. (1998). SMART, a simple modular architecture research tool: identification of signaling domains. *Proc. Natl. Acad. Sci.* **95**, 5877-5864.

SPSS Inc. (2011). Statistical Package for Social Sciences Study. SPSS for Windows, Version 20. Chicago SPSS Inc.

Szklarczyk D., Franceschini A., Wyder S., Forslund K., Heller D., Huerta-Cepas J., Simonovic M., Roth A., Santos A., Tsafou K.P., Kuhn M., Bork P., Jensen L.J. and von Mering C. (2015). STRING v10: protein-protein interaction networks, integrated over the tree of life. *Nucleic Acids. Res.* **43**, 447-452.

Tee A.R., Manning B.D., Roux P.P., Cantley L.C. and Blenis J. (2003). Tuberous sclerosis complex gene products, Tuberin and Hamartin, control mTOR signaling by acting as a GTPase-activating protein complex toward Rheb. *Curr. Biol.* **13**, 1259-1268.

Tohidi nezhad F., Mohammadabadi M.R., Esmailizadeh A.K. and Najmi Noori A. (2015). Comparison of different levels of Rheb gene expression in different tissues of Raini Cashmir goat. *J. Agric. Biotechnol.* **6**, 35-50.

Yamagata K., Sanders L.K., Kaufmann W.E., Yee W., Barnes C.A., Nathans D. and Worley P.F. (1994). Rheb, a growth factor-and synaptic activity-regulated gene, encodes a novel Ras-related protein. *J. Biol. Chem.* **269**, 16333-16339.

Yang W., Tabancay A.P., Urano J. and Tamanoi F. (2001). Failure to farnesylate Rheb protein contributes to the enrichment of G0/G1 phase cells in the Schizosaccharomyces pombe farnesyltransferase mutant. *Mol. Microbiol.* **41**, 1339-1347.

Zheng X., Yang J.F., Wang X.J., Liang Y., Wu M.L., Shi J.J., Zhang T., Yin Q., Li S.Y. and Hao X.Y. (2011). Molecular characterization and expression pattern of Rheb gene in Inner Mongolia Cashmere goat (*Capra hircus*). *Agric. Sci. China.* **10**, 1452-1458.

15

In vitro Assessment of Adsorbents to Counteract Lead Toxicity in Ruminal Fermentation

S. Azadbakht[1], A.A. Khadem[1] and M.A. Norouzian[1*]

[1] Department of Animal Science, College of Abouraihan, University of Tehran, Tehran, Iran

*Correspondence E-mail: manorouzian@ut.ac.ir

ABSTRACT

An *in vitro* gas production technique was used to evaluate the effects of different adsorbents on the gas production parameters of lead-exposed diet. Ruminal fluid, obtained from fistulated sheep (3-4 years of age and 40±5 kg BW), and experimental diets samples, 200 ± 0.2 mg DM, were incubated in 100 mL glass bottles. To each bottle, one of the following treatments was applied: 1) control (no Pb and adsorbent), 2) 15 mg/kg DM Pb as Pb acetate and no adsorbent, 3) 15 mg/kg DM Pb as Pb acetate and 3% activated carbon, 4) 15 mg/kg DM Pb as Pb acetate and 3% nanoclay and 5) 15 mg/kg DM Pb as Pb acetate and 3% bentonite. Supplementation of ruminal medium with 15 ppm Pb significantly (P<0.01) decreased total and rate of gas production at all incubation times. Rate and total gas production were significantly (P<0.05) higher for nanoclay compared to control and the other adsorbents treatments. Lead inclusion decreased organic matter digestibility (OMD), net and metabolizable energy contents and short chain fatty acids production (P<0.01). Among the adsorbents, nanoclay showed to be more effective on ruminanl fermentation parameters and improved rate and total gas production.

KEY WORDS adsorbent, lead, rumen fermentation, sheep.

INTRODUCTION

Both primary and secondary lead–zinc smelters have been implicated in producing lead toxicity in cattle, buffalo and goat reared around these industrial units (Radostits *et al.* 2000; Swarup *et al.* 2006). The higher lead levels in animals reared around such industrial activities are mainly due to ingestion of pasture contaminated with lead as well as inhalation of lead particles (Mohajeri *et al.* 2014). These toxic trace elements hinder rumen fermentation, thereby decreasing the productivity of ruminants. Forsberg (1977) showed that the elements can causing 50% inhibition of *in vitro* rumen fermentation. Ions of heavy metals may also inhibit activity of some ruminal enzymes (Faixova and Faix, 2002). Despite of this fact that the effect of heavy

metal contamination on rumen fermentation has been previously reported (Faixova and Faix, 2002; Forsberg, 1977), based on our survey it appears no literature on the effect of adsorbents supplementation on ruminal fermentation parameters of sheep chronically exposed to lead contamination. Therefore, the present study was aimed to assess the effect of different adsorbent supplements on *in vitro* ruminal fermentation parameters.

MATERIALS AND METHODS

In this study an *in vitro* gas production technique (Menke *et al.* 1979) was used to estimate fermentation parameters, organic matter digestibility (OMD), net energy (NE) and metabolizable energy (ME) contents and short chain fatty acids (SCFA) production. Ruminal fluid obtained from 4 fistulated sheep (3-4 years of age and 40±5 kg BW) 3 h after morning feeding was mixed and strained through 4 layers of cheesecloth into a pre-warmed thermos and transported to the laboratory. The lambs were fed a total mixed ration (60:40 forage:concentrate; DM basis) and 0.6% mineral and vitamin premix. The lambs were fed twice daily at 0700 and 1900 hours and had free access to water.

Incubation medium was prepared as described by Menke et al. (1979). Sample from experimental diets, each of 200 ± 0.2 mg DM, were incubated in 100 mL glass bottles in which 30 mL of the incubation medium was added. To each bottle, one of the following experimental treatments was applied: 1) control (no Pb and adsorbent), 2) 15 mg/kg DM Pb as Pb acetate and no adsorbent, 3) 15 mg/kg DM Pb as Pb acetate and 3% activated carbon, 4) 15 mg/kg DM Pb as Pb acetate and 3% nanoclay and 5) 15 mg/kg DM Pb as Pb acetate and 3% bentonite.

Samples were incubated in triplicate and cumulative gas production was monitored at 2, 4, 6, 8, 10, 12, 15, 19, 24, 30, 36, 48, 72 and 96 h post-incubation. Three bottles with incubation medium only, were used as blanks to correct the gas production values for gas release from the rumen contents. The gas production data were fitted to the following model introduced by France et al. (2000):

$$A = b \times [1 - e^{-c(t-L)}]$$

Where:
A: volume of gas production at time t.
b: asymptotic gas production (based on mL/200 mg DM).
c: rate of gas production per hour from the slowly fermentable feed fraction b.
Time lag (L): discrete lag time prior to gas production.

The rate of gas production (RGP) at 4 and 6 h was calculated from recorded volumes of gas produced before and after these times (Vázquez-Armijo et al. 2011). For example, RGP at 4 h was calculated as:

RGP 4 h [(mL/g DM)/h)]= (volume of gas produced at 6 h-volume of gas produced at 2 h) / (4×sample weight (mg))

OMD, ME (Menke et al. 1979) and net energy (NE) (Menke and Steingass, 1988) contents were estimated according to the equations given below:

OMD (%)= 14.88 + 0.8893 $IVGP_{24}$ + 0.448 CP (%DM) + 0.651 A (% DM)
ME (MJ/kg DM)= 2.20 + 0.136 $IVGP_{24}$ (mL/200 mg DM) + 0.057 CP (% DM)

NE (MJ/kg DM)= 0.101 $IVGP_{24}$ + 0.051 CP (% DM) + 0.11 EE (% DM)

Where:
$IVGP_{24}$: 24 h in vitro gas production volume.
CP: crude protein.
A: ash.
EE: ether extract contents of the feed sample.

SCFA were estimated by the equation of Makkar (2005):

SCFA (mmol/g DM)= 0.0222 (mL gas at 24 h) - 0.00425

Data were analyzed as a completely randomized design using the MIXED procedure of SAS (2004). Duncan multiple range test was used to detect statistical significance between treatments using a significance level of 0.05.

RESULTS AND DISCUSSION

Rate of gas production (mL/200 mg DM) in different incubation times are presented in Table 1. Lead-polluted non-supplemented batch (group 2) had lower rate of gas production compared to the groups supplemented with adsorbents and the control (P< 0.01).

Gas production volumes (mL/200 mg DM) in different incubation times of experimental groups are shown in Table 2. Supplementation of ruminal medium with 15 ppm Pb has significantly (P<0.01) decreased total gas production at all incubation times. Gas production volumes were significantly (P<0.05) higher for nanoclay compared to control and other adsorbents containing groups (Table 2 and Figure 1).

Gas production parameters (b, c) and calculated amounts of OMD, ME, NE and SCFA are presented in Table 3. Supplementation of lead adsorbents improved gas production parameters (P<0.01). Among the different adsorbents, the nanoclay was the most effective one on ruminanl fermentation parameters. However, there were no significant differences between the different treatments regarding the fractional rate of gas production (c). Lead inclusion led to reduction in OMD, NE, ME contents and SCFA production (P<0.01).

Heavy metals prevent rumen fermentation, thereby decreasing the productivity of ruminants. Forsberg (1977) believed that these elements may be inhibitory to both the fermentative activity and the growth of microorganisms present in the rumen.

In this study, inclusion of lead to ruminal fermentation bottleshad an reducing effect on the measured fermentation parameters.

Table 1 Effect of adsorbents addition to lead-exposed diets on rate of gas production at different incubation times

Incubation time	Treatment[1]					SEM	P-value
	1	2	3	4	5		
RGP 4 h	11.4[c]	9.6[d]	12.2[bc]	14.1[a]	11.9[bc]	2.2	< 0.01
RGP 6 h	9.9[c]	8.5[d]	10.6[b]	12.8[a]	10.1[b]	2.4	< 0.01
RGP 8 h	8.9[c]	8.4[c]	11.2[ab]	12.5[a]	10.0[b]	2.1	< 0.01
RGP 12 h	6.5[c]	7.0[c]	9.7[ab]	10.8[a]	9.0[ab]	1.9	< 0.01
RGP 24 h	2.9[c]	2.1	3.1[ab]	3.4[a]	2.9[ab]	0.4	< 0.01
RGP 48 h	2.9[c]	2.8[d]	4.0[a]	3.9[a]	3.7[b]	0.5	< 0.01
RGP 72 h	0.5[b]	0.6[b]	1.0[a]	1.0[a]	0.9[a]	0.1	< 0.01

[1] Treatment: 1) control (no Pb and adsorbent); 2) 15 mg/kg DM Pb as Pb acetate and no adsorbent; 3) 15 mg/kg DM Pb as Pb acetate and 3% activated carbon; 4) 15 mg/kg DM Pb as Pb acetate and 3% nanoclay and 5) 15 mg/kg DM Pb as Pb acetate and 3% bentonite.
RGP: rate of gas production in milliliters per 200 mg DM per hour at 4, 6, 8, 12, 24, 48 and 72 h after incubation.
The means within the same row with at least one common letter, do not have significant difference (P>0.05).
SEM: standard error of the means.

Table 2 Effect of adsorbents addition to lead-exposed diets on total gas production (mL) at different incubation times

Incubation time	Treatment[1]					SEM	P-value
	1	2	3	4	5		
GP 2 h	9.1[b]	7.0[c]	8.9[b]	12.5[a]	9.6[b]	2.3	< 0.01
GP 4 h	15.2[b]	11.9[c]	15.3[b]	19.5[a]	16.1[b]	2.8	< 0.01
GP 8 h	23.7[b]	19.1[c]	24.1[b]	30.2[a]	24.5[b]	3.5	< 0.01
GP 12 h	30.5[c]	26.0[d]	34.0[b]	40.6[a]	32.8[b]	4.5	< 0.01
GP 16 h	35.0[c]	30.9[d]	40.4[b]	48.3[a]	39.6[b]	5.2	< 0.01
GP 24 h	44.2[c]	36.2[d]	48.4[b]	58.1[a]	47.7[b]	5.8	< 0.01
GP 48 h	55.2[b]	45.5[c]	61.1[b]	71.2[a]	59.6[b]	6.4	< 0.01
GP 72 h	59.6[c]	50.6[d]	68.5[b]	78.0[a]	66.3[b]	6.4	< 0.01
GP 96 h	60.6[c]	52.2[d]	71.7[b]	81.5[a]	69.1[b]	6.5	< 0.01

[1] Treatment: 1) control (no Pb and adsorbent); 2) 15 mg/kg DM Pb as Pb acetate and no adsorbent; 3) 15 mg/kg DM Pb as Pb acetate and 3% activated carbon; 4) 15 mg/kg DM Pb as Pb acetate and 3% nanoclay and 5) 15 mg/kg DM Pb as Pb acetate and 3% bentonite.
GP: gas production (in milliliters per 200 mg at 2, 4, 8, 12, 16, 24, 48, 72 and 96 h after incubation).
The means within the same row with at least one common letter, do not have significant difference (P>0.05).
SEM: standard error of the means.

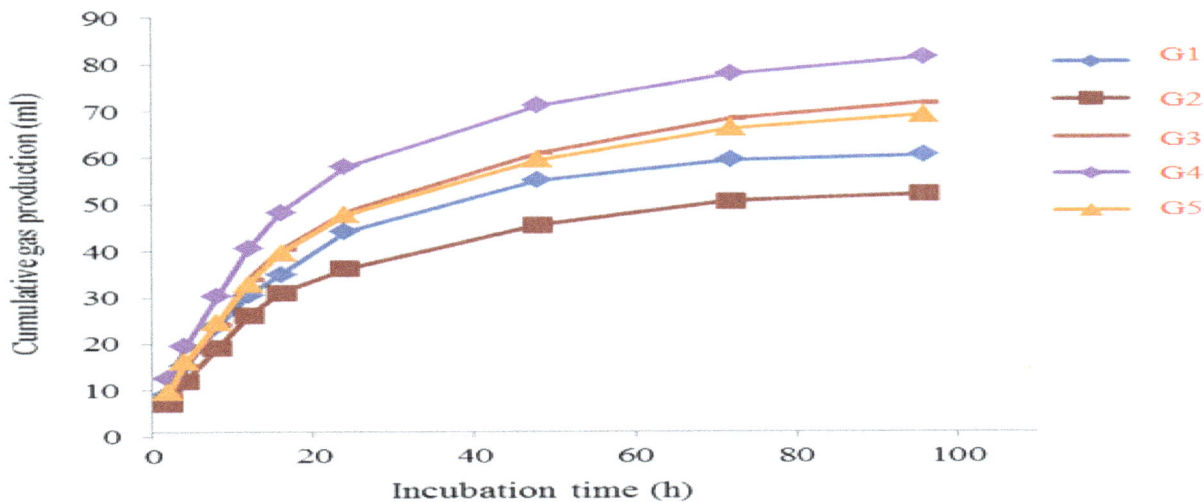

Figure 1 Gas production of lead-exposed diets supplemented with adsorbents: G1) control (no Pb and bentonite); G2) 15 mg/kg DM Pb as Pb acetate and no adsorbent; G3) 15 mg/kg DM Pb as Pb acetate and 3% activated carbon; G4) 15 mg/kg DM Pb as Pb acetate and 3% nanoclay and G5) 15 mg/kg DM Pb as Pb acetate and 3% bentonite

Table 3 Effect of adsorbents addition to lead-exposed diets on fermentation parameters and energy and fatty acids production

Factor	Treatment[1]					SEM	P-value
	1	2	3	4	5		
b (mL/g DM)	60.3[c]	50.9[c]	69.7[b]	79.3[a]	67.7[b]	5.8	< 0.01
c (h)	0.05[9]	0.057	0.053	0.059	0.054	0.006	0.74
OMD (%)	61.4[b]	54.2[c]	65.1[b]	73.7[a]	64.5[b]	5.2	< 0.01
ME (MJ/kg DM)	9.0[c]	7.9[d]	9.6b[c]	10.9[a]	9.5b[c]	0.8	< 0.01
NE (MJ/kg DM)	6.1[b]	5.1[c]	6.5[b]	7.7[a]	6.5[b]	0.6	< 0.01
SCFA (mmol/g DM)	0.97[b]	0.80[c]	1.07[b]	1.28[a]	1.05[b]	0.13	< 0.01

[1] Treatment: 1) control (no Pb and adsorbent); 2) 15 mg/kg DM Pb as Pb acetate and no adsorbent; 3) 15 mg/kg DM Pb as Pb acetate and 3% activated carbon; 4) 15 mg/kg DM Pb as Pb acetate and 3% nanoclay and 5) 15 mg/kg DM Pb as Pb acetate and 3% bentonite.
b: asymptotic gas production (in milliliters per 200 mg DM); c: fractional rate of gas production (per hour); OMD: organic matter digestibility (%); ME: metabolizable energy (in MJ per kilogram DM); NE: net energy yield (in MJ per kilogram DM) and SCFA: short chain fatty acid (millimole per g DM).
The means within the same row with at least one common letter, do not have significant difference (P>0.05).
SEM: standard error of the means.

In agreement with our results, prior studies showed that exposure to Pb affected rumen metabolism by forming lipid soluble organometallic compounds that inhibit the growth and respiration of micro-organisms (Chaudhary *et al.* 2006). Surveys conducted by Phillips *et al.* (2011) have shown that exposure to Pb led to a reduction in the digestion of long particles because of a toxic effect on rumen-micro-organisms. Regarding our study the main finding was a much lower production of total gas and volatile fatty acids (VFA) in the rumen fluid of contaminated with Pb compared to the control and adsorbents supplemented groups. In this regards, Varadyova *et al.* (2006a) observed depressed the gas and methane production, production of VFA and altered the fermentation pattern in ruminal fermentation of sheep grazing in an area polluted. Marounek and Joch (2014), observed decreased molar proportion of acetate with increasing concentrations of lead in rumen cultures. In this study production of VFA in cultures containing Pb, Hg, Cd and As at 50 µg/mL was decreased by 15.9, 40.0, 29.1 and 35.6%, respectively. The elements in increasing order of toxicity based on the inhibition of VFA production at 50 µg/mL were Pb, Cd, As and Hg. The corresponding ranking of trace elements toxicity, based on the inhibition of gas evolution reported by Forsberg (1977) was Pb, As, Cd and Hg. Probably as a consequence of inhibition the microbial activity and the methanogenic population appeared not to be fully able to consume hydrogen produced in bacterial and protozoan communities (Varadyova *et al.* 2006b). In anaerobic digestion medium, Chen *et al.* (2008) demonstrated that the toxic effect of heavy metals is attributed to disruption of enzyme function and structure by binding of the metals with thiol and other groups on protein molecules or by replacing naturally occurring metals in enzyme prosthetic groups.

Change in fermentation pattern (Varadyova *et al.* 2007) and protozoan counts (Hristov *et al.* 2001) by clay sorbent has been documented previously.

In this study addition of different absorbents ameliorated toxic effects of lead and improved ruminal fermentation parameters. With consistent to our results, in study of Varadyova *et al.* (2006a), addition of bentonite to fermentation bottles containing lead tended to enhance production of total and rate of gas production.

Supplementation with 1.5% of dietary DM clay decreased the adverse effects of lead contamination. These results are in agreement with those reported in ruminants by Khalifeh *et al.* (2012) who found that, addition of bentonite led to a significant increase in gas production rate constant and organic matter digestibility. Also, activated carbon has been used as an adsorbent for removal of heavy metal pollutants from wastewater and has proved to be effective (Gueu *et al.* 2007; Okpareke *et al.* 2009).

However, in our study with using different adsorbents, nanosized clay, was more effective for removal of Pb from ruminal medium and has better improvement for fermentation parameters than natural adsorbents. This is partly because of their large surface areas and high activities caused by the size-quantization effect (Hua *et al.* 2012). With agreement with our results, Muhammad and Munawar (2007) mentioned that nano zeolite Y is an effective adsorbent for Pb removal. In another study by Rasouli *et al.* (2012) low silica nano-zeolite X be much suitable adsorbent for the removal of Pb from aqueous solution. Recent studies suggested that many nanosized clay exhibit very favorable sorption to heavy metals in terms of high capacity and selectivity, which would result in deep removal of toxic metals to meet increasingly strict regulations (Deliyanni *et al.* 2009).

CONCLUSION

In conclusion exposure to Pb affected rumen function, reducing rate and total gas production and energy and fatty acids production. Among the different adsorbents, nanosized clay was found to be much suitable adsorbent for the removal of Pb from ruminal medium.

ACKNOWLEDGEMENT

The authors would like to acknowledge the financial support of University of Tehran for this research.

REFERENCES

Chaudhary A., Agarwal M. and Singh R.V. (2006). Organotin (IV) and organolead (IV) complexes as biocides and fertility regulators, synthetic, spectroscopic and biological studies. *Appl. Organomet. Chem.* **20,** 295-303.

Chen Y., Cheng J.J. and Creamer K.S. (2008). Inhibition of anaerobic digestion process: a review. *Bioresource. Technol.* **99,** 4044-4064.

Deliyanni E.A., Peleka E.N. and Matis K. (2009). Modeling the sorption of metal ions from aqueous solution by iron-based adsorbents. *J. Hazard. Mater.* **172,** 550-558.

Faixova Z. and Faix S. (2002). Influence of metal ions on ruminal enzyme activities. *Acta. Vet. Brno.* **71,** 451-455.

Forsberg C.W. (1977). Effects of heavy metals and other trace elements on the fermentative activity of the rumen microflora and growth of functionally important rumen bacteria. *Canadian J. Microbiol.* **24,** 298-306.

France J., Dijkstra J., Dhanoa M.S., López S. and Bannink A. (2000). Estimating the extent of degradation of ruminants feeds from a description of their gas production profiles observed *in vitro*: derivation of models and other mathematical considerations. *British. J. Nutr.* **83,** 43-150.

Gueu S., Yao B., Adouby K. and Ado G. (2007). Kinetics and thermodynamics study of lead adsorption on to activated carbons from coconut and seed hull of the palm tree. *Int. J. Environ. Sci. Technol.* **4,** 11-17.

Hristov A.N., Ivan M., Rode L.M. and McAllister T.A. (2001). Fermentation characteristics and ruminal ciliate protozoal populations in cattle fed medium-or high-concentrate barley-based diets. *J. Anim. Sci.* **79,** 515-524.

Hua M., Zhang S., Pan B., Zhang W., Lv L. and Zhang Q. (2012). Heavy metal removal from water / wastewater by nanosized metal oxides: a review. *J. Hazard. Mater.* **212,** 317-331.

Khalifeh M.J., Mohammadabadi T., Chaji M., Salari S. and Khalil M. (2012). The effect of different levels of sodium bentonite on *in vitro* fermentation and digestibility of soybean meal. Pp. 23-27 in Proc.15th AAAP Anim. Sci. Congr. Thammasat University, Rangsit Campus, Thailand.

Makkar H.P.S. (2005). *In vitro* gas methods for evaluation of feeds containing phytochemicals. *Anim. Feed Sci. Technol.* **123,** 291-302.

Marounek M. and Joch M. (2014). Effects of heavy metals and arsenate on the Ovine rumen fermentation *in vitro*. *Agric. Trop. Subtrop.* **47,** 106-108.

Menke K.H., Raab L., Salewski A., Stingass H., Fritz D. and Schneide W. (1979). The estimation of the digestibility and metabolizable energy content of ruminant feedstuffs from the gas production when they are incubated with rumen liquor *in vitro*. *J. Agric. Sci.* **93,** 217-222.

Menke K.H. and Steingass H. (1988). Estimation of the energetic feed value obtained from chemical analysis and *in vitro* gas production using rumen fluid. *Anim. Res. Dev.* **28,** 47-55.

Mohajeri G., Afzalzadeh A., Norouzian M.A. and Mohseni M. (2014). Changes in blood metals, hematology and hepatic enzyme activities in lactating cows reared in the vicinity of a lead–zinc smelter. *Bull. Environ. Contam. Toxicol.* **92,** 693-697.

Muhammad S. and Munawar E. (2007). Nanocrystalline zeolite Y: synthesis and heavy metal removal. *J. Rekayasa Kimia dan Lingkungan.* **6,** 55-62.

Okpareke O.C., Agha I. and Ejikeme P.M. (2009). Removal of Cu (II), Cd (II) and Hg (II) ions from simulated wastewater by Brachystagea Eury coma seed pod. Pp. 23-27 in Proc. 32nd Int. Conf. Chem. Soc. Nigeria. Bauchi, Nigeria.

Phillips C.J.C., Mohamed M.O. and Chiy P.C. (2011). Effects of duration of exposure to dietary lead on rumen metabolism and the accumulation of heavy metals in sheep. *Small Rumin. Res.* **100,** 113-121.

Radostits O.M., Blood D.C., Gay C.C. and Hinchcliff H.E. (2000). Veterinary Medicine. A Text Book of Disease of Cattle, Sheep, Pigs, Goats and Horses. WB Saunders, London, United Kingdom.

Rasouli M., Yaghobi N., Hafezi M. and Rasouli M. (2012). Adsorption of divalent lead ions from aqueous solution using low silica nano-zeolite X. *J. Ind. Eng. Chem.* **18,** 1970-1976.

SAS Institute. (2004). SAS®/STAT Software, Release 9.1. SAS Institute, Inc., Cary, NC. USA.

Swarup D., Patra R.C., Naresh R., Kumar P., Shekhar P. and Alagangatharathilagar P. (2006). Lowered blood copper and cobalt contents in goats reared around lead–zinc smelter. *Small Rumin. Res.* **63,** 309-313.

Varadyova Z., Mihalikova K., Kisdayova S. and Javorsky P. (2006a). Fermentation pattern of the rumen and hindgut inocula of sheep grazing in an area polluted from the non-ferrous metal industry. *Czech J. Anim. Sci.* **51,** 66-72.

Varadyova Z., Isidayova K., Mihalikova S. and Baran K. (2006b). Influence of natural magnesium sources on the *in vitro* fermentation and protozoan population in the rumen fluid collected from sheep. *Small Rumin. Res.* **61,** 63-71.

Varadyova Z., Styriakova I. and Kisidayova S. (2007). Effect of natural dolomites on the *in vitro* fermentation and rumen protozoan population using rumen fluid and fresh faeces inoculum from sheep. *Small Rumin. Res.* **73,** 58-66.

Vázquez-Armijo J.F., Martínez-Tinajero J.J., López D., Salem A.Z.M. and Rojo R. (2011). *In vitro* gas production and dry matter degradability of diets consumed by goats with or without copper and zinc supplementation. *Biol. Trace Elem. Res.* **144,** 580-587.

16

The Effects of Different Levels of Saturated and Unsaturated Fats and Their Composition in Growing and Finishing Periods on Productive Performance and Blood Lipids of Broilers

G.F. Baighi[1] and A. Nobakht[1*]

[1] Department of Animal Science, Maragheh Branch, Islamic Azad University, Maragheh, Iran

*Correspondence E-mail: anobakht20@iau-maragheh.ac.ir

ABSTRACT

This experiment was conducted to investigate the effects of different levels of saturated and unsaturated fats and their composition in growing and finishing periods on performance, carcass traits and blood lipids level in broiler chickens. In this experiment, 432 Ross 308 broilers were used from 11 up to 42 days in 9 treatments, 4 replicates and 12 birds in each replicate in growing (11-24 days) and finishing (25-42 days) periods in a completely randomized design. Treatments included: 1) control group (without fat), 2) 2% canola oil, 3) 4% Canola oil, 4) 2% beef tallow, 5) 4% beef tallow, 6) 2% canola oil + 2% beef tallow, 7) 4% canola oil + 2% beef tallow, 8) 2% canola oil + 4% beef tallow and 9) 4% canola oil + 4% beef tallow. In the growing period, different levels of fat sources had no significant effects on the performance of broilers. In the finishing period, the higher amounts of daily weight gain and final live weight were obtained in group 8 (P<0.05). In the whole feeding period, the best feed conversion ratio and final live weight were observed in group 8 (P<0.05). The lowest amount of intestine, abdominal, gizzard and liver weights belonged to control group (P<0.05). However, a difference was observed between the control group and group 8 in these respects. Using different levels of saturated and unsaturated fat in growing and finishing periods had no significant effect on the blood lipids level of broilers (P>0.05). The overall results indicated that in broilers, using mixing of saturated and unsaturated fats have beneficial effects on their performance.

KEY WORDS beef tallow, blood lipids, broiler chickens, canola oil, performance.

INTRODUCTION

Fats and oils can be used as an alternative energy source in place of prime energy feed ingredients in broiler production. The dietary metabolizable energy (ME) increased through supplementation of fat in broiler rations and significantly increased their body weight gain (Leeson and Atteh, 1995). The cholesterol content of the food products especially from animal sources becomes the prime area of consumer's concern because of the increased awareness on higher dietary cholesterol and the incidence of coronary heart diseases. In poultry and other monogastric animals, the fatty acid composition of tissue lipids depends on the dietary fatty acids (Lopez Ferre *et al.* 1999). One of the major concerns related to fat usage is the actual ME value that should be assigned to each fat source. This number is often difficult to determine in a practical sense and may have little practical value in diet formulation. The type of fat added to the diet has a significant influence on the profile of fatty acids of the abdominal fat. It was evident that inclusion of saturated fats produces the higher accumulation of intramuscular, mesenteric and abdominal fat in broilers (Sanz *et al.* 1999; Crespo and Esteve-Garcia, 2002). Many factors influence the absorption of fats. The chemical character of the fat itself is important. Absorption will be influenced by whether the fatty acids are free fatty acids (FFA) or triglycerides (Garrett and Young, 1975). The position of the fatty acids in the triglyceride molecule affects absorption (Renner and Hill, 1961a). The digestibility of free fatty acids decreases with the increasing length of the carbon chain and saturation (Renner and Hill, 1961b). An increase in the content of unsaturated fats in relation to saturated fats the absorption of the saturated fatty acids increased (Young and Garrett, 1963). Oils added to the rations of animals are effective on the fatty acid composition and amount of abdominal fat. In fact, fatty acids composition of oils used in poultry rations are reflected in the animal products because dietary fatty acids are incorporated with little change into the bird body fats (Scaife et al. 1994). Thus, the type of fat used in the feed, influence the composition of broiler body lipids. Abdominal fat is a good indicator of chicken body fats because it is very sensitive to changes in dietary fatty acid composition (Yau et al. 1999). In this context, reported that broiler chickens fed with diets enriched with polyunsaturated fatty acids have less abdominal fat or total body fat deposition than do broiler chickens fed diets containing saturated fatty acids (Sanz et al. 2000a). There are different kinds of fats in the market. Canola oil and beef tallow are the most important of them. Canola oil has been recognized as adequate mixture of essential fatty acids, unsaturated fatty acids such as α-linolenic acid (C18:3) that can improve broiler performance, also linolenic acids can be converted to longer chain omega-3 fatty acids (Sim et al. 1990; Young et al. 1963) that is an important factor in animal feeding and is for promote of health (Bezard et al. 1994). Adding 3% of canola oil and poultry fat resulted in significant improvement in body weight and better feed conversion ratio than other groups, no significant different were found in liver, breast and thigh weights between groups fed lipid in comparison with the control group. Addition 6% poultry fat caused significant increasing on abdominal fat and gizzard weight was significantly higher in control group (Shahryar et al. 2011). It has been accepted that dietary canola oil is excellent supplement for commercial fish such as salmon (Huang et al. 2007).

On the other hand, canola oil contains less than 2% of erucic acid (docosanoic acid, C22:1, (-9) in relation to the total fatty acids and less than 30 moles of glucosinolates per gram of free oil on seed dry matter basis. In birds, the adverse effects of adding erucic acid to the diets are reflected on intake, feed growth performance and the apparent digestibility of total lipid and individual fatty acids (Leeson and summers, 2001). Using 2% of canola oil in broiler diets positively improved their performance and carcass traits (Nobakht et al. 2012). In broiler diets inclusion 5% canola oil improved their weight gain and increased blood triglyceride (Kiani et al. 2016). Using canola oil up to 5% in native turkeys had no effects on blood lipids concentration (Salamatdoust Nobar et al. 2010). In laying hens, incorporating 2% canola oil in diets improved their performance and reduced their egg and blood cholesterol level (Ismail et al. 2013). Inclusion up to 5% beef tallow to broiler diets had positive effects on their performance (Nasiri Moghaddam et al. 2000). In laying hens using 4% beef tallow could not affect their performance, egg traits and blood parameters (Safamehr et al. 2011).

As in young birds, the digestive tract not fully complicated and the amount of secreted enzymes and other juices are not considerable, it thought that in this period, the birds have several difficulties in digestion and absorption of some feed ingredients such as saturated fats. In the current study, the effects of different levels of saturated and unsaturated fats and several compositions of them in growing and finishing breeding periods of broilers on their performance, carcass traits and blood lipids had been investigated.

MATERIALS AND METHODS

In this experiment, 432 Ross 308 broilers were used from 11 up to 42 days in 9 treatments, 4 replicates and 12 birds in each replicate in growing (11-24 days) and finishing (25-42 days) periods in a completely randomized design. Treatments included: 1) control group (without fat), 2) 2% canola oil, 3) 4% Canola oil, 4) 2% beef tallow, 5) 4% beef tallow, 6) 2% canola oil + 2% beef tallow, 7) 4% canola oil + 2% beef tallow, 8) 2% canola oil + 4% beef tallow and 9) 4% canola oil + 4% beef tallow. The diets were formulated to meet the requirements of birds established by the Aviagen (2014) for broilers in grower (11-24 days) and finisher (25-42 days) periods are shown in Tables 1 and 2.

In all experimental periods, the diets and water were provided *ad libitum* for birds. The lighting program during the experimental periods consisted of a period of 23 hours light and 1 hour of darkness. House temperature was gradually decreased from 33 °C to 25 °C on day 21 and was then kept constant. Body weight, feed intake and feed conversion ratio were determined at the end of each experimental period on bird bases. Mortality was also recorded if it occurred. At the end of the experiment, two birds from each replicate were randomly chosen for blood collection and approximately 5 mL blood samples were collected from the brachial vein of randomly chosen birds. The blood was centrifuged to obtain serum for determining the blood lipids which included cholesterol, triglyceride, low density lipoprotein (LDL) and high density lipoprotein (HDL). Kit packages (Pars Azmoon Company; Tehran, Iran) were used for determining the blood biochemical parameters using Anision-300 auto-analyzer system (Nazifi, 1997).

Table 1 The composition of broiler diets in growing period (11-24 days)

Feed ingredients	0%	2% canola oil	4% canola oil	2% beef tallow	4% beef tallow
Corn	62.83	49.83	44.46	50.10	45.02
Soybean meal (42% CP)	33.17	42.31	43.70	42.35	43.57
Fat	0.00	2.00	4.00	2.00	4.00
Inert (sand)	0.27	0.56	3.55	1.33	3.10
Oyster shell	0.25	0.23	0.21	0.24	0.21
Bone meal	2.11	2.65	2.68	2.65	2.68
Salt	0.40	0.41	0.41	0.41	0.41
Vitamin premix[1]	0.25	0.25	0.25	0.25	0.25
Mineral premix[2]	0.25	0.25	0.25	0.25	0.25
DL-methionine	0.28	0.31	0.31	0.31	0.31
L-lysine hydrochloride	0.19	0.20	0.18	0.21	0.20
Calculated composition					
Metabolizable energy (kcal/kg)	3000	3000	3000	3000	3000
Crude protein (CP) (%)	20.63	20.63	20.63	20.63	20.63
Calcium (%)	0.80	0.80	0.80	0.80	0.80
Available phosphorus (%)	0.39	0.39	0.39	0.39	0.39
Sodium (%)	0.19	0.19	0.19	0.19	0.19
Lysine (%)	1.02	1.02	1.02	1.02	1.02
Methionine + cysteine (%)	0.81	0.81	0.81	0.81	0.81
Tryptophan (%)	0.19	0.19	0.19	0.19	0.19

[1] Vitamin premix per kg of diet: vitamin A (retinol): 2.7 mg; vitamin D_3 (cholecalciferol): 0.05 mg; vitamin E (tocopheryl acetate): 18 mg; vitamin K_3: 2 mg; Thiamine: 1.8 mg; Riboflavin: 6.6 mg; Panthothenic acid: 10 mg; Pyridoxine: 3 mg; Cyanocobalamin: 0.015 mg; Niacin: 30 mg; Biotin: 0.1 mg; Folic acid: 1 mg; Choline chloride: 250 mg and Antioxidant 100 mg.
[2] Mineral premix per kg of diet: Fe ($FeSO_4.7H_2O$, 20.09% Fe), 50 mg; Mn ($MnSO_4.H_2O$, 32.49% Mn), 100 mg; Zn (ZnO, 80.35% Zn), 100 mg; Cu ($CuSO_4.5H_2O$), 10 mg; I (K_1, 58% I), 1 mg; Se ($NaSeO_3$, 45.56% Se): 0.2 mg.

Table 2 The composition of broiler diets in finishing period (25-42 days)

Feed ingredients	0%	2% canola oil	4% canola oil	2% beef tallow	4% beef tallow
Corn	66.38	61.10	56.76	61.68	55.93
Soybean meal (42% CP)	30.22	33.53	33.86	32.97	34.74
Fat	0	2.00	4.00	2.00	4.00
Inert (sand)	0	0.04	2.05	0	2.00
Oyster shell	0.24	0.27	0.24	0.26	0.24
Bone meal	2.07	2.04	2.07	2.05	2.07
Salt	0.37	0.39	0.39	0.39	0.39
Vitamin premix[1]	0.25	0.25	0.25	0.25	0.25
Mineral premix[2]	0.25	0.25	0.25	0.25	0.25
DL-methionine	0.18	0.13	0.14	0.15	0.13
Calculated composition					
Metabolisable energy (kcal/kg)	3050	3050	3050	3050	3050
Crude protein (CP) (%)	19.54	19.54	19.54	19.54	19.54
Calcium (%)	0.81	0.81	0.81	0.81	0.81
Available phosphorus (%)	0.40	0.40	0.40	0.40	0.40
Sodium (%)	0.17	0.17	0.17	0.17	0.17
Lysine (%)	1.01	1.01	1.01	1.01	1.01
Methionine + cysteine (%)	0.78	0.78	0.78	0.78	0.78
Tryptophan (%)	0.24	0.24	0.24	0.24	0.24

[1] Vitamin premix per kg of diet: vitamin A (retinol): 2.7 mg; vitamin D_3 (cholecalciferol): 0.05 mg; vitamin E (tocopheryl acetate): 18 mg; vitamin K_3: 2 mg; Thiamine: 1.8 mg; Riboflavin: 6.6 mg; Panthothenic acid: 10 mg; Pyridoxine: 3 mg; Cyanocobalamin: 0.015 mg; Niacin: 30 mg; Biotin: 0.1 mg; Folic acid: 1 mg; Choline chloride: 250 mg and Antioxidant 100 mg.
[2] Mineral premix per kg of diet: Fe ($FeSO_4.7H_2O$, 20.09% Fe), 50 mg; Mn ($MnSO_4.H_2O$, 32.49% Mn), 100 mg; Zn (ZnO, 80.35% Zn), 100 mg; Cu ($CuSO_4.5H_2O$), 10 mg; I (K_1, 58% I), 1 mg; Se ($NaSeO_3$, 45.56% Se): 0.2 mg.

Also, at 42 days of age, two birds from each replicate randomly chosen based on the average weight of the group and sacrificed. Dressing yield was calculated by dividing eviscerated weight by live weight. Abdominal fat, gizzard, liver, spleen, breast and thigh were collected, weighed and calculated as a percentage of carcass weight.

The data were subjected to one-way analysis of variance procedures as completely randomized design using the General Linear Model procedures of SAS (2005). Means were compared using the Duncan test (Valizadeh and Moghaddam, 1994). Statements of statistical significance were based on ($P<0.05$).

RESULTS AND DISCUSSION

Performance

The effects of different levels of saturated and unsaturated fats on the performance of broilers in growing period are shown in Table 3. Different levels and composition of saturated and unsaturated fats in growing period had no significant effects on performance of broilers (P>0.05).

The effects of different levels of saturated and unsaturated fats on the performance of broilers in the finishing period are summarized in Table 4. In the finishing period, different levels of saturated and unsaturated significantly changed the amounts of daily weight gain and final live weight (P<0.01). The highest amounts of daily weight gain (71.07 g) and final live weight 2346.33 g were observed in group 8 (contained 2% canola oil+4% beef tallow), whereas the lowest value of them, 63.02 g and 2182.33 g belonged to group 4 with 4% beef tallow. The values of daily feed intake and feed conversion ratio were not significantly different between treatments (P>0.05).

The effects of different levels of saturated and unsaturated fats on performance of broilers in whole feeding period are shown in Table 5.

Different levels and composition of saturated and unsaturated fats in whole feeding period significantly affected the performance of broilers (P>0.01). In whole feeding period, the highest amounts of daily feed intake and weight gain were observed in control group, whereas the amounts of these parameters not different between other treatments (P>0.05). In this period such as finishing period the final live weight belonged to group 8. There were not any significant difference between treatment about feed conversion ratio, livability, and production index (P>0.05).

Carcass traits

The effects of different levels of saturated and unsaturated fat on carcass traits of broilers are shown in Table 6. Using different levels of saturated and unsaturated fats in growing and finishing periods significantly affected the carcass traits of broilers (P<0.05). The lowest values for abdominal fat, gizzard and liver percentages were observed in control group, however there were not any significant difference between control group and group 8 about this organs values (P>0.05). Carcass, breast and thigh percentages were not affected by using different levels of saturated and unsaturated fats (P>0.05).

Blood lipids

The effects of different levels of saturated and unsaturated fats in growing and finishing periods of blood lipids in broilers are summarized in Table 7.

Different levels of saturated and unsaturated fats in growing and finishing periods had no significant effects on blood lipids of broilers (P>0.05).

In growing period in contrast to finishing period, the amount of feed intake is low, this caused the performance not to be change significantly. Whereas in finishing period the amount of daily feed intake is high, so, it caused the amounts of daily weight and final live weight changes significantly, the highest amount is occurred in group 8 with mixing of 2% canola oil + 4% beef tallow. High performance in group 8 may be related to positive corporation combination between saturated and unsaturated fat sources.

Canola oil more than energy is a rich source of essential fatty acids and fat soluble vitamins. Using it in corporation with beef tallow by supplying optimum levels of nutrients supported broiler performance and improved the amount of their weight gain.

The current results are in line with the reports of Nobakht and Mehmannavaz (2012) and Poorghasemi *et al.* (2013) who indicated that a combination of vegetable and animal fat sources in laying hens and broiler diets supported positively of their performance. Achieving to the high amount of daily feed intake and daily weight gain in group without fat sources (control group), may be related to the absent of fat sources, as the fats can easily supply considerable amount of energy (Leeson and summers, 2001), in contrast, other energy sources can reduce the amount of feed intake and vest versa. In this relation, as other groups contain fats, so the amounts of daily feed intake and weight gain, had not been changed significantly, whereas the highest amount of live final weight belonged to group 8 with combination 2% canola oil + 4% beef tallow. This result is in agree with the experiment results of Nobakht and Mehmannavaz (2012) and Poorghasemi *et al.* (2013) reported that combination of vegetable and animal fat sources in laying hens and broilers diets supported positively their performance. Changes in the amounts of daily feed intake and weight gain, could not alter feed conversion ratio, that it is not in agree with Jeffri *et al.* (2010) reported that by increasing fat sources to broiler diet, the feed conversion ratio was improved.

The results showed that no significant changes took place on feed conversion ratio in this experiment. There are conflicting reports on the effect of fat supplementation on feed conversion ratio of broiler.

Al Athari and Watkins (1988) found no difference in the feed conversion ratio of broiler diets containing 5% added saturated fat or soybean oil. In contrast, Pinchasov and Nir (1992) and Zollitsch *et al.* (1996) reported an improved feed conversion ratio in broiler when dietary polyunsaturated fatty acid intake increased.

Table 3 The effects of saturated and unsaturated fats on performance of broilers in growing period (11-24 days)

Treatments	Weight gain (g/d/h)	Feed intake (g/d/h)	Feed conversion ratio	Final weight (g)
1) control (without fat)	69.19	97.58	1.42	1168.67
2) 2% canola oil	65.85	95.54	1.46	1177.33
3) 4% canola oil	61.68	93.28	1.52	1173.33
4) 2% beef tallow	64.25	96.58	1.51	1155.67
5) 4% beef tallow	63.86	94.01	1.48	1174.33
6) 2 canola oil + 2% beef tallow	64.27	92.65	1.44	1183.67
7) 4 canola oil + 2% beef tallow	62.41	92.93	1.50	1171.33
8) 2 canola oil + 4% beef tallow	63.27	91.66	1.45	1182.67
9) 4 canola oil + 4% beef tallow	62.96	90.56	1.44	1141.67
SEM	2.14	2.47	0.2	10.80
P-value	0.4174	0.5581	0.0881	0.2105

SEM: standard error of the means.

Table 4 The effects of saturated and unsaturated fats on the performance of broilers in finishing period (25-42 days)

Treatments	Weight gain (g/d/h)	Feed intake (g/d/h)	Feed conversion ratio	Final weight (g)
1) control (without fat)	65.65[bcd]	121.39	1.85	2243.33[de]
2) 2% canola oil	64.43[d]	120.69	1.88	2213.00[de]
3) 4% canola oil	64.93[de]	124.86	1.93	2212.33[de]
4) 2% beef tallow	64.22[d]	119.41	1.86	2188.67[de]
5) 4% beef tallow	63.02[d]	121.30	1.93	2182.33[e]
6) 2 canola oil + 2% beef tallow	66.16[bcd]	125.52	1.90	2242.00[de]
7) 4 canola oil + 2% beef tallow	68.98a[bc]	127.39	1.85	2282.67[bc]
8) 2 canola oil + 4% beef tallow	71.07[a]	128.45	1.81	2346.33[a]
9) 4 canola oil + 4% beef tallow	69.72[ab]	128.45	1.85	2297.67[ab]
SEM	1.27	4.92	0.6	17.22
P-value	0.0026	0.8288	0.8324	0.0001

SEM: standard error of the means.
The means within the same column with at least one common letter, do not have significant difference (P>0.05).

Sanz *et al.* (2000b) confirmed that the source and level of different fats and the rates of use did not affect the feed conversion of broilers. Not improving in feed conversion ratio in the current study may be related to proportion between daily feed intake and weight gain.

The lowest amount of abdominal fat was observed in control group (diet without fat) and the highest it related to group with 4% canola oil. As fats are the reach sources of energy in diets, so using them (especially at high levels), not only can supply enough energy for supporting of performance, but also more of it can be changed to body fat and save in different forms such as abdominal fats (Leeson and summers, 2001), as oils in contrast to fats have highly amount of unsaturated fatty acids, the digestion and supplying of energy is in high amount, so by feeding 4% of canola oil, not only the amount of energy requirement for growth can be available, but also, more than it, can be changed to body fat and reserve different forms such as abdominal fat, whereas in control group the digestible energy supplied by other ingredients without fat sources mainly supported performance and tissue growth. In contrast to the present study, high amount of abdominal fat was reported by using 4% tallow in broiler diet (Duraisamy *et al.* 2013).

Such as abdominal fat, the highest and the lowest amounts of gizzard and liver related to control group and group contained 4% canola oil. As the gizzard and liver contain highly amount of fat, the size of them can be related to the amount of abdominal fat, as the amount of abdominal fat is low and high in groups 1 and 3, it could affect the size of these organs. Our results in the present study in consist with Crespo and Esteve-Garcia (2000) reports, that using 4% sunflower oil in broiler diets reduced the amount of abdominal fat. There suggested that reduction of abdominal fat in broilers fed a diet supplemented with oil seems to be a consequence of higher lipid oxidation despite the higher synthesis of endogenous fatty acids. Among available fat sources in poultry nutrition, it has been recognized that canola oil is an adequate combination of essential fatty acids, unsaturated fatty acids (such α-linolenic acid) that can improve broiler performance, and also linolenic acid can be converted to longer chain omega-3 fatty acids that are important factors in animal feeding as health promoter (Bezard *et al.* 1994).

Shahryar *et al.* (2011) found that the addition 6% animal fat caused a significant increase in abdominal fat weight compared to control groups.

Table 5 The effects of saturated and unsaturated fats on performance of broilers in whole feeding period (11-42 days)

Treatments	Weight gain (g/d/h)	Feed intake (g/d/h)	Feed conversion ratio	Final weight (g)	Livability (%)	Production index
1) control (without fat)	67.52[a]	109.21[a]	1.62	2243.33[dc]	97.22	320.46
2) 2% canola oil	50.44[b]	80.75[b]	1.60	2213.00[de]	97.22	319.88
3) 4% canola oil	49.59[b]	81.19[b]	1.65	2212.33[de]	94.45	302.67
4) 2% beef tallow	49.86[b]	80.42[b]	1.62	2188.67[de]	97.22	313.98
5) 4% beef tallow	49.48[b]	80.08[b]	1.62	2182.33[e]	97.22	312.03
6) 2 canola oil + 2% beef tallow	50.52[b]	81.21[b]	1.61	2242.00[dc]	97.22	322.74
7) 4 canola oil + 2% beef tallow	50.87[b]	81.70[b]	1.61	2282.67[bc]	94.45	319.00
8) 2 canola oil + 4% beef tallow	51.62[b]	81.71[b]	1.59	2346.33[a]	94.45	332.00
9) 4 canola oil + 4% beef tallow	50.88[b]	81.02[b]	1.60	2297.67[ab]	94.45	329.06
SEM	0.80	1.99	0.03	17.22	3.21	8.81
P-value	0.0001	0.0001	0.9219	0.0001	0.9853	0.4464

SEM: standard error of the means.
The means within the same column with at least one common letter, do not have significant difference (P>0.05).

Table 6 The effects of saturated and unsaturated fats on carcass traits (carcass %) of broilers (42 days)

Treatments	Carcass	Abdominal fat	Gizzard	Liver	Breast	Thigh
1) control (without fat)	69.77	3.62[c]	4.04[c]	2.90[c]	34.57	27.85
2) 2% canola oil	73.85	4.03[abc]	4.40[abc]	3.07[bc]	34.23	28.35
3) 4% canola oil	68.42	4.41[a]	4.75[a]	3.47[a]	35.96	30.36
4) 2% beef tallow	70.95	4.19[ab]	4.53[abc]	3.32[ab]	34.43	28.57
5) 4% beef tallow	69.64	4.08[abc]	4.60[ab]	3.34[ab]	36.21	30.00
6) 2 canola oil + 2% beef tallow	71049	3.68[bc]	4.01[c]	3.04[bc]	33.87	28.02
7) 4 canola oil + 2% beef tallow	69.66	3.93[abc]	4.30[abc]	3.15[abc]	34.99	29.22
8) 2 canola oil + 4% beef tallow	70.35	4.13[abc]	4.42[abc]	3.08[bc]	34.75	28.62
9) 4 canola oil + 4% beef tallow	70.97	3.82[bc]	4.18[bc]	3.09[bc]	33.34	27.53
SEM	1.82	0.16	-	0.10	1.37	1.52
P-value	0.6734	0.0497	-	0.0187	0.8747	0.9063

SEM: standard error of the means.
The means within the same column with at least one common letter, do not have significant difference (P>0.05).

Table 7 The effects of saturated and unsaturated fats on blood lipids (mg/dL) of broilers (42 days)

Treatments	Cholesterol	Triglyceride	LDL	HDL
1) control (without fat)	115.75	87.09	57.67	40.39
2) 2% canola oil	142.59	62.41	46.12	75.79
3) 4% canola oil	124.75	98.95	45.40	76.97
4) 2% beef tallow	136.73	94.06	65.33	56.90
5) 4% beef tallow	139.57	85.77	51.00	74.34
6) 2 canola oil + 2% beef tallow	126.53	89.85	40.90	59.63
7) 4 canola oil + 2% beef tallow	155.39	94.04	40.78	68.41
8) 2 canola oil + 4% beef tallow	136.03	93.63	57.67	66.86
9) 4 canola oil + 4% beef tallow	125.23	95.03	49.73	69.72
SEM	11.43	25.08	10.54	12.11
P-value	0.4219	0.8160	65.43	0.5211

SEM: standard error of the means.
LDL: low density lipoprotein and HDL: high density lipoprotein.

Moreover, adding 3% of canola oil and poultry fat mixture resulted in significant enhancement in organs' weight, whereas no differences were found in liver, breast and thigh weights between groups fed fat in comparison with control group.

Blood lipids composition had not changed in experimental groups. Not changes in the blood lipids in our study can have some reasons such as the length of experimental diets using, the amounts and composition of fats in diets, other diets ingredients, birds strain, performance and health status of birds.

The present results is in line with Salamatdoust Nobar et al. (2010) report that using canola oil up to 5% of native turkeys diets had no effects on their blood lipids, whereas on the base of Ismail et al. (2013) reported using 2% canola oil in laying hens diets reduced their egg and blood cholesterol content.

CONCLUSION

The overall results indicated that using combination of oil and fat such as group 8 (2% canola oil+4% beef tallow) in growing and finishing diets of broilers can improve their performance.

ACKNOWLEDGEMENT

This study has been supported by Islamic Azad University of Maragheh Branch. The author likes to appreciate Dr Mehmannvaz for his supporting during experimental period in poultry farm.

REFERENCES

Al Athari A.K. and Watkins B.A. (1988). Distribution of trans and cis 18:1 fatty acid isomers in chicks fed different fats. *Poult. Sci.* **67**, 778-786.

Aviagen. (2014). Ross 308: Broiler Nutrition Specification.. Aviagen Ltd., Newbridge, UK.

Bezard J., Blond J.P., Bernard A. and Clouet P. (1994). The metabolism and availability of essential fatty acids in animal and human tissues. *Reprod. Nutr. Dev.* **34**, 539-568.

Crespo N. and Esteve-Garcia E. (2002). Dietary polyunsaturated fatty acids decrease fat deposition in separable fat depots but not in the remainder carcass. *Poult. Sci.* **81**, 1533-1542.

Duraisamy K., Senthilkumar M. and Mani K. (2013). Effect of saturated and unsaturated fat on the performance, serum and meat cholesterol level in broilers. *Vet. World.* **10**, 159-162.

Garrett R.L. and Young R.J. (1975). Effect of micelle formation on the absorption of neutral fat and fatty acids by the chicken. *J. Nutr.* **105**, 827-838.

Huang S.S.Y., Hggis D.A., Brauner C.J. and Satoh S. (2007). Effect on the dietary canola oil level on the growth performance and fatty acid composition of Jurenile red sea bream, Pagrus major. *Aquaculture.* **271**, 420-431.

Ismail I.B., Al-Busadah K.A. and El-Bahr S.M. (2013). Effect of dietary supplementation of canola oil on egg production, quality and biochemistry of egg yolk and plasma of laying hens. *Int. J. Biol. Chem.* **7**, 27-37.

Jeffri D., Firman H. and Kamyab A. (2010). Comparison of soybean oil with an animal/vegetable blend at four energy levels in broiler rations from hatch to market. *Int. Poult. Sci.* **9**, 1027-1030.

Kiani A., Sharifi S.D. and Ghazanfari S. (2016). Effect of graded levels of canola oil and lysine on performance, fatty acid profile of breast meat and blood lipids parameters of broilers. *Anim. Sci. Res. J.* **26(2)**, 109-121.

Leeson S. and Atteh J.O. (1995). Utilization of fats and fatty acids by turkey poults. *Poult. Sci.* **74**, 2003-2010.

Leeson S. and Summers J.D. (2001). Nutrition of the Chicken. Published by University Books, Ontario, Canada.

Lopez Ferrer S., Baucells M.D., Barroeta A.C. and Grashorn M.A. (1999). Influence of vegetable oil sources on quality parameters of broiler meat. *Arch. Fiir. Geflugelkunde.* **63**, 29-35.

Nasiri Moghaddam H., Razian S.H.R. and Khajeh Ali F. (2000). The effect different fat sources on performance of broiler. *Sci. Thchnol. Agric. Nat. Res.* **3(1)**, 63-72.

Nazifi S. (1997). Hematology and Clinical Biochemistry of Birds.

Shiraz University Publication, Shiraz, Iran.

Nobakht A., Ariyana A. and Mazlum F. (2012). Effect of different levels of canola oil with vitamin E on performance and carcass traits of broilers. *Int. Res. J. Appl. Bas. Sci.* **3(5)**, 1059-1064.

Nobakht A. and Mehmannavaz Y. (2012). Effects of saturated and unsaturated fats in starter and grower feeds on performance and carcass traits of broilers. *J. Bas. Appl. Sci. Res.* **2(2)**, 967-997.

Pinchasov Y. and Nir I. (1992). Effect of dietary polyunsaturated fatty acid concentration on performance, fat deposition and carcass fatty acid composition in broiler chickens. *Poult. Sci.* **71**, 1504-1512.

Poorghasemi M.R., Seidavi A.R. and Tufarelli V. (2013). Influence of dietary fat source on growth performance responses and carcass traits of broiler chicks. *Asian Austerilain J. Anim. Sci.* **26(5)**, 705-710.

Renner R. and Hill F.W. (1961a). Factors affecting the absorbability of saturated fatty acids in the chick. *J. Nutr.* **74**, 254-258.

Renner R. and Hill F.W. (1961b). Utilization of fatty acids by the chicken. *J. Nutr.* **74**, 259-264.

Safamehr A.R., Thagavi E. and Nobakht A. (2011). Influence of tallow, soybean and sunflower oils on performance, eggshell quality, biochemical and immune parameters of laying hens. *Vet. Res.* **7**, 19-27.

Salamatdoust Nobar R., Gorbani A., Nazeradl K., Ayazi A., Hamidiyan A., Fani A., Aghdam Shahryar H., Giyasi ghaleh kandi J. and Ebrahim Zadeh Attari V. (2010). Beneficial effects of canola oil on breast fatty acids profile and some of serum biochemical parameters of Iranian native turkeys. *J. Cell. Anim. Biol.* **4(8)**, 125-130.

Sanz M., Flores A. and Lopez Bote C.J. (2000). The metabolic use of energy from dietary fat in broilers is affected by fatty acid saturation. *Br. Poult. Sci.* **41**, 61-68.

Sanz M., Flores A., Perez D.E., Ayala P. and Lopez Bote C.J. (1999). Higher lipid accumulation in broilers fed on saturated fats than in those fed unsaturated fats. *Br. Poult. Sci.* **40**, 95-101.

Sanz M., Lopez-Bote C.J., Monoyo D. and Bautista J.M. (2000b). Abdominal fat deposition and fatty acid synthesis are lower and fl- oxidation is higher in broiler chickens fed diets containing unsaturated rather than saturated fat. *J. Nutr.* **130**, 3034-3037.

SAS Institute. (2005). SAS®/STAT Software, Release 9.1 SAS Institute, Inc., Cary, NC. USA.

Scaife J.R., Moyo J., Galbraith H., Michie W. and Campbell V. (1994). Effect of different dietary supplemental fats and oils on the tissue fatty acid composition and growth of female broilers. *Br. Poult. Sci.* **35**, 107-118.

Shahryar H.A., Salamatdoustnobar R., Lak A. and Lotfi A.R. (2011). Effect of dietary supplemented canola oil and poultry fat on the performance and carcass characterizes of broiler chickens. *Curr. Res. J. Biol. Sci.* **3**, 388-392.

Sim J.S. (1990). Flax seed as a high energy/protein/ omega-3 fatty acid feed ingredient for poultry. Pp. 65-72 in Proc. 53rd Flax Inst. Fargo, North Dakota, USA.

Valizadeh M. and Moghaddam M. (1994). Experimental Designs

in Agriculture. Pishtaz Elem Publication, Tehran, Iran.

Yau J.C., Denton J.H., Bailey C.A. and Sams A.R. (1991). Customizing the fatty acid content of broiler tissues. *Poult. Sci.* **70,** 167-170.

Young R.J. and Garrett R.L. (1963). Effect of oleic and linoleic acid on the absorption of saturated fatty acids in the chick. *J. Nutr.* **81,** 321-329.

Zollitsch W., Kmaus W., Aichinger F. and Lettner F. (1996). Effects of different dietary fat sources on performance and carcass characteristics of broilers. *Anim. Feed Sci. Technol.* **66,** 63-73.

17

Effects of Pre-Pubertal Plane of Nutrition on Skeletal Growth, Lamb Mortality, IGF-1 Concentrations, Quantity and Quality of Colostrum Production in Kurdish Female Lambs

S. Menatian[1*], H.R. Mirzaei Alamouti[1], F. Fatahnia[2] and R. Masoumi[1]

[1] Department of Animal Science, Faculty of Agriculture, University of Zanjan, Zanjan, Iran
[2] Department of Animal Science, Faculty of Agriculture, Ilam University, Ilam, Iran

*Correspondence E-mail: menatian@znu.ac.ir

ABSTRACT

This experiment was conducted to investigate the effect of pre-pubertal plane of nutrition on the skeletal growth, lamb mortality, insulin-like growth factor 1 (IGF-1) concentrations, quantity and quality of colostrum produced in ewe lambs. A total of 40 clinically health Kurdish female lambs (30 ± 8.6 d and weighing 10.2 ± 3.4 kg) were randomly allocated to one of two experimental diets in pre-weaning period: high quality diet (HQD, 2.50 Mcal ME/kg dry matter (DM) and 148 g CP/kg DM) or low quality diet (LQD, 2.02 Mcal ME/kg DM and 87 g CP/kg DM). At weaning, one half of lambs from each group was randomly separated and assigned to HQD or LQD. So there were four treatment groups in post-weaning period: H-H (HQD pre- and post-weaning); H-L (HQD pre-weaning and LQD post-weaning); L-H (LQD pre-weaning and HQD post-weaning) and L-L (LQD pre and post-weaning, control group). Weekly DM intake was determined. Serum IGF-1 concentrations was determined by ELISA method. The HQD treatment increased DM intake and BW (body Weight) compared with the LQD treatment during pre-weaning period (P<0.01). At 210 d of age, animals fed LQD during the pre-weaning period and HQD during the post-weaning period (L-H sequence) had greater body length (BL), wither height (WH), hip height (HH) and hip width (HW) than animals on the H-L sequence. No interaction of the two periods was detected for heart girth, HW, WH, BL at 210 d of age (P>0.05). HW at weaning time was not affected by quality of diet. However, lambs of H-H and L-H sequences had a higher increase in HW during post-weaning period compared with lambs of H-L and L-L sequences. Quantity and quality of colostrum was not influenced by pre-pubertal plane of nutrition (P>0.05). Based on the results of current study the authors' suggestion is that skeletal compensatory growth during post-weaning period can compensate losses caused by poor nutrition during pre-weaning period. And also, the results indicated that skeletal size is a better indication for first-lactation colostrum yield than body weight (BW).

KEY WORDS colostrum, IGF-1, Kurdish ewes, lamb mortality, pre-pubertal diets, skeletal growth.

INTRODUCTION

Small ruminants are a common of many traditional farming systems near the Zagros Mountains in the west of Iran. Despite advances in animal science husbandry during previous decades, selling lambs are the main source of incomes for small holder in this area. Profitability in sheep husbandry in Iran is closely related to the ability of breeding ewes to raise lambs at the maximum economically feasible level. One approach to economic feasibility sheep production in a closed space is manipulating the age of puberty. In order to accelerate the maturity of the ewe lamb, the traditional plane of nutrition must be changed. However, due to the high positive correlation between BW and lamb mortality it is necessary to apply the right strategies. All of factors that affect the rate of growth pre- and post-weaning are important determinants of age at puberty. If a ewe lamb fails to achieve puberty in its first autumn, it will be delayed until the following breeding season (Kenyon et al. 2014).

Studies on animals revealed that both prenatal and neonatal programming of skeletal system development may be induced with the use of nutritional manipulation (Harrison et al. 2004; Tatara et al. 2007; Andersen et al. 2008). This suggests that the opportunity for increasing skeletal growth rate is greater during the time prior to puberty. Body weight fails to indicate the composition of the animal, therefore, measurements of the animal's frame can be considered indirect indicators in determining meat leanness (Greyling and Taylor, 1999).

Abnormal or difficulty in giving birth can be led to lamb mortality. There are two types of factors that lead to incidence of dystocia in ewe. Firstly, the fetal factors which include oversized fetus, lamb malpresentation, malposition, postural defects, and congenital abnormalities. Secondly, the maternal factors which include over feeding of dam during pregnancy, uterine inertia in polytocous ewes, and small diameter of pelvic canal (Pugh and Baird, 2012).

After a safe parturition, another challenge for livestock producer is to produce healthy newborn lambs. Ruminant neonates rely entirely on colostrum and milk from their dam for survival (Stelwagen et al. 2009). Newborn ruminants require a sufficient amount of colostrum within 48 h post-partum to survive (Stelwagen et al. 2009). The colostrum contains an important antibody which provides a defense mechanism for newborn ruminants until their own immune system is established (Ahmad et al. 2000; Yilmaz and Kaşikçi, 2013). Furthermore, nutritional supplementation of dams can enhance the erythropoietic response and therefore improve offspring survival (Ahmad et al. 2000).

The objective of this study was to compare the effects of diet quality fed during the pre- and post-weaning periods and existence of potential interactions between pre- and post-weaning diets on serum IGF-1 concentrations, skeletal size, lamb mortality, quantity and quality of colostrums production in Kurdish ewe lambs.

MATERIALS AND METHODS

Hormonal drugs
Controlled internal drug release (CIDR) with 300 mg of progesterone, a progestagen analogue (InterAg, Hamilton, New-Zealand), PMSG (folligon; Intervet International B.V., Boxmeer, the Netherlands), IGF-1 (LDN. Germany. LOT:150702) and progesterone (DiaMetra. Italy. LOT N:4026) were used.

Locations, animals and treatment schedule
This study was performed at Nomadic Management Department, Ilam Province, Iran (33° 5′ N, 46° 27′ E) from January 2013 to December 2015. All procedures involving animal care and management were approved by the University of Zanjan Animal Care Committee (proposal no. 1169739). A total of 40 clinically health Kurdish female lambs (30±8.6 d and weighing 10.2±3.4 kg) were used in this experiment.

At 30 d of age, lambs were randomly housed together with twice daily access to their mother milk and to one of two supplemental dietary treatments to achieve either high or low rates of BW gain during two consecutive periods of 30 to 120 (pre-weaning period) and from 121 to 210 d of age (post-weaning period). They were kept in individual pens (1×2 m) for 3 consecutive days every 2 weeks for recording dry matter intake (DMI).

In pre-weaning period the lambs fed high quality diet (HQD, n=20) or low quality diet (LQD, n=20) and at the weaning time HQD and LQD fed lambs were re-randomized. So that one half of lambs from each group randomly allocated to HQD or LQD. So there were four treatment groups (n=10) in post-weaning period: HQD pre- and post-weaning (H-H); HQD pre- weaning and LQD post- weaning (H-L); LQD pre- weaning and HQD post-weaning (L-H) and LQD pre- and post- weaning (L-L, control group).

The HQD and LQD were formulated according to nutrient requirements for small ruminants (NRC, 2007) recommendations covered the energy and protein needs for a 20 kg growing lamb with an average daily gain of 200 and 100 g/d, respectively. The HQD and LQD contained 2.50 and 2.02 Mcal ME/kg DM and 14.9 and 8.9% CP (DM basis), respectively. Rations were totally hand-mixed for each pen and offered in equal proportions twice daily at 09:00 and 16:00 in pre- and post-weaning period. Ingredients and chemical composition of the experimental diets are shown in Table 1.

Estrous synchronization and pregnancy diagnosis
When ewe lambs reached 210-d-old, estrus was induced and synchronized by CIDR. Animals were treated with CIDR for 14 d and were injected with 500 IU PMSG at the time of CIDR withdrawal. Twenty four hours after CIDR withdrawal, all of ewe lambs were monitored for estrus detection by 5 intact fertile rams and were ultimately naturally bred. The rams remained with the ewe lambs until the termination of estrous signs.

After serving, all ewe lambs were kept together in the same nutritional and managerial conditions and reared in pasture until 2 weeks before expected parturition. Pregnancy diagnosis was determined by using of trans-abdominal ultrasound (Piemedical, Falco 100; Netherlands) at 60 d after serving.

Table 1 Ingredients and chemical composition of the experimental diets

Ingredients (%)	Pre- and post-weaning diets	
	HQD	LQD
Alfalfa hay	445.1	-
Wheat straw	-	513.7
Ground barley	445.1	428.1
Soybean meal	59.3	-
Calcium carbonate	5.9	6.8
Salt	5.0	5.0
Mineral and vitamin premix[1]	39.6	46.4
Composition		
DM (%)	916.0	919.0
CP (%)	148.0	87.0
EE (%)	58.0	22.0
NDF (%)	285.0	450.0
NFC (%)	466.0	371.0
ME (Mcal/kg)	2.50	2.02

HQD: high quality diet; LQD: low quality diet; DM: dry matter; CP: crude protein; EE: ether extract; NFC: non-fiber carbohydrates and ME: metabolite energy.
[1] Each kg (DM basis) of mineral and vitamin premix contained: Ca 180 g ; P: 70 g; K: 35 g; Na: 50 g; Cl: 58 g; Mg: 30 g; S: 32 g; Mn: 5 g; Fe: 4 g; Zn: 3 g; Cu: 300 mg; I: 100 mg; Co: 100 mg; Se: 20 mg; vitamin A: 400000 IU; vitamin D$_3$: 100000 IU and vitamin E: 245 IU.

Data collection and calculation

The body weight (BW), body length (BL), heart girth (HG), wither height (WH), hip height (HH) and hip width (HW) were measured every 2 weeks from 30 to 210 d of age. WH and HH were measured by using of vertical graduated rod, BL, HG and HW by tape measure. BW was measured every 2 weeks from 30 to 210 d of age. Feed offered and feed refusals of individual pens were weighed and recorded daily and DM content of total mix ration (TMR) and orts were determined to estimate DMI. ME and CP intake were calculated as DMI from each diet multiplied by their ME and CP contents, respectively (NRC, 2007). DM, CP and ether extract (EE) of experimental diets were measured according to the methods of AOAC (1995). The neutral detergent fiber (NDF) was measured according to the method described by Van Soest et al. (1991) without α-amylase and sodium sulfite and was expressed exclusive of residual ash. Non-fibrous carbohydrates (NFC) content was calculated according to NRC (2001) dairy cattle model as: 100 - (CP+NDF+EE+ash).

Milk intake by ewe lambs was measured by the weigh-suckle-weigh method (WSW) in 3 consecutive days every 2 weeks from the start of study to weaning (30-120 d). At the start of WSW method at each suckling occasion (twice daily), ewe lambs were weighed, allowed to suckle the udder of their dams and weighed again immediately after

suckling. The difference between pre- and post-suckling weights was defined as milk intake. After lambing, ewe lambs were hand milked twice daily throughout lactation and milk yield was recorded at each milking for the entire lactation (two months). On each milking occasion, ewes were milked by hand after intravenous injection of 1 IU synthetic oxytocin. Milk samples of dams and ewe lambs in subsequent lactation were collected in 3 consecutive days every 2 weeks and analyzed for fat, protein and lactose by using of Milk-O-Scan 133B (Foss Electric, Hillerod, Denmark). Milk protein, fat and lactose yields were calculated by multiplying milk yield from the respective day by protein, fat and lactose contents of the milk for each ewe. Milk gross energy (GE) was calculated as: GE= ((0.0547×CP %)+(0.0929×fat %)+(0.0395×lactose %)) according to NRC (2001). The mean metabolize ability of the ewe milk GE is 0.94 (Treacher and Caja, 2002), therefore, milk ME content was calculated as GE × 0.94.

Energy corrected milk (ECM) and fat corrected milk (6.5% FCM) were calculated as ECM= (0.327×kg milk) + (12.95×kg fat) + (7.2×kg protein) and FCM= milk yield × (0.37+(0.097×fat %)).

Blood sampling

Before the first meal of the day, blood samples (5 mL) were collected by jugular venipuncture from each lamb every 2 weeks from 90 d of age until puberty (age at puberty was assessed by serum concentrations of progesterone when 2 consecutive blood samples contained at least 1 ng of progesterone/mL). Hence, samples were centrifuged for 15 min (3000 rpm), sera were separated into 1.5 mL micro tubes and then placed in freezer (-20 °C). Serum samples were tested for progesterone and IGF-1 by ELISA method. Standard commercial kits were used for analysis and the procedures were adopted as recommended by the manufacturer of these kits.

Statistical analyses

The data of pre-weaning parameters were subjected to statistical analysis by using of completely randomized design (CRD). Data were analyzed as a CRD in factorial arrangement (2×2) by using of the mixed model procedure of SAS (2003) with fixed effects of treatment and random effects of lamb nested in treatments.

(1) $Y_{ik} = \mu + D_i + L_k(D_i) + \varepsilon_{ik}$

Where:

Y_{ij}: dependent variable.

μ: mean.

D_i: fixed effect of dietary treatment I.

$L_k(D_i)$: effect of lamb k nested in the dietary treatment.

ε_{ik}: error.

For repeated measure data, the model was:

(2) $Y_{ijk} = \mu + D_i + Time_j + D_i \times Time_j + L_k(D_i) + \varepsilon_{ijk}$

Where:

$Time_j$: effect of time j as a fixed effect.

Measurements obtained before administration of dietary treatments were used as covariates. The covariates were removed from the model one at a time, starting with the least significant. LSM, SEM and P-values are reported. Statistical differences were considered significant when (P<0.05) and trends are discussed when (P<0.01).

RESULTS AND DISCUSSION

Accelerating the growth of sheep has the potential to increase the profitability by reducing the time need from birth to first lambing, subsequently reducing feed, labor, housing, and other costs associated with raising replacement animals. Kurdish ewe is the most popular indigenous dairy breed of sheep in west of Iran. Its main characteristics are high prolificacy and high milk yield. Considering the high genetic potential of Kurdish sheep it is important to ensure that appropriate management practices are implemented in their intensive production systems.

Intake and growth

Table 2 shows feed intake and skeletal growth (as measured by body weight (BW), body length (BL), heart girth (HG), wither height (WH), hip height (HH) and hip width (HW)) measurements by treatment between 30 and 210 d age. The HQD treatment increased DMI and BW compared with the LQD treatment during pre-weaning period (P<0.01). Lambs of H-H sequence had higher DMI, ME and CP compared with lambs of H-L, L-H and L-L treatments at post-weaning period with no interaction of periods (P>0.05).

BL was increased by 39 and 31 cm by HQD and LQD treatments, respectively and HG was increased by 24 cm by HQD and 19 cm by LQD treatment, during pre-weaning period (P<0.01). At 120 d of age, ewe lambs fed the HQD treatment had greater WH and HH than lambs on the LQD treatment (P<0.01).

There were no differences among treatments in HW at weaning time. At 210 d of age, animals fed LQD during the pre-weaning period and HQD during the post-weaning period (L-H sequence) had greater BL, WH, HH and HW than animals on the H-L sequence. Also at the end of experiment, lambs on the L-H treatment had greater BL, HG, WH and HH than L-L sequence (P<0.01).

Ewe lambs on the H-H and L-H treatments were taller at 210 d of age than lambs on the H-L and L-L treatments (P<0.01). By comparing the pre- and post-weaning periods in the H-H sequence, BL increased by 35 cm during pre-weaning period, while it raised only 12 cm during post-weaning period. A part from the HW, the trend was the same for other measured parameters. Pre-weaning skeletal growth rate was more than three times the growth rate post weaning. Some parameters like shoulder height and shoulder width grow at a slower rate than body length, but again these measurements had highly linear correlation with live body weight (Greyling and Taylor, 1999). Results of the present study indicated that the ewe lambs fed the HQD would gain faster than ewe lambs fed the LQD in both BW and skeletal size. Skeletal growth rate of H-L group was lower than L-L group during post-weaning period, but final skeletal size of H-L sequence was higher than L-L sequence, showing the importance of pre-weaning plane of nutrition. Our experiment results showed that more than 80% of pre-pubertal skeletal growth to maturity can be related to the pre-weaning period. With respect to higher skeletal growth rate, responses to diet quality will be dependent on several factors including the capacity of the animal for skeletal growth, the quality of the diet and protein with respect to meeting the animal's mineral and amino acid needs for increased bone synthesis. Mature skeletal size is determined by genetic potential, but quality of diet and feeding plane can result in animals achieving that genetic potential earlier or being retarded in growth (Owens *et al.* 1993). However, in some experiments (Radcliff *et al.* 1997; Van Amburgh *et al.* 1998) increased energy or a combination of energy and protein decreased skeletal size in ruminants. Since the rate of protein and mineral deposition decreases with age, the response will diminish as the animal matures. Based on the results of the current study the HQD diet was more effective during the early parts of the study than the mature lambs. As expected, we observed that that skeletal compensatory growth during post-weaning period can be compensating for losses caused by poor nutrition during this period. The most important functions of IGF-1 in relation to the skeletal system include its stimulating effect on longitudinal bone growth, proliferation and differentiation of chondrocytes in the growth plate, cortical bone formation, proliferation and differentiation of osteoblasts and type I collagen synthesis (Kanbur *et al.* 2005). Tatara (2008) was reported that the anabolic response of biochemical markers of bone formation at both these developmental stages of animals was analogous to the changes in serum concentration of IGF-1 induced by the pre-weaning treatment of lambs with HQD, great importance of somatotrophic axis function is in the regulation of bone metabolism and skeletal system development.

Table 2 Effect of pre- and post-weaning plane of nutrition on intake and skeletal growth of ewe lambs (30-210 d of age)

Item	Pre-weaning treatments		Item	Post-weaning treatments			
				HQD		LQD	
	HQD	LQD		H-H	H-L	L-H	L-L
n	20	20	**n**	10	10	10	10
Intake			**Intake**				
DM (kg/d)	0.97	0.64	DM (kg/d)	1.54	1.21	1.31	0.87
Fresh milk intake (kg/d)	1.11	1.18	ME (Mcal/d)	3.85	2.42	3.27	1.76
ME intake (diet+milk, Mcal/d)	3.49	2.44	CP (g/d)	228	104	194	76
CP intake (diet+milk, g/d)	187.4	103	Puberty age (d)	123	245	168	267
Body weight			**Body weight**				
30 d (kg)	10.04	10.2	-	-	-	-	-
120 d (kg)	31.20	22.50	121-210 d (g/d)	138	31	153	57
30-120 d (g/d)	235	136	210 d (kg)	43.8	33.9	36.3	26.6
Body length			**Body length**				
30 d (cm)	93.7	90.6	-	-	-	-	-
120 d (cm)	132.9	122	121-210 d (cm/d)	0.13	0.003	0.19	0.03
30-120 d (mm/d)	4.4	3.5	210 d (cm)	145	133.2	138.9	124.8
Heart girth			**Heart girth**				
30 d (cm)	55.6	53.3	-	-	-	-	-
120 d (cm)	80	72.5	121-210 d (cm/d)	0.101	0.016	0.09	0.03
30-120 d (mm/d)	2.7	2.1	210 d (cm)	89.1	81.5	80.5	74.9
Wither height			**Wither height**				
30 d (cm)	44.3	43.9	-	-	-	-	-
120 d (cm)	59.4	55	121-210 d (cm/d)	0.07	0.03	0.11	0.04
30-120 d (mm/d)	1.67	1.2	210 d (cm)	66.5	63	65	59.2
Hip height			**Hip height**				
30 d (cm)	46.4	45.6	-	-	-	-	-
120 d (cm)	61.8	57.2	121-210 d (cm/d)	0.07	0.03	0.11	0.04
30-120 d (mm/d)	1.71	1.3	210 d (cm)	68.1	64.8	67.4	60.4
Hip width			**Hip width**				
30 d (cm)	16.8	15.8	-	-	-	-	-
120 d (cm)	22	21.8	121-210 d (cm/d)	0.06	0.02	0.05	0.01
30-120 d (mm/d)	0.6	0.6	210 d (cm)	27.5	23.9	26.2	22.8

HQD: high quality diet; LQD: low quality diet; H-H: HQD pre and post-weaning; H-L: HQD pre-weaning and LQD post-weaning; L-H: LQD pre-weaning and HQD post-weaning and L-L: LQD pre and post-weaning (control).
DM: dry matter; CP: crude protein and ME: metabolite energy.

Reproductive performance

The L-L treatment was removed from the statistical analysis of reproduction performance, because of eight animals of this treatment were not pregnant. The rate of pregnancy H-H, H-L, L-H and L-L treatments were 100, 50, 70 and 20%, respectively (Table 3). Normal gestation length for sheep is between 144 and 152 days, and also according to results of Echternkamp and Gregory (1999), factors linked to gestation length were retained placenta, age of the dam, and sex of the lamb. There were differences in lamb birth weights between three groups (Table 3). Age at puberty, age at first lambing, safe and successful parturition at a younger age, stress, optimal nutrition, breed and geographical region are important traits concerning overall reproductive performance.

However, 3.6 kg by L-H sequence is relevant to 7 lambs and 2.7 and 2.8 kg by H-H and H-L groups are relevant to 10 and 5 lambs, respectively (Table 3). In general, at the mating time, animals that have sufficient weight, struggle

less to reach their mature weight and bring the nutrients to the fetus. Lambs birth weight of H-H sequence was lower than in other treatments (Table 3). It seems that more than 80% of mature weight by the age of 7 months has a negative effect on lambs' birth weight. The incidence of lambs mortality was higher in L-H sequence (43%) compared with H-H (20%) and H-L (40%) treatments, which may be due to the higher birth weight of lambs of the treatment. In spite of good HW of dams, animals of H-H sequence showed dystocia and lower lambs' birth weight (Tables 2 and 3). Small pelvises in ewes are associated with high incidences of dystocia, high prenatal ewe and lamb mortality rates and poor lifetime rearing performance of ewes (Hartwig, 2002). High birth weights have been associated with increased dystocia in ewe bearing single lambs and young ewes are more susceptible to lambing problems than mature ewes that have lambed previously (Anderson, 1992; Hartwig, 2002). Growing animals on a low nutrient diet have clearly resulted in an increase in dystocia.

Table 3 Effect of pre-and post-weaning plane of nutrition on reproduction performance and colostrum production of ewe lambs (30-210 d of age)

Item	Post-weaning treatments			
	HQD		LQD	
	H-H	H-L	L-H	L-L
Number of pregnant ewe (n=10)	10	5	7	2
Pregnancy days (d)	149.1	147.8	146.6	-
Lamb birth weight (kg)	2.7	2.8	3.6	-
Lamb weaning weight at day 60 (kg)	15.2	14.8	16.5	-
Lamb mortality	2 (10)	2 (5)	3 (7)	2 (2)
Stillbirth	0 (10)	2 (5)	0 (7)	2 (2)
Colostrum (mL)	223	152	213	-
Fat (%)	12.29	10.97	12.19	-
Protein (%)	16.80	15.03	16.10	-

HQD: high quality diet; LQD: low quality diet; H-H: HQD pre and post-weaning; H-L: HQD pre-weaning and LQD post-weaning; L-H: LQD pre-weaning and HQD post-weaning and L-L: LQD pre and post-weaning (control).

Overfeeding animals causes internal fat deposits which obstruct the pelvic canal. All managers, however, must maintain a balance between achieving maximum frame growth without allowing excessive fat deposits. Fat animals will have high incidences of dystocia just as severely as underdeveloped animals (Wilson and Rossi, 2006).

The mortality rates in H-H, H-L, L-H and L-L sequences were 0%, 40%, 0% and 100% respectively (Table 3). High mortality of L-L and H-L groups may be related to their difficult parturition. Environmental factor such as season of birth is responsible for approximately 55% of dystocia (Anderson, 1992). Survival lambs were less in L-L and H-L groups and this situation may be due to poor mother ewes and low colostrum produced of these mothers (Table 3). Newborn lambs from H-L group with a lower birth weight tended to be weaker and therefore to have more trouble suckling adequate amounts of colostrum to provide sufficient levels of antibodies in their blood for initial immune protection, which is in accordance with report of Ahmad *et al.* (2000) in Pak-Karakul sheep.

In their study, they found that maternal body status is more effective than the parturition position on lamb survival, so that dystocia had little impact on H-H and L-H groups lamb survival.

The provision of an adequate amount of high quality maternal colostrum is essential for the health and survival of neonatal lambs. Quantity of colostrum and fat percentage in colostrum did not respond to pre-pubertal plane of nutrition (P>0.05), but colostrum protein percentages in H-H sequence showed tendency to rise (Table 3) which might be related to higher concentrations of serum total protein (the results have not reported) in ewe lambs fed H-H sequence. The quantity and quality of colostrum can be influenced by various factors including breed, lactation number, age, health status, nutrition, body condition score at parturition

and genetics, as well as environmental factors (Hart *et al.* 2009).

CONCLUSION

Based on the results of the current study, most skeletal growth in Kurdish lambs was taken place in pre-weaning period and strongly influenced by pre-pubertal plane of nutrition. The results also showed that adequate skeleton size in Kurdish ewes is needed in minimizing the lambs' mortality associated with parturition. However, further researches are needed to evaluate the effects of nutrition and mobility on dystocia and lambs' mortality in first parturition.

ACKNOWLEDGEMENT

The authors wish to acknowledge Mr. Ali Salimi, Farshad Yasemi, Mohammad Rashid Taheri, Behzad Abdolahi and Dr Misban and Nomadic Management Department of Ilam, Iran for their coorporation. In addition, the authors wish to thank Nomadic Management Department of Ilam, Iran for funding this study.

REFERENCES

Ahmad R., Khan A., Javed M.T. and Hussain I. (2000). The level of immunoglobulin's in relation to neonatal lamb mortality in Pak-Karakul sheep. *Vet.* **3,** 129-139.

Andersen N.K., Tatara M.R., Krupski W., Majcher P. and Harrison A.P. (2008). The long-term effect of alpha-ketoglutarate, given early in postnatal life, on both growth and various bone parameters in pigs. *J. Anim. Physiol. Anim. Nutr.* **92,** 519-528.

Anderson P. (1992). Minimizing calving difficulty in beef cattle. MS Thesis. University of Minnesota, Minnesota, USA.

AOAC. (1995). Official Methods of Analysis. Vol. I. 16th Ed. Association of Official Analytical Chemists, Arlington, VA, USA.

Echternkamp S.E. and Gregory K.E. (1999). Effect of twinning in gestation length, retained placenta and dystocia. *J. Anim. Sci.* **77,** 39-47.

Greyling J.P. and Taylor G.J. (1999). The effect of the anabolic agent, nandrolone laurate, on certain production and reproduction parameters in ram lambs, under intensive and extensive feeding regimes. *South African J. Anim. Sci.* **29,** 130-188.

Harrison P., Tygesen M.P., Sawa-Wojtanowicz B., Husted S. and Tatara M.R. (2004). α-ketoglutarate treatment early in postnatal life improves bone density in lambs at slaughter. *Bone.* **35,** 204-209.

Hart K.W., Contou C., Blackberry M. and Blache D. (2009). Merino ewes divergently selected for calm temperament have a greater concentration of immunoglobulin G in their colostrum than nervous ewes. *Proc. Adv. Anim. Breed. Genet.* **18,** 576-579.

Hartwig N. (2002). Sheep Health. Iowa State University of Science and Technology Press, USA.

Kanbur N.O., Derman O. and Kinik E. (2005). The relationships between pubertal development, IGF-1 axis and bone formation in healthy adolescents. *J. Bone. Miner. Metab.* **23,** 76-83.

Kenyon P.R., Thompson A.N. and Morris S.T. (2014). Breeding ewe lambs successfully to improve lifetime performance. *Small Rumin. Res.* **118,** 2-15.

NRC. (2001). Nutrient Requirements of Dairy Cattle. 7th Ed. National Academy Press, Washington, DC, USA.

NRC. (2007). Nutrient Requirements of Small Ruminants: Sheep, Goats, Cervids and New World Camelids. National Academy Press, Washington, DC, USA.

Owens F.N., Dubeski P. and Hanson C.F. (1993). Factors that alter the growth and development of ruminants. *J. Anim. Sci.* **71,** 3138-3150.

Pugh D.G. and Baird N.N. (2012). Sheep and goat medicine. Elsevier Health Sciences, Amsterdam, Netherlands.

Radcliff R.P., Vandehaar M.J., Skidmore A.L., Chapin L.T., Radke B.R., Lloyd J.W., Stanisiewski E.P. and Tucker H.A. (1997). Effects of diet and bovine somatotropin on heifer growth and mammary development. *J. Dairy Sci.* **80,** 1996-2003.

SAS Institute. (2003). SAS®/STAT Software, Release 8.0. SAS Institute, Inc., Cary, NC. USA.

Stelwagen K., Carpenter E., Haigh B., Hodgkinson A. and Wheeler T.T. (2009). Immune components of bovine colostrum and milk. *J. Anim. Sci.* **87,** 3-9.

Tatara M.R. (2008). Neonatal programming of skeletal development in sheep is mediated by somatotrophic axis function. *Exp. Physiol.* **93,** 763-772.

Tatara M.R., Śliwa E. and Krupski W. (2007). Prenatal programming of skeletal development in the offspring: effects of maternal treatment with β-hydroxy-β-methylbutyrate (HMB) on femur properties in pigs at slaughter age. *Bone.* **40,** 1615-1622.

Treacher T.T. and Caja G. (2002). Nutrition during lactation. Pp. 101-110 in Sheep Nutrition. M.H. Freer and H. Dove, Eds. CSIRO Publishing, Australia.

Van Amburgh M.E., Galton D.M., Bauman D.E., Everett R.W., Fox D.G., Chase L.E. and Erb H.N. (1998). Effects of three prepubertal body growth rates on performance of Holstein heifers during first lactation. *J. Dairy Sci.* **81,** 527-538.

Van Soest P.J., Robertson J.B. and Lewis B.A. (1991). Methods for dietary fiber, neutral detergent fiber and nonstarch polysaccha-rides in relation to animal nutrition. *J. Dairy Sci.* **74,** 3583-3597.

Wilson T.W. and Rossi J. (2006). Factors affecting calving difficulty. *Bulletin.* **943,** 4.

Yilmaz O. and Kaşikçi G. (2013). Factors affecting colostrum quality of ewes and immuno stimulation. *Turkish J. Vet. Anim. Sci.* **37,** 390-394.

18

Nutritional Value, Fourier Transform Infrared Spectroscopic Molecular Structures, Mycotoxines and Heavy Metals Concentration of Un-Ripe, Ripe and Sun- Dried Fruit from 'Sultana' Grapevine for Ruminants

M. Yari[1,2*], M. Manafi[1], M. Hedayati[1], R. Karimi[2,3], R. Valizadeh[4] and
A. Jonker[5]

[1] Department of Animal Science, College of Agriculture, Malayer University, Malayer, Iran
[2] Iranian Grape and Raisin Institute, Malayer University, Malayer, Iran
[3] Department of Landscape Engineering, College of Agriculture, Malayer University, Malayer, Iran
[4] Department of Animal Science, Faculty of Agriculture, Ferdowsi University of Mashhad, Mashhad, Iran
[5] Grasslands Research Centre, AgResearch Limited, Palmerston North, New Zealand

*Correspondence E-mail: m.yari@malayeru.ac.ir

ABSTRACT

Grapes and grape-derived products have worldwide importance due to its consumption by human, however, they may also be used for ruminant feeding when their price or quality is low. The objectives of current study were to determine the nutritive value, in terms of chemical composition, *in vitro* rumen gas production kinetics and predicted nutrient supply, Fourier Transform Infrared Spectroscopic (FTIR) molecular structures and mycotoxin and heavy metal contaminants in un-ripe, ripe and sun-dried (raisin) fruits of 'Sultana' grapevine for ruminants. Un-ripe fruit had higher nitrogen to total carbohydrate ratio and total phenol and tannin concentration (P<0.05) than ripe and sun-dried grapevine, and tended to have higher vibration spectroscopy peak area related to phenolic compounds and related to structural carbohydrates than ripe and sun dried fruit. Raisin had higher *in vitro* cumulative gas production at 24 h of incubation (P<0.05). Aflatoxin B1, B2, G1 and G2 and Ochratoxin A were not detectable in un-ripe and ripe grapevine fruit, while low concentrations were found in raisin (P<0.05). The lead concentration was lower in un-ripe fruit and raisin than in ripe fruit (P<0.05). In conclusion, phenolic compounds and tannins, mycotoxins and heavy metal concentration in un-ripe and ripe fruits and raisin of Sultana' grapevine were lower than toxic level for animal nutrition and nutrient profile and availability of the grapevine products make it a suitable feed to replace forage in the diet of ruminants.

KEY WORDS energy value, mycotoxines, FTIR vibration spectroscopy, grapevine products, *in vitro* gas production kinetics, ruminants.

INTRODUCTION

Grapes were among the first fruit species to be domesticated and today is the most economically important fruit crop in the world (Keller, 2010). Grapevines are planted on approximately 7.3 million hectares worldwide producing approximately 67 million metric tons of fruit in 2007. Iran is one of the major grapevine growers in the world with 215000 ha producing 2.15 million tons of fruit (FAO, 2009) and Malayer is the leading grapevine growing areas (Hamedan province) of Iran, with about 11200 ha grapevines (Karimi and Ershadi, 2014). 'Sultana' (synonym White Kishmish) is one of main grapevine cultivars planted in Malayer, which produces seedless fruit and raisins that are mainly exported as raisins or locally sold fresh for human consumption (Karimi and Ershadi, 2014). The oversupply of grapevine products (e.g. fruit) or low quality or contaminated (e.g. mycotoxins and heavy metals) grapevine and raisin products not suitable for the human market are usually used for domestic ruminant feeding (Besharati Taghizadeh, 2009). However, no information was found in literature regarding nutritive value, plant secondary compounds and contaminants in un-ripe, ripe and sun-dried (raisin) 'Sultana' grapevines for ruminants. This information is essential for ration formulation and for safe feeding of these products to ruminants. Nutritive value and plant secondary compounds in grapevine products likely change with phenological stage, berry development and fruit ripening. Berry development and ripening in grapevine consist of three main phases being early fruit development, lag phase and berry ripening. The duration of the early fruit development phase is specific to individual cultivars and this phase ends simultaneously with the end of the herbaceous phase followed by a lag phase without fruit growth. Then a second growth phase takes place, with the onset of ripening indicated by a change in grape berry skin colour called 'véraison' in French. The largest changes in grape berry composition occur during this ripening phase (Boss and Davies, 2001). Chemical composition, molecular features and nutritive value of feeds for ruminants can be analysed by wet-chemical methods, *in vitro* gas production technique and Fourier Transform Infrared (FTIR) vibration spectroscopy.

Contamination of fruit with mycotoxins (secondary metabolites of moulds) and heavy metals can occur during berry development, ripening and sun drying (to produce raisins) through management practices, weather conditions and ground contamination, which lower the product quality for local consumption and export. Mycotoxins and heavy metals can be hazardous to animal health (Besharati Taghizadeh, 2009), therefore, their levels need to be determined. The objectives of current study were to determine the nutritive value, in terms of chemical composition, *in vitro* gas production kinetics, predicted nutrient supply, FTIR molecular structures, and mycotoxin and heavy metal contaminants in un-ripe, ripe and sun-dried (raisin) fruits of 'Sultana' grapevine. Our hypothesis was that the chemical composition, nutritive value, FTIR features and presence of mycotoxines and heavy metals in un-ripe, ripe and sun-dried fruits of 'Sultana' grapevine would change with phenological stage and during product processing for raisin production.

MATERIALS AND METHODS

Plant materials and grapevine sampling
In this study, un-ripe and ripe fruit of grapevine were sampled from 14 years old own-rooted 'Sultana' grapevines (*Vitis vinifera*) grown at four different vineyards (blocks) in the Malayer region (Malayer, Iran; lat. 34° 30′ N, long. 48° 85′ E, alt. 1550 m). All vines were spaced 2.5 m apart in north–south-orientated rows that were 3 m wide. The vines were pruned when fruit spurs were at 5-6 buds, which occurred in the middle of March, and vines were irrigated every two weeks.

The sample collection was performed during the 2012 growing season (dates and weather conditions in Table 1) at berry véraison (colour change and maturation nascent; i.e. un-ripe fruit), grape harvest (grape maturity; i.e. ripe fruit) and postharvest after sun drying (i.e. raisin). In each vineyard, 3 rows with 12 vines were used as replicates and about 36 grapevine clusters (at least 3 representative clusters per vine) were randomly collected from different parts of grapevine: un-ripe ($4\times3\times12\times3$), ripe ($4\times3\times12\times3$) and sun-dried ($4\times3\times12\times3$) harvest. The grapevine clusters from vines within each row were pooled to produce 3 samples (3 rows) per vineyard. Fresh samples were packed in polyethylene bags and placed on ice in styrofoam boxes and shipped to the Laboratory.

Raisin production process
For raisin preparation, part of the ripe grapevine clusters harvested were dipped into a solution of potassium carbonate and olive oil (90 g/kg K_2CO_3+1.5 g/kg olive oil). This operation preserves the vitamins and minerals in the grape, accelerates the grape drying process and gives the raisins the golden yellow colour. The treated grapevines were spread directly on the ground in the sun to dry for approximately 7-12 days according to the heat intension (Pala *et al.* 1993). After drying, raisins were separated from the stalks and stored in fruit boxes.

Chemical composition analysis
Before chemical composition analysis and *in vitro* ruminal fermentation measurements, all samples (12=4 vineyards×3 rows within vineyard) were dried at 50 °C to constant weight and ground to pass a 1 mm screen (Ghods miller, Ghods Company, Iran). Standard procedures described by the Association of Official Analytical Chemists (AOAC, 10) were used to determine dry matter (DM; method 930.15), ash (AOAC method 942.05), crude protein (CP; AOAC method 984.13) and ether extract (EE; AOAC, method 954.02).

Neutral detergent fibre (NDF), assayed with heat stable alpha-amylase, and acid detergent fibre (ADF) were determined according to Van Soest *et al.* (1991) with the ANKOM A200 Filter Bag technique (Ankom Technology, Fairport, NY, USA). Sodium sulfate was used for NDF and ADF determination to remove nitrogen attached to cell wall structure.

Table 1 Weather conditions during sampling of un-ripe, ripe and sun-dried (raisin) Sultana grapevine in 2012

Items	Grapevine product[1]		
	Un-ripe (n=12)	Ripe (n=12)	Sun-dried (n=12)
Sampling date	July 26	Sep 5	Sep20
Max. temperature (°C)	35.00	32.00	27.00
Min. temperature (°C)	18.00	12.00	10.00
Relative humidity (%)	10.00	15.00	25.00
Day length (h)	14:01	13:01	12:14

[1] n= 12; (4 vineyards×3 rows within vineyard).

Acid detergent lignin was determined by soaking the ADF filter bag residue in 72% sulphuric acid for 3 h followed by washes with water (method 973.18; 10). All chemical analyses were performed in duplicate and repeated if error was higher than 5%. Non-fiber carbohydrates (g/kg DM; NFC= 1000 - (NDF+CP+EE+Ash)) and total carbohydrates [g/kg DM; CHO=1000-(CP+EE+Ash)] were calculated according to NRC (2001).

Before total phenolics and tannin analysis, samples (4 vineyards×3 rows within each vineyard=12) were ground through a 0.5 mm screen (Makkar, 2000). Total phenolic compounds were extracted from 200 mg dried samples in 10 ml aqueous acetone water (700:300, v/v) at 4 °C overnight.

The samples were then centrifuged at 3000 g at 4 °C for 15 min, and the supernatant (i.e. extract) used in the colorimetric Folin–Ciocalteu assay, as described by Singleton and Rossi (1965) with tannic acid (Merck GmbH, Darmstadt, Germany) as a standard. Total tannin was estimated indirectly after binding tannins in the supernatant of the phenolic extraction to insoluble polyvinyl-polypyrrolidone followed centrifugation at 3000 g at 4 °C for 15 min, and supernatant again used in the Folin–Ciocalteu assay. Concentration of total tannin was then calculated by subtracting phenolic compounds remaining in the supernatant after the polyvinyl-polypyrrolidone precipitation of tannins from total phenolic compounds (Singleton and Rossi, 1965).

In vitro gas production and predicted nutrient supply in ruminants

In vitro rumen incubations were performed using the semi-automated gas production technique (Theodorou *et al.* 1994; Rogerio *et al.* 1999) with buffered rumen fluid prepared according to Menke and Steinglass (1988). Rumen fluid was collected before the morning feeding from four ruminally fistulated steers (482.5±22.5 kg, body weight) fed 9 kg DM/d (in g/kg DM; total mixed ration with 556 g barley silage, 300 g alfalfa hay, and 144 g dairy cow concentrate) twice daily in equal portions at the experimental farm of the Ferdowsi University of Mashhad (Mashhad, Iran) as described by Yari *et al.* (2014).

Feeding and animal husbandry of the steers were according to procedures of the Iranian Council on Animal Care (ICAC; 1995) guidelines. After collection, ruminal contents were strained through four layers of cheese cloth, to eliminate large feed particles, and transported to the laboratory at Ferdowsi University of Mashhad in a pre-warmed thermos.

Each sample [3 samples per vineyard × 4 vineyards × 3 products (un-ripe, ripe and sun-dried berries)= 36] was incubated in triplicate vials (125 mL) with 10 mL of rumen liquid and 20 mL of buffer (Menke and Steinglass, 1988) for 96 h at 37.5 °C.

Three vials with buffered rumen medium, without sample, were incubated to correct for gas release from the inoculum.

Gas accumulated in the head-space of the vial was determined using a pressure transducer (Razi Instruments, Mashhad, Iran) and head-space gas volume (Gp) was predicted by Boyle's Gas Law from pressure measurements as:

$$GP= (Vh/Pa) \times Pt$$

Where:
Vh: represents head-space volume (95 ml).
Pa: represent atmospheric pressure (14.692 psi; Meteorological Office, Mashhad, Iran).
Pt: represents pressure transducer reading (psi) (Theodorou *et al.* 1994; Rogerio *et al.* 1999).

Head-space pressure readings were taken at 0, 2, 4, 8, 12, 20, 24, 48, 72 and 96 h after the start of incubation. All incubations were repeated in two runs. The rate and extent of gas production were determined for each sample by fitting gas production data over time to a nonlinear equation:

$$Y= b(1-\exp^{-ct})$$

Where:
Y: volume of gas produced at time t.
b: asymptotic gas production.
c: fractional rate of gas production (Ørskov and McDonald, 1979).

Parameters b and c were calculated using the NLIN (nonlinear) procedure of SAS using iterative least-squares regression (SAS, 2003). Linear regression equations (Menke and Steinglass, 1988) were used to estimate organic matter digestibility (OMD, g kg^{-1})= 14.88 + 0.8893 × gas + 0.0448 × CP + 0.0651 × ash and net energy for lactation (NE$_L$; MJ kg DM^{-1}= -0.22 + 0.1062 × gas + 0.0048 × CP + 0.0132 × fat based on cumulative gas volume at 24 h of incubation (gas) and sample CP, fat and ash concentrations.

Molecular spectroscopic study and spectral features

All samples (4 vineyards×3 rows within vineyard=12 per grapevine product) were scanned by FTIR vibration spectroscopy using a JASCO FTIR-ATR-4200 (JASCO Corporation, Tokyo, Japan). Spectra were generated in the mid-infrared spectra range from 4000 to 700 cm^{-1} in transmission mode. Quantitative analyses (peak area and height) of the molecular spectral were performed using JASCO software. Typical spectra of the three grapevine products with different functional groups is presented in Figure 1. The spectral peak area, height and centre were identified for phenolic compounds at 3000-3730 cm^{-1} and 1700-1830 cm^{-1}, (Fernaandez and Agosin, 2007; Musingarabwi, 2015), for lipids at *ca.* 2790-3000 cm^{-1}, for proteins at 1546-1703 cm^{-1}, for total carbohydrates at 920-1188 cm^{-1}, and for structural carbohydrates at 1188-1496 cm^{-1} (Wetzel *et al.* 1998; Yu *et al.* 2004).

Mycotoxines and heavy metals analysis

Mycotoxines, aflatoxin B1, B2, G1 and G2 were determined according to ISIRI (No. 6872; Ref. 2012; ISIRI, 2012a) and for Ocratoxin A (No. 9238; 2012; ISIRI, 2012b) according to AOAC (1995) and Romer *et al.* (1978) using HPLC method and immunoaffinity column clean up-Test method. Heavy metals were analysed according to procedures in ISIRI standards (No. 12968; Ref. 2012; ISIRI, 2012) based on AOAC (1995). Number of samples used for mycotoxines and heavy metal analysis were 12 (4 vineyards×3 rows within each vineyard).

Statistical analysis

Data was analysed using PROC MIXED of SAS (2003) with the following statistical model:

$$Y_{ij}= \mu + T_i + B_j + e_{ij}$$

Where:

Y_{ij}: observation of the dependent variable ij.

μ: fixed effect of population mean for the variable.

T_i: fixed effect of treatment (i=3; un-ripe, ripe and sun dried fruit of grapevine).

B_j: random effect of block (vineyard; j=4).

e_{ij}: random error associated with the observation ij.

For gas production values, triplicate vials per run per samples were averaged and the effect of run was included in the model as random effect. For all analysis, experimental replicates were three samples per block (i.e. vineyard) for each treatment (3 samples×4 vineyards×3 treatments=36). The Fisher's protected least significant difference test was used for multiple-treatment comparisons using the LSMEAN statement of SAS. For the different statistical tests, significance was declared at P ≤ 0.05 and trend at P ≤ 0.15.

RESULTS AND DISCUSSION

Chemical composition

Raisin samples had higher dry matter (DM) concentration than ripe grapevine, which both had higher DM concentration than un-ripe grapevine (P<0.01) (Table 2). Nitrogen to CHO ratio (N:CHO), N to organic matter ratio (N:OM), total phenolics and total tannin (P<0.05) were higher in un-ripe grapevine compared with ripe grapevine, and these constituent were lowest in raisin(P<0.05) (Table 2). None-fibre carbohydrate and CHO concentrations were lower and ADF and acid detergent lignin (ADL) concentration were higher in un-ripe and ripe grapevine compared with raisin (P<0.05) (Table 2). Ripe grapevine had higher NDF concentration than un-ripe grapevine, which had higher NDF than raisin (P<0.01). Ether extract was similar among grapevine products.

In vitro gas production and estimated nutrient supply in ruminants

In vitro potential gas production (i.e. b) and rate of gas production (i.e. c) were similar among the three grapevine products (Table 3).

Cumulative gas production after 24 h of incubation was higher for un-ripe grapevine than for raisins, which were both higher than for ripe grapevine (P<0.05). Unripe grapevine tended (P≤0.10) to have a lower predicted organic matter digestibility and net energy for lactation than ripe and sun-dried grapevine.

FTIR spectroscopic features

The three grapevine products had similar FTIR molecular structures related to amides, aromatic compound bonds and total carbohydrate (Table 4). Ripe grapevine had lower peak area related to CH$_2$ and CH$_3$ symmetric and anti-symmetric stretching compared with raisin and un-ripe grapevine (P<0.05). Un-ripe grapevine tended to have higher area related to phenolic compounds and tannins (P=0.15) compared with ripe grapevine.

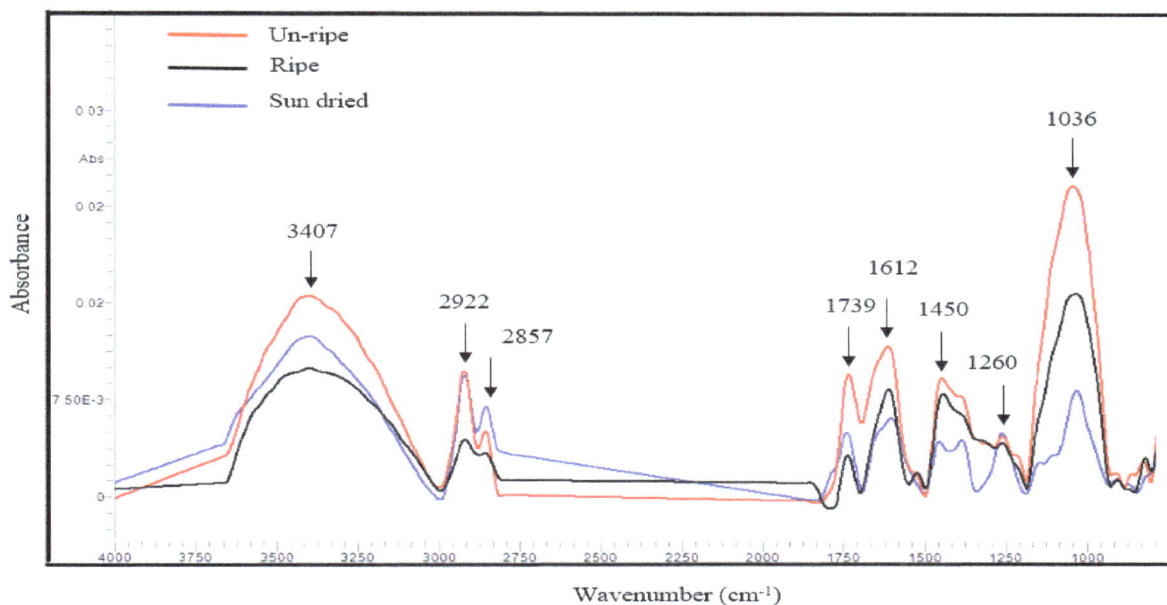

Figure 1 Typical FTIR graph for un-ripe, ripe and sun-dried grape vine after baseline correction and normalization scale

Table 2 Chemical composition of un-ripe, ripe and sun-dried (raisin) fruits Sultana grapevine

Items	Grapevine product[1]			SEM	P-value
	Un-ripe (n=12)	Ripe (n=12)	Sun-dried (n=12)		
Basic chemical composition (% of dry matter)					
Dry matter	13.45[c]	18.16[b]	84.95[a]	0.401	< 0.01
Crude protein	7.65[a]	4.94[b]	3.72[c]	0.188	< 0.01
Ether extract	2.60	2.02	1.92	0.622	0.71
Acid detergent fiber	30.23[a]	33.64[a]	3.75[b]	1.72	0.01
Aacid detergent lignin	29.92[a]	32.74[a]	2.3[b]	1.84	< 0.01
Neutral detergent fiber	45.75[b]	59.43[a]	4.8[c]	2.05	< 0.01
Non-fiber carbohydrates	37.02[b]	26.01[b]	85.02[a]	2.36	< 0.01
Total carbohydrates	82.77[b]	85.44[b]	89.82[a]	0.757	0.03
Nitrogen to carbohydrate (CHO) and organic matter ratio (g/kg)					
N:carbohydrates	14.78[a]	9.26[b]	6.62[c]	0.41	< 0.01
N:organic matter	13.15[a]	8.56[b]	6.24[c]	0.399	< 0.01
Phenolic compounds (g/100 g of dry matter)					
Total phenol	6.51[a]	2.85[b]	0.87[c]	0.414	0.02
Total tannin	5.33[a]	2.34[b]	0.57[c]	0.232	< 0.01

[1] n= 12; (4 vineyards×3 rows within vineyard).
The means within the same row with at least one common letter, do not have significant difference (P>0.05).
SEM: standard error of the means.

Un-ripe grapevine tended to have higher peak height related to structural carbohydrate compared with raisin (P=0.15).

FTIR molecular ratios were similar among grapevine products, except for CH_2 and CH_3 symmetric and antisymmetric functional groups related to lipids to phenolic compounds and tannins ratio, which tended to be higher for raisin compared with un-ripe and ripe grapevine (P=0.07) (Table 4).

Mycotoxines and heavy metals
Concentration of aflatoxin B1, B2, G1 and G2 and Ocratoxin A were lower in un-ripe and ripe grapevine than in raisins (P<0.01) (Table 5).

All three grapevine product had similar cadmium concentrations, while lead concentration was lower in un-ripe grapevine and raisin than in ripe grapevine (P<0.05).

Chemical composition
Un-ripe and ripe grapevine and raisin are mostly used for human consumption, however, they may be used for animal feeding if fruit is physically damaged, contaminated with mycotoxins or other pollutant, or if the price is low. The chemical composition of 'Sultana' grapevine differed greatly among un-ripe, ripe and sun dried grapevine in this study.

Table 3 Kinetics of ruminal fermentation of un-ripe, ripe and sun-dried (raisin) fruit of sultana grapevine measured by the *in vitro* gas production technique

Items	Grapevine treatment[1]			SEM	P-value
	Un-ripe (n=12)	Ripe (n=12)	Sun-dried (n=12)		
Kinetics of gas production[1]					
b (mg/0.2 g DM)	230.80	245.25	256.30	7.668	0.26
c (/h)	0.1591	0.1409	0.1040	0.0088	0.09
gas 24 (mg/0.2 g DM)	39.73[c]	44.60[b]	47.28[a]	0.780	< 0.01
Potential of nutrient supply in ruminants					
OMD (% DM)	58.19	61.70	61.56	0.754	0.06
NE_L (MJ/kg DM)	4.71	5.02	5.23	0.090	0.10

[1] n= 12; (4 vineyards×3 rows within vineyard).
b: total potential gas production and c: rate of gas production.
DM: dry matter; OMD: organic matter digestibility and NE_L: net energy for lactation.
The means within the same row with at least one common letter, do not have significant difference (P>0.05).
SEM: standard error of the means.

Table 4 Fourier Transform Infrared Spectroscopic (FTIR) spectroscopy molecular structures in un-ripe, ripe and sun-dried (raisin) fruit of sultana grapevine[1]

Items		Grapevine products[2]			SEM	P-value
		Unripe (n=12)	Ripe (n=12)	Sun-dried (n=12)		
FTIR molecular structures						
Total carbohydrates 920 to 1188 cm[-1]	Height	0.021	0.013	0.015	0.0037	0.48
	Area	2.32	2.97	1.80	0.590	0.50
Structural carbohydrates 1188 to 1496 cm[-1]	Height	0.0097	0.010	0.0075	0.00102	0.35
	Area	2.03	1.38	1.17	0.209	0.15
Amides 1546 to 1703 cm[-1]	Height	0.01	0.01	0.01	0.0001	0.95
	Area	0.981	0.976	0.855	0.1772	0.89
Aromatic combination bands 1700 to 1830	Height	0.012	0.007	0.010	0.0022	0.47
	Area	0.513	0.426	0.450	0.1191	0.90
CH_2 and CH_3 symmetric and anti-symmetric 2790 to 3000	Height	0.012	0.008	0.010	0.001541	0.42
	Area	1.261[a]	0.630[b]	1.165[a]	0.1232	0.05
O-H stretching and C-H stretching vibrations (phenolic compounds and tannins)	Height	0.0195	0.0133	0.0150	0.00312	0.51
	Area	8.394	4.713	6.725	0.9945	0.15
Ratio between FTIR molecular structures						
Amides: total carbohydrate		0.505	0.833	0.750	0.1634	0.49
Amides: structural carbohydrate		1.03	1.00	1.50	0.2044	0.35
Structural carbohydrate: total carbohydrate		0.489	0.833	0.500	0.1253	0.18
Lipids: phenolic compounds		0.1495	0.1266	0.1798	0.01173	0.07
Lipids: structural carbohydrate		1.1228	0.7000	1.5000	0.2534	0.20
Lipids: total carbohydrate		0.566	0.700	0.750	0.1833	0.80
Lipids: amides		1.100	0.700	1.000	0.1414	0.25

[1] Lipids, CH_2 and CH_3 symmetric and anti-symmetric 2790 to 3000 cm[-1].
[2] n= 12; (4 vineyards×3 rows within vineyard).
The means within the same row with at least one common letter, do not have significant difference (P>0.05).
SEM: standard error of the means.

The CP concentration decreased with increasing berry ripening, which was consistent with Vivin *et al.* (2003) who suggested that N containing compounds were diluted by the increased concentration of soluble carbohydrates with advancing berry ripening.

Further, with advancing berry ripening, concentration of N commences to decline while it increases in magnitude in rachises with their lateral branches, peduncles and shoots. This translocation of N occurs to replenish the N pool of woody parts of grapevine for the re-growth of next year (Wermelinger, 1991).

The CP of grapevine products in the current study were in a similar range as found for raisin industry by-products (Yari *et al.* 2015a), but more than 50% lower than in wine industry rest-products (Alipour and Rouzbehan, 2007; Spanghero *et al.* 2009).

Changes in FTIR molecular structures related to proteins (amid I+amid II) were previously found to be related to digestibility of proteins in ruminants (Damiran and Yu, 2011; Yari *et al.* 2013).

The FTIR proteins molecular structures were similar among grapevine products, which suggests that the type of protein and digestibility would be similar for the three grapevine products.

This is different than for alfalfa hay where FTIR proteins molecular structures changed with advancing maturity of the forage harvested (Yari et al. 2013). The N:CHO and N:OM ratios declined from the un-ripe to ripe and sun-dried grapevine stages as a result of reduction in N and increase in both non-structural and total CHO, and these ratios were in a similar range as found for raisin industry by-products (Yari et al. 2015a).

Fibre fractions (NDF and ADF) were lower and NFC higher in raisin compared with un-ripe and ripe grapevine. Fibre fractions and NFC in un-ripe and ripe grapevine were in a similar range as for wine industry rest-products (Alipour and Rouzbehan, 2007; Spanghero et al. 2009), while NDF and ADF were lower and NFC higher in raisin of the current study compared with values for raisin industry by-products (Yari et al. 2015a).

Plant secondary metabolites

Phenolics like coumaric, caffeic, ferulic and vanillic acids are plant secondary compounds that are relatively simple, while others have more complex polymeric structures such as condensed tannins, which strongly contribute to the mouth feel, antioxidant activity, diet digestibility, ruminal N availability and anti-parasitic activity of grape products fed to animals (Corrales et al. 2008; Waghorn, 2008). Total phenolics and total tannins by wet chemistry and FTIR molecular structures related to phenolic compounds and tannins reduced with maturation of grapevine, and with sun drying to generate raisins, in the current study, which was consistent with Vivin et al. (2003). Total phenolics and total tannins concentration in un-ripe berries were similar to values previously reported for raisin by-products (Besharati and Taghizadeh, 2009; Yari et al. 2015a) and winery rest products (Alipour and Rouzbehan, 2007; Spanghero et al. 2009), while these were higher than in ripe berries and raisins in this study. Small berries have higher tannin concentration than large berries because the skin constitutes (fraction which contains most of the tannins) a larger percentage of the berry mass (Wermelinger, 1991), which might explain the higher tannin concentration in un-ripe (i.e. smaller berries) grapevines.

Gas production kinetics and predicted nutrient supply in ruminants

Total gas production during in vitro incubation provides an indication of the fermentablity, digestibility and energy value of a diet for ruminants (Menke and Steinglass, 1988).

Asymptotic gas production was numerically highest for raisin (Table 3 and Figure 2) among grapevine by-products and rate of gas production tended to be higher for un-ripe and ripe grapevine than for raisin. These values were in a similar range as previously found for raisin by-products (Yari et al. 2015b).

The higher gas production in raisin (Figure 2) at different times of incubation and at 24 h may be due to much higher NFC, which is more fermentable than NDF (Menke and Steinglass 1988; Tavendale et al. 2005; Tefera et al. 2008), than in un-ripe and ripe grapevine. The lower 24 h gas production of un-ripe fruit compared to ripe fruit might have resulted from their higher total phenolic and tannin concentration, which can impair fermentation and gas production. In vitro gas production during different times of incubation (Figure 2) where similar for ripe and sun-dried grapevine, while both were higher than in un-ripe grapevine. The FTIR spectroscopy features of ripe and sun-dried grapevine, also, were similar in terms of most FTIR functional groups while being more different compared with unripe grapevine (Figure 1).

Energy values (OMD and NE$_L$) grapevine products in the current study were higher than of raisin by-products (Yari et al. 2015b) and winery rest products (Alipour and Rouzbehan, 2007; Spanghero et al. 2009). The energy content of un-ripe grapevine was, however, lower than in ripe grapevine and raisin, which might be due to the higher NDF and total phenolics and tannins concentrations (which all in general reduce feed digestibility and therefore energy value) in un-ripe grapevine. Energy values of the grapevine products in the current study were in a similar range as for alfalfa hay and barley silage (NRC, 2001), which are forages that are commonly fed to ruminants in Iran. Therefore, grapevine products tested in the current study can likely replace part of alfalfa hay or barley silage in the diet of ruminants.

Mycotoxines and heavy metals

Many fungi can grow and infect berries in the vineyard depending on the weather conditions (Pitt, 1993). Fungi most commonly grow on berries when the moisture of air is high, when the ambient temperature ranges between 20 to 30 °C (Serra et al. 2006) and during maturation of fruit. Temperature and moisture conditions during growth of grapevine in the current study were not ideal for common fungi growth, which might explain why mycotoxines were not detected in un-ripe and ripe grapevine. This is different to Serra et al. (2006) who detected mycotoxin ochratoxin A (OTA) concentration in pea berries and OTA were found to differ among different grape cultivars and with advancing maturation stage.

Table 5 Mycotoxines and heavy metal concentrations in un-ripe, ripe and sun dried (raisin) fruits of sultana grapevine

Items	Treatments			SEM	P-value
	Un-ripe (n=12)	Ripe (n=12)	Sun-dried (n=12)		
Mycotoxines (mg/kg of dry matter; ppm)					
Afl-B1	ND	ND	1.775	0.00433	-
Afl-B2	ND	ND	0.2355	0.00087	-
Afl-G1	ND	ND	1.775	0.00433	-
Afl-G2	ND	ND	0.2355	0.00087	-
OTA	ND	ND	1.775	0.00433	-
Heavy metals (mg/kg of dry matter)					
Cadmium	0.01863	0.01347	0.001469	0.01133	0.48
Lead	0.2054[b]	0.5966[a]	0.08548[b]	0.06817	0.0001

[1] n= 12; (4 vineyards×3 rows within vineyard).
Afl: aflatoxine; OTA: Ochratoxin A and ND: not detected.
The means within the same row with at least one common letter, do not have significant difference (P>0.05).
SEM: standard error of the means.

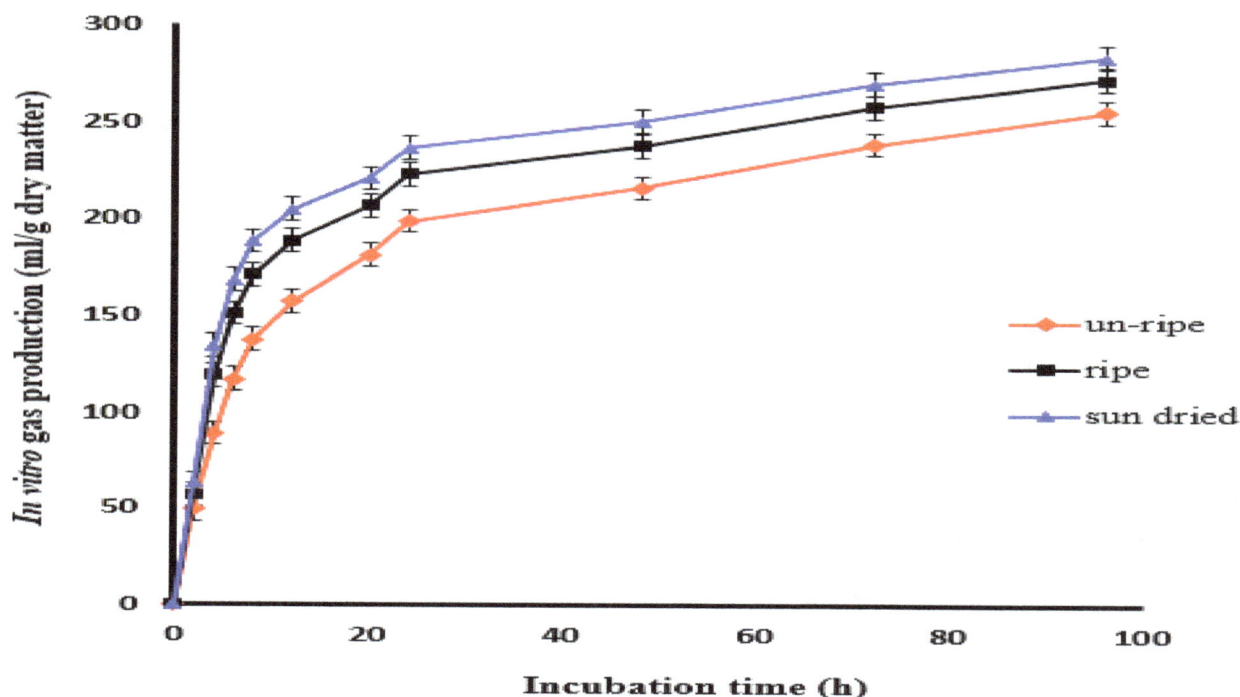

Figure 2 *In vitro* gas production pattern of un-ripe and ripe fruits and sun-dried (raisin) of Sultana grapevine

Accessibility of fruit nutrient due to physical damage of berries during harvest might provide good conditions for fungal growth, which might explain why higher mycotoxines concentrations were found on raisins after sun-drying. Aspergillus and Penicillium species are commonly found on dried fruits, including grape products (Palumbo *et al.* 2011). Raisin samples in the current study had an average OTA concentration of 1.875 ppb, which was lower than average OTA level of over 2 ppb and maximum value up to 100 ppb reported for worldwide dried vine fruits (e.g., sultanas, raisins) (Palumbo *et al.* 2011) and lower than maximum tolerable levels of mycotoxins in barley, maze and legumes for ruminants [respectively, 10, 5 and 5 ppb for aflatoxin B1; 10 ppb for the sum of aflatoxin B1, B2, G1 and G2; 50, 30 and 20 ppb for OTA (ISIRI 2002 confirmed by FAO, 2003; FDA, 2011)]. Mycotoxins are considered less toxic in ruminants than in simple stomach animals because microorganisms in the rumen can partially detoxify mycotoxins.

Contamination of food with toxic heavy metals, such as cadmium, lead and mercury, can occur through for example rainfall, irrigation water, soil, traffic density, atmospheric dust etc., which can be absorbed through the leaf blades (Zaidi *et al.* 2005; Sobukola *et al.* 2010). The concentration of toxic heavy metals differs among plants species, which mainly depends on soil conditions and the ability of plants to selectively accumulate some metals (Divrikli *et al.* 2006). Levels of lead in the current grapevine samples was in similar range as previously reported for fruits and vegetables bought on the open market (Radwan and Salama, 2006), while cadmium levels were in general higher in samples of the current study than found by Radwan and Salama (2006), but still far below the maximum allowable cadmium level of 1 mg/kg set by the Joint FAO/WHO Expert Committee on Food Additives (FAO/WHO, 1999).

CONCLUSION

Crude protein, non-fibre carbohydrates, structural carbohydrate, lignin, ash, total phenolics, total tannin content and the FTIR spectroscopic molecular structures such as phenolic compounds and tannins, structural carbohydrates and lipids (CH_2 and CH_3 symmetric and antisymmetric functional groups) differed among un-ripe, ripe and sun-dried (raisin) 'Sultana' grapevine. Mycotoxines contamination was detected only in raisin samples, which were lower than hazardous levels set for ruminant feeding. Ripe and sun-dried grapevine (raisins) had higher predicted nutritive value than un-ripe grapevine.

ACKNOWLEDGEMENT

The authors gratefully thank the Iranian Grape and Raising Institute, Malayer University for financial support of this project. The authors also would like to thank Jahad Keshavarzi Malayer for helping to collect samples.

REFERENCES

Alipour D. and Rouzbehan Y. (2007). Effects of ensiling grape pomace and addition of polyethylene glycol on *in vitro* gas production and microbial biomass yield. *Anim. Feed. Sci. Technol.* **137,** 138-149.

AOAC. (1995). Official Methods of Analysis. Vol. I. 15th Ed. Association of Official Analytical Chemists, Arlington, VA, USA.

Besharati M. and Taghizadeh A. (2009). Evaluation of dried grape by-product as a tanniniferous tropical feedstuff. *Anim. Feed. Sci. Technol.* **152,** 198-203.

Boss P.K. and Davies C. (2001). Molecular biology of sugar and anthocyanin accumulation in grape berries. Pp. 1-33 in Molecular Biology and Biotechnology of the Grapevine. K.A. Roubelakis-Angelakis, Eds. Kluwer Academic Publishers, Dordrecht, The Netherlands.

Corrales M., Toepf S., Butz P., Knorr D. and Tauscher B. (2008). Extraction of anthocyanins from grape by-products assisted by ultrasonics, high hydrostatic pressure or pulsed electric fields: a comparison. *Innov. Food Sci. Emerg. Technol.* **9,** 85-91.

Damiran D. and Yu P. (2011). Molecular basis of structural makeup of hulless barley in relation to rumen degradation ki-

netics and intestinal availability in dairy cattle: a novel approach. *J. Dairy Sci.* **94,** 5151-5159.

Divrikli U., Horzum N., Soylak M. and Elci L. (2006). Trace heavy metal contents of some spices and herbal plants from western Anatolia, Turkey. *Int. J. Food Sci. Technol.* **41,** 712-716.

Fernaandez K. and Agosin E. (2007). Quantitative Analysis of Red Wine Tannins Using Fourier-Transform Mid-Infrared Spectrometry. *J. Agric. Food Chem.* **55,** 7294-7300.

FAO. (2009). Food and Agriculture Organization of the United Nations (FAO), Rome, Italy.

FAO. (2003). Food and Agriculture Organization of the United Nations (FAO), Rome, Italy.

FAO/WHO. (1999). Expert Committee on Food Additives. Pp. 1-10 in Proc. 53rd Meet. Summ. Concl. Rome, Italy.

FDA. (2011). National Grain and Feed Association. Regulatory Guidance for Mycotoxins. Available at: https://www.ngfa.org/wp-content/uploads.

ICAC. (1995). Iranian Council of Animal Care. Guide to the Care and Use of Experimental Animals. Isfahan University of Technology, Isfahan, Iran.

ISIRI. (2002). Institute of Standards and Industrial Research of Iran. Food and Feed-Mycotoxin-Maximum Tolerated Level. ISIRI No. 5925. Karaj, Iran.

ISIRI. (2012a). Institute of Standards and Industrial Research of Iran. Food and Feed Stuffs: Determination of Aflatoxins B and G by HPLC Method Using Immunoaffinitycolumn Clean up-Test Method. ISIRI No. 6872. Karaj, Iran.

ISIRI. (2012b). Institute of Standards and Industrial Research of Iran. Foodstuffs-Cereal and Cereals Products: Determination of Ochratoxin A by HPLC Method and Immunoaffinity Column Clean up-Test Method. ISIRI No. 9238. Karaj, Iran.

ISIRI. (2012). Institute of Standards and Industrial Research of Iran. Food and Feed-Maximum Limit of Heavy Metals. ISIRI No. 12968. Karaj, Iran.

Karimi R. and Ershadi A. (2015). Role of exogenous abscisic acid in adapting of 'Sultana' grapevine to low temperature stress. *Acta Physiol. Plan.* **37,** 151-162.

Keller M. (2010). The Science of Grapevines: Anatomy and Physiology. Academic Press, Burlington, Massachusetts.

Makkar H.P.S. (2000). Quantification of Tannins in Tree Foliage. A Laboratory Manual for the FAO/IAEA Co-Ordinated Research Project on "Use of Nuclear and Related techniques to Develop Simple Tannin Assays for Predicting and Improving the Safety and Efficiency of Feeding Ruminants on Tanniniferous Tree Foliage" Joint FAO/IAEA, FAO/IAEA of Nuclear Techniques in Food and Agriculture, Animal Production and Health Sub-Programme, FAO/IAEA Working Document, International Atomic Energy Agency (IAEA), Vienna, Austria.

Menke K.H. and Steinglass H. (1988). Estimation of the energetic feed value obtained from chemical analysis and *in vitro* gas production using rumen fluid. *Anim. Res. Devel.* **28,** 47-55.

Musingarabwi D.M. (2015). Characterisation of grapevine berry samples with infrared spectroscopy methods and multivariate data analyses tools. MS Thesis. Stellenbosch Univ., Stellenbosch, South Africa.

NRC. (2001). Nutrient Requirements of Dairy Cattle. 7th Ed. Na tional Academy Press, Washington, DC, USA.

Ørskov E.R. and McDonald I. (1979). The estimation of protein degradability in the rumen from incubation measurements weighted according to rate of passage. *J. Agric. Sci.* **92**, 499-503.

Pala M., Saygi Y.B. and Sadikoglu H.A. (1993). Study on the drying of sultana grapes by different techniques and effective parameters. Pp. 437-444 in In Developments in Food Science 32: Food Flavors, Ingredients and Composition. G. Charalambous, ed. Elsevier Science Publishers, Amsterdam, the Netherlands.

Palumbo J.D., O'Keeffe T.L., Vasquez S.J. and Mahoney N.E. (2011). Isolation and identification of ochratoxin A-producing *Aspergillus* section Nigri strains from California raisins. *Lett. Appl. Microbiol.* **52**, 330-336.

Pitt R.E. (1993). A descriptive model of mold growth and aflatoxin formation as affected by environmental conditions. *J Food Prot.* **2**, 139-146.

Radwan M.A. and Salama A.K. (2006). Market basket survey for some heavy metals in Egyptian fruits and vegetables. *Food Chem. Toxicol.* **44**, 1273-1278.

Rogerio M., Mauricio F.L., Moulda M.S., Dhanoa E.O., Kulwant S. and Michael K.T. (1999). A semi-automated *in vitro* gas production technique for ruminant feedstuff evaluation. *Anim. Feed Sci. Technol.* **79**, 321-330.

Romer T.R., Boling T.M. and Macdonald J.L. (1978). Gas liquid chromatographic determination of T-2 toxin and diacetoxy-scirpenol in corn and mixed feeds. *J. Assoc. Off. Anal. Chem.* **61**, 801-807.

SAS Institute. (2003). SAS®/STAT Software, Release 9.2. SAS Institute, Inc., Cary, NC. USA.

Serra R., Mendonça C. and Venâncio A. (2006). Ochratoxin A occurrence and formation in Portuguese wine grapes at various stages of maturation. *Int. J. Food Microbiol.* **111**, 35-39.

Singleton V.L. and Rossi J.A. (1965). Colorimetry of total phenolics with phosphomolybdic phosphotungstic acid reagents. *Am. J. Enol. Viticcult.* **16**, 144-158.

Sobukola O.P., Adeniran O.M., Odedairo A.A. and Kajihausa O.E. (2010). Heavy metal levels of some fruits and leafy vegetables from selected markets in Lagos, Nigeria. *African J. Food Sci.* **4**, 389-393.

Spanghero M., Salem A.Z.M. and Robinson P.H. (2009). Chemical composition, including secondary metabolites, and rumen fermentability of seeds and pulp of Californian (USA) and Italian grape pomaces. *Anim. Feed Sci. Technol.* **152**, 243-255.

Tavendale M.H., Meagher L.P., Pacheco D., Walker N., Attwood G.T. and Sivakumaran S. (2005). Methane production from *in vitro* rumen incubations with *Lotus pedunculatus* and *Medicagosativa*, and effects of extractable condensed tannin fractions on methanogenesis. *Anim. Feed Sci. Technol.* **123**, 403-419.

Tefera S., Mlamboa V., Dlamini B.J., Dlamini A.M., Koralagama K.D.N. and Mould F.L. (2008). Chemical composition and *in vitro* ruminal fermentation of common tree forages in the semi-arid rangelands of Swaziland. *Anim. Feed Sci. Technol.* **142**, 99-110.

Theodorou M.K., Williams B.A., Dhanoa M.S., McAllan A.B. and France J.A. (1994). Simple gas production method using a pressure transducer to determine the fermentation kinetics of ruminant feeds. *Anim. Feed Sci. Technol.* **48**, 185-197.

Van Soest P.J., Robertson J.B. and Lewis B.A. (1991). Methods for dietary fiber, neutral detergent fiber, and non starch polysaccharides in relation to animal nutrition. *J. Dairy Sci.* **74**, 3583-3597.

Vivin P., Castelan-Estrada M. and Gaudillere J.P. (2003). Seasonal changes in chemical composition and construction costs of grape vine tissues. *Vitis.* **42**, 5-12.

Waghorn G. (2008). Beneficial and detrimental effects of dietary condensed tannins for sustainable sheep and goat production-Progress and challenges. *Anim. Feed Sci. Technol.* **147**, 116-139.

Wermelinger B. (1991). Nitrogen dynamics in grapevine: Physiology and modelling. Pp. 23-31 in Proc. Int. Symp. Nitr. Grap. Wine. American Society for Enology and Viticulture, Davis, USA.

Wetzel D.L., Eilert A.J., Pietrzak L.N., Miller S.S. and Sweat J.A. (1998). Ultraspatially-resolved synchrotron infrared microspectroscopy of plant tissue *in situ*. *Cell. Mol. Boil.* **44**, 145-168.

Yari M., Valizadeh R., Naserian A.A., Jonker A. and P. Yu. (2013). Protein molecular structures in alfalfa hay cut at three stages of maturity and in the afternoon and morning and relationship with nutrient availability in ruminants. *J. Sci. Food Agric.* **93**, 3072-3080.

Yari M., Manafi M., Hedayati M., Khalaji S., Mojtahedi M., Valizadeh R. and Hosseini Ghaffari M. (2015a). Nutritive value of several raisin by-products for ruminants evaluated by chemical analysis and *in situ* ruminal degradability. *Res. Opin. Anim. Vet. Sci.* **5**, 198-204.

Yari M., Manafi M., Hedayati M., Khalaji S., Valinejad S., Valizadeh R. and Hosseini-Ghaffari A. (2015b). Prediction of energy contents and potential nutrient supply of raisin by-products for ruminants using National Research Council feeding system and *in vitro* gas production method. *Res. Opin. Anim. Vet. Sci.* **5**, 284-289.

Yari M., Valizadeh R., Naserian A.A., Jonker A., Azarfar A. and Yu P. (2014). Effects of including alfalfa hay cut in the afternoon or morning at three stages of maturity in high concentrations on dairy cows performance, diet digestibility and feeding behaviour. *Anim. Feed Sci. Technol.* **192**, 62-72.

Yu P., McKinnon J.J., Christensen C.R. and Christensen D.A. (2004). Imaging molecular chemistry of Pioneer corn. *J. Agric. Food Chem.* **52**, 7345-7352.

Zaidi M.I., Asrar A., Mansoor A. and Farooqui M.A. (2005). The heavy metal concentrations along roadsides trees of Quetta and its effects on public health. *J. Appl. Sci.* **5**, 708-711.

19

In Silico Prediction of B-Cell and T-Cell Epitopes of Protective Antigen of *Bacillus anthracis* in Development of Vaccines Against Anthrax

M. Tahmoorespur[1*], N. Nazifi[1] and Z. Pirkhezranian[1]

[1] Department of Animal Science, Faculty of Agriculture, Ferdowsi University of Mashhad, Mashhad, Iran

*Correspondence E-mail: tahmoores@um.ac.ir

ABSTRACT

Protective antigen (PA), a subunit of anthrax toxin from *Bacillus anthracis*, is known as a dominant component in subunit vaccines in protection against anthrax. In order to avoid the side effects of live attenuated and killed organisms, the use of linear neutralizing epitopes of PA is recommended in order to design recombinant vaccines. The present study is aimed at determining the dominant epitopes based on multi-parameter and multi-method analysis. The epitopes were identified by the well-known online bioinformatics server and then they were selected and compared based on the highest score and the highest repetition rate. Further analysis on predicted epitopes has been carried out by online VaxiJen 2.0 and Protein Digest server. Among the selected epitopes, those with the highest antigenicity score (>0.9 threshold) and less susceptibility to gastrointestinal tract proteases, were selected as final epitopes. Final B-cell predicted epitopes were amino acid residues 292-308, 507-521 and 706-719; residues 17-31, 315-329 and 385-400 which were determined as the best major histocompatibility complex I (MHCI) class of T-cells epitopes; in addition, residues 455-464 and 661-669 were also considered the best MCHII class of T-cells epitopes. Since random coil structure had a high probability of protein forming of antigenic epitope, the results of secondary structure analysis of the final PA epitopes have shown that all these epitopes form a 100% random coil structure.

KEY WORDS *anthracis*, epitope prediction, protective antigen (PA).

INTRODUCTION

Anthrax is known as an epizootic and zoonotic disease in domestics which could spread through spore transmission via ingestion, inhalation or an open skin wound; it could also affect the humans who are in contact with the infected animals and their contaminated products (Leppla *et al.* 2002; Inglesby *et al.* 2002). *Bacillus anthracis* is a gram-positive, facultatively anaerobic and rod-shape pathogen with two different plasmids named pXO1 and pXO2 (Brey, 2005). PXO1 plasmid encodes toxin factors including protective antigen (PA) and two other catalytically active components; lethal factor (LF) and edema factor (EF). Exotox-

ins production was mediated by binary combinations of these three regions (Stanley and Smith, 1961). Using proteolytic cleavage of PA into a 20-kDa amino-terminal fragment and a 63-kDa polypeptide via furin (Gordon *et al.* 1995) along with the formation of heptameric oligomers, *B. anthracis* can bind with the cellular receptors via PA which later translocate LF and EF into the cytosol with enzymatic activity. With the use of zinc ion, LF inactivates mitogen-activated protein kinases (MAPKKs) which cause toxic shock and death (Vitale *et al.* 1998). EF factor, through high converting intracellular ATP into cAMP as an adenylate cyclase, stimulate rate of the intracellular cAMP levels and finally leading to edema (Leppla, 1982). PXO2's encoded capsules enhance virulence in vivo by inhibiting phagocytosis of the organism (Little and Ivins, 1999). It has been proved that a truncated recombinant of PA could stimulate a protective immune response to anthrax (Abboud and Casadevall, 2008; Flick-Smith et al. 2002). In animal studies it has been demonstrated that protective immunity against anthrax is associated with the induction of neutralizing anti-PA antibodies (Farchaus et al. 1998; Little et al. 1997; Pitt et al. 2001). In recent researches, the purified recombinant PA (rPA) has been reported as an advanced anthrax vaccine. Therefore, PA with its four distinct domains could be considered the best choice for epitope prediction. Epitopes are specific sites of antigens as antigenic determinant which are classified into two major gropes; B-cell (continuous and discontinuous) and T-cell (major histocompatibility complex I (MHCI) and major histocompatibility complex II MHCII)) (Zhang et al. 2012). B- and T-cell epitopes of antigens can be identified and predicted using computational tools in order to design recombinant vaccines which are important in stimulation of antibodies. These predictor tools are as in-silico environment which are advantageous since they are inexpensive and noninfectious in vaccine designing, whereas viruses or bacteria could be harmful during experimental process. In contrast with the experimental methods which are costly and time-consuming, these tools are cheap and available (Ponomarenko and Van, 2009). The present study is aimed to identify B-cell and T-cell epitopes of PA antigen in vaccine designing in order to counter against anthrax using molecular biology software which could reveal the dominant epitopes of the protective antigen of B. anthracis.

MATERIALS AND METHODS

Amino acid sequence of the protective antigen protein of B. anthracis

Amino acid sequence of protective antigen protein of B. anthracis (Accession number: CAL49462) was obtained from GeneBank (http://www.ncbi.nih.gov/genbank/). PA protein is composed of 735 amino acid residues.

Prediction of the secondary and tertiary structure of the PA antigen

Different conformational states (helices, sheets, turns and coils) of PA antigen of B. anthracis protein were analyzed to predict the secondary structures using the improved self-optimized prediction method (SOPMA) software (http://npsa pbil.ibcp.fr/cgi bin/npsa_automat.pl?page=/NPSA/npsa_sopma.html) (Geourjon and Deléage, 1995).

The next step of augury of tertiary structure of PA antigen was accomplished using 3DLigandSite ligand banding sit prediction Server (http://www.sbg.bio.ic.ac.uk) (Wass et al. 2010).

Servers and software used for epitope prediction

Using in-silico softwares which have been listed in Table 1, the process of epitope prediction of PA antigen was carried out using antigen primary sequence. These softwares are designed for B-cell or T-cell epitopes and they use liner sequences of amino acids to determine antigenicity of hot spot regions, known as epitope region, and also to report them with different antigencity scores.

In T-cell prediction softwares, parameters of each server were adjusted as follows; 'MHC alleles (A-0101, A0201 and B-2705 alleles for MHCI class and DRB1-0101, DRB1-0401 and DRB1-0401 alleles for MHCII class)' and 'desired length' of related epitopes. The remaining parameters did not change.

Characterization of epitopes

The final predicted epitopes of B and T-cell were evaluated using VaxiJen 2.0 server, an alignment-independent prediction of protective antigens. VaxiJen server classifies antigens according to physicochemical properties of proteins without having recourse to sequence alignment (http://www.ddgApharmfac.net/vaxijen/VaxiJen/VaxiJen.html).

The study on enzymatic digestion of final predicted epitopes PA protein has been done using (http://db.systemsbiology.net:8080/proteomicsToolkit/proteinDigest.html) server. Mass (Da) and point of isoelectric (pI) of each predicted epitopes were determined using this Protein Digest server.

Prediction of the secondary structure of the PA protein

The secondary structure of PA protein encompasses four conformational states; helices, sheets, turns and coils, which were analyzed by the improved self-optimized prediction method (SOPMA) software (https://npsa-prabi.ibcp.fr/cgibin/npsa_automat.pl?page=/NPSA/npsa_sopma.html) (Geourjon and Deléage, 1995).

Required parameters for prediction of the secondary structure such as threshold and window width were set to 8 and 17, respectively.

Tertiary structure was conducted using iterative threading ASSEmbly refinement (I-TASSER) site (http://zhanglab.ccmb.med.umich.edu/I-TASSER) which is a hierarchical approach to protein structure and function prediction.

And finally using PyMOLV1 Viewer software, the primary structure of the studied template from protein database (PDB) format was analyzed and viewed.

RESULTS AND DISCUSSION

Prediction of the secondary structure of PA protein

SOPMA software was used in order to identify some details of antigenic property of PA protein in its secondary structure. As it is shown in Figure 1, the number of extended strand, random coil and alpha helix were the most dominant region in PA protein (26.67, 35.65 and 26.26 respectively).

The existence of the extended strands and random coils in protein is the leading cause of the probability of protein formation as an antigenic epitope.

Prediction of B-cell and T-cell epitopes for PA antigen

B-cell and T-cell epitopes' prediction of PA antigen has been done using online software listed in Table 1. The predicted epitopes were selected based on the highest score and the highest repetition rate from all the softwares' outputs (data shown in supplementary file).

Pre-final B-cell epitopes which had the most conserved sequences among all proposed epitopes are being listed in Table 3. It is worth mentioning that the software utilizes different scoring systems. Using different online software the predicted epitopes of MHCI (A-0101, A0201 and B-2705 alleles) and MHCII (DRB1-0101, DRB1-0401 and DRB1-0401 alleles), as two separate classes of T-cell, were listed in Table 3. The high-scored regions which had a high potential, were selected as epitope region compared to all the other utilized softwares. The pre-final selection among T-cell epitopes was based on some sequences of epitopes which were presented in all MHCI as well as MHCII class alleles.

Antigenicity and characterization of Protein Digest selected epitopes

Further analysis to assign the best epitopes for PA antigen continued using the final results of the previously mentioned online software.

```
SOPMA :
    Alpha helix       (Hh) :   193 is   26.26%
    3₁₀ helix         (Gg) :     0 is    0.00%
    Pi helix          (Ii) :     0 is    0.00%
    Beta bridge       (Bb) :     0 is    0.00%
    Extended strand   (Ee) :   196 is   26.67%
    Beta turn         (Tt) :    84 is   11.43%
    Bend region       (Ss) :     0 is    0.00%
    Random coil       (Cc) :   262 is   35.65%
    Ambiguous states  (?)  :     0 is    0.00%
    Other states           :     0 is    0.00%
```

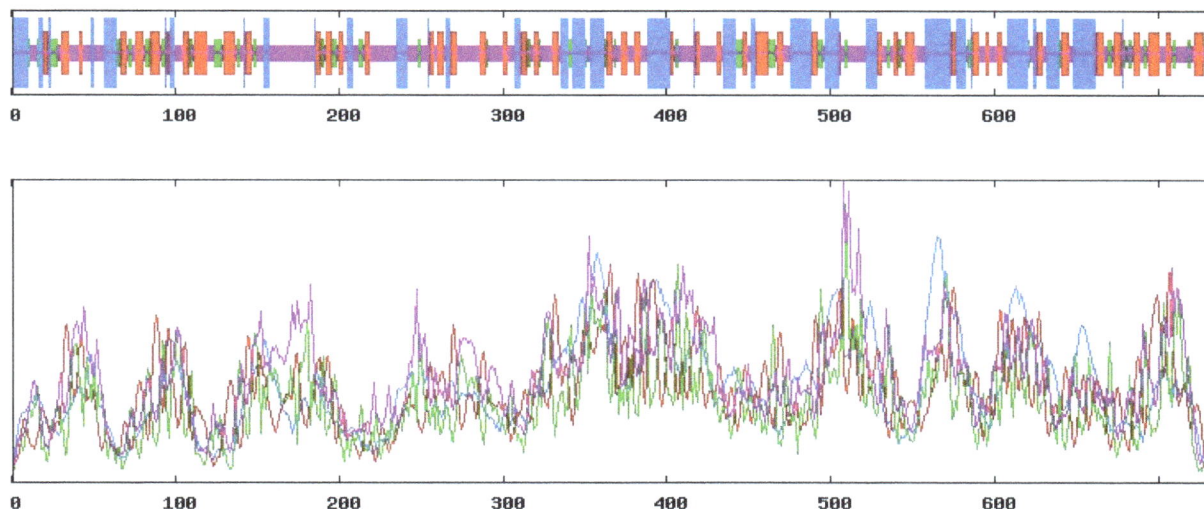

```
Parameters :
    Window width        :  17
    Similarity threshold :   8
    Number of states    :   4
```

Figure 1 Secondary structure prediction results for the PA protein. An increased number of extended strands and random coils in the protein corresponded with an increased likelihood of the protein forming an antigenic epitope

Lines in different colors represent different secondary structures: blue, α helix; green, β turn; red, extended strand; and purple, random coil

Lines in different colors represent different secondary structures: blue, α helix; green, ß turn; red, extended strand; and purple, random coil

Table 1 Bioinformatics software that used in present study

Servers	Description	Link
T-cell epitopes prediction		
IEDB	SVM and ANN-based method for prediction	http://tools.immuneepitope.org/mhci
SYFPEITH	A database of MHC ligands and peptide motifs; predictive server for MHC binding peptide	http://www.syfpeithi.de
NetMHC	ANN-based method for prediction of HLA	http://www.cbs.dtu.dk/services/NetMHC
NetCTL	Prediction of T-cell epitope	http://www.cbs.dtu.dk/services/NetCTL
PropredI	Predict MHC class I binding peptides	http://www.imtech.res.in/raghava/propred1
Propred	Predict MHC class II binding peptides	http://www.imtech.res.in/raghava/propred
B-cell epitopes prediction		
Bcepred	Physio-chemical properties of amino acids based predictive server for linear B-cell epitope	http://www.imtech.res.in/raghava/bcepred
ABCpred	ANN based predictive server	http://www.imtech.res.in/raghava/abcpred
BepiPred	Predictor of linear B-cell epitopes using a combination of a hidden Markov model and a propensity scale method	http://www.cbs.dtu.dk/services/BepiPred
BCPred	Prediction of linear B-cell epitopes using amino acid pair antigenicity scale and string kernels	http://ailab.cs.iastate.edu/bcpreds
SVMTrip	Predictor of linear B-cell epitopes using Support Vector Machine (SVM)	http://sysbio.unl.edu/SVMTriP
LEPS	Prediction of linear B-cell epitopes using Support Vector Machine classification and Amino Acid Propensity	http://leps.cs.ntou.edu.tw
IEDB	Physio-chemical properties of amino acids based predictive server for linear B-cell epitope	http://tools.immuneepitope.org/tools/bcell/iedb_input

Table 2 Training bioinformatics software that used in present study[*]

Antigen	Predicted epitopes	Experimental epitopes	Reference
GroEL[1] of *Yarsinia*	28-42, 78-92, 178-185, 275-290, **315-336**, 430-440, 526-545	316-326	Yamaguchi *et al.* (1996)
Dnak[2] of *Brucella*	40-67, 78-92, 210-227, 357-370, 523-537, **609-640**	617-637	Vizcaino *et al.* (1997)
Omp31[3] of *Brucella*	25-28, **46-73**, 122-127, 175-182	48-74	Wang *et al.* (2014) Cassaratro *et al.* (2005)
SOD[4] of *Brucella*	44-50, **70-86, 134-153, 147-165**	75-86, 136-150, 149-162	Tabatabai *et al.* (1994)

[1] Heat shock protein 60.
[2] Heat shock protein 70.
[3] Outer membrane 31.
[4] Sodium oxide dismutase.
[*] Similar epitopes between predicted epitopes using bioinformatics tools and experimental studies were specified by bold and were underline.

Antigenicity of the selected epitopes was determined along with the enzyme digestion position of each epitope using VaxiJen 2.0 and Protein Digest servers.

Epitopes mentioned in Table 3 were first filtered by the VaxiJen software with 0.5 threshold, and then scored more than 0.5 threshold and also checked for determination of their mass (Da), pI and enzymatic digestion (Table 3). Final B-cell predicted epitopes, based on the high score of antigenicity and maximum number of undigested enzymes, have been highlighted and ranged in 292-308, 507-521 and 706-719 amino acids (Table 3).

More investigations were carried out through determining the characterization of final MHCI and MHCII T-cell predicted epitopes. As it has been shown in Table 3, amino acids sequences in 17-31, 315-329 and 385-400 regions are the most suitable epitopes for MCHI class, and amino acids sequences in 455-464 and 661-669 range are the best ones for MCHII class of T-cells, according to their high VaxiJen score and the maximum number of undigested enzymes.

Prediction of the tertiary structure of PA protein

The tertiary structure of PA protein is predicted by I-TASSER server (http://zhanglab.ccmb.med.umich.edu/I-TASSER) and viewed by PyMOL Viewer software. This result demonstrated that our final predicted epitopes are exposed and located on the surface of the protein as it is shown in Figure 3; the red spheres represent the potential B-cell epitopes, the green spheres indicate stronger potential of MHCI class epitopes of T-cells, the blue spheres is related to MHCII class epitopes of T-cells and finally the white spheres represent the remainder of the protein.

Vaccine production based on live attenuated or killed organisms has long been criticized for its side effects; however, with the use of newly developed technique in fourth generation of recombinant DNA technology, vaccine production has been conferred by using some immunogenic component pathogens.

Table 3 Antigenicity ability of predicted epitopes using Vaxijen server and Protein Digestion analysis of pre-final B-and T-cell epitopes*

Final B-cell predicted epitopes				
Sequence	V.J score	Mass	pI	Undigested enzyme
$_3$KQENRLLNESESSSQG$_{19}$	1.0122	1902.99	4.79	Chymotrypsin, Cyanogen_Bromide, IodosoBenzoate, Proline_Endopept, AspN
$_{35}$VTSSTTGDLSIPSSELENIP$_{55}$	0.7880	2047.20	3.57	Trypsin, Chymotrypsin, Clostripain, IodosoBenzoate, Trypsin_K, Trypsin_R,
$_{70}$KVKKSDEYTFATSADNHVTM$_{90}$	0.9250	2272.51	6.75	Clostripain, Cyanogen_Bromide, IodosoBenzoate, Proline_Endopept, Trypsin_R
$_{99}$KASNSNKIRLEKGRLYQIKI$_{118}$	1.1331	2359.80	10.56	Cyanogen_Bromid, IodosoBenzoate, Proline_Endopept, AspN
$_{115}$QIKIQYQREDPTEKGL$_{131}$	0.9824	1946.19	6.18	Cyanogen_Bromid, IodosoBenzoate
$_{155}$ELKQKSSNSKKRRS$_{168}$	2.0535	1675.91	11.17	Chymotrypsin, Cyanogen_Bromide,IodosoBenzoate, Proline_Endopept, AspN
$_{167}$RSTSAGPTVPDRDNDG$_{187}$	0.8526	1644.68	4.43	Chymotrypsin, Cyanogen_Bromide, IodosoBenzoate, Staph_Protease, Trypsin_K, Chymotrypsin(modified)
$_{241}$GRIDKNVSPEARHPLVAAYP$_{261}$	0.6977	2190.49	8.60	Clostripain, Cyanogen_Bromide, IodosoBenzoate, Proline_Endopept, Trypsin_R
$_{274}$NEDQSTQNTDSQTRTISKNTSTSRTHTSEVHGNAE$_{308}$	1.6135	3862.91	5.48	Chymotrypsin, Cyanogen_Bromide, M IodosoBenzoate, Proline_Endopept, Chymotrypsin (modified)
$_{292}$**NTSTSRTHTSEVHGNA**$_{308}$	1.7064	1698.73	6.92	Chymotrypsin, Cyanogen_Bromide, IodosoBenzoate, Proline_Endopept, Trypsin_K, AspN, Chymotrypsin (modified),
$_{422}$NAQDDFSSTPITMNYN$_{438}$	0.6708	1817.90	3.56	Trypsin, Clostripain, IodosoBenzoate, Staph_Protease, Trypsin_K, Trypsin_R
$_{465}$ENGRVRVDTGSNWSEVLPQI$_{485}$	0.5150	2256.46	4.68	Cyanogen_Bromide, Trypsin_K,
$_{507}$**AVNPSDPLETTKPDM**$_{521}$	0.5455	1614.79	4.03	Trypsin, Chymotrypsin, Clostripain, Cyanogen_Bromide, IodosoBenzoate, Trypsin_K, Trypsin_R
$_{511}$SDPLETTKPDMTLKEALKIA$_{531}$	0.8427	2201.56	4.78	Chymotrypsin, Clostripain, IodosoBenzoate, Trypsin_R,
$_{530}$AAVNPSDPLETTKPDMTLKE$_{550}$	0.8128	2157.42	4.54	Chymotrypsin, Clostripain, IodosoBenzoate, Trypsin_K, Trypsin_R
$_{542}$YQGKDITEFDFNFDQQTSQN$_{561}$	1.4173	2425.51	3.84	Clostripain, Cyanogen_Bromide, IodosoBenzoate, Proline_Endopept, Trypsin_R,
$_{688}$YISNPNYKVNVYAVTKENTI$_{707}$	0.7403	2330.62	8.38	Clostripain, Cyanogen_Bromide, IodosoBenzoate, Trypsin_R, AspN
$_{706}$**TIINPSENGDTSTN**$_{719}$	1.2134	1462.49	3.67	Trypsin, Chymotrypsin, Clostripain, Cyanogen_Bromide, IodosoBenzoate, Trypsin_K, Trypsin_R, Chymotrypsin (modified)
$_{705}$NTIINPSENGDTSTNGIKRI$_{725}$	0.8542	2144.33	6.07	Chymotrypsin, Cyanogen_Bromide, IodosoBenzoate, Chymotrypsin (modified)

Final T-cell predicted epitopes (MHCI class)				
Sequence	V.J score	Mass	pI	Undigested enzyme
$_1$EVKQENRLLNESESSS$_{16}$	0.7739	1848.94	4.49	Chymotrypsin, Cyanogen_Bromide, IodosoBenzoate, Proline_Endopept
$_{17}$**GLLGYYFSDLNFQA**$_{31}$	0.8413	1607.78	3.80	Trypsin, Clostripain, Cyanogen_Bromide, IodosoBenzoate, Proline_Endopept, Staph_Protease, Trypsin_K, Trypsin_R
$_{111}$GRLYQIKIQY$_{121}$	1.8157	1281.52	9.70	Cyanogen_Bromide, IodosoBenzoate, Proline_Endopept, Staph_Protease, AspN
$_{123}$EDPTEKGLDFKLYWTDS$_{129}$	1.1810	2044.20	4.11	Clostripain, Cyanogen_Bromide, Trypsin_R,
$_{153}$LPELKQKSSNSKKRRS$_{168}$	2.0389	1886.18	11.17	Chymotrypsin, Cyanogen_Bromide, IodosoBenzoate, AspN,
$_{163}$SKKRRSTSAGPTVPDRD$_{179}$	0.6143	1858.04	10.90	Chymotrypsin, Cyanogen_Bromide, IodosoBenzoate, Staph_Protease, Chymotrypsin (modified)
$_{268}$NIILSKNEDQSTQNTD$_{283}$	1.0858	1819.90	4.27	Chymotrypsin, Clostripain, Cyanogen_Bromide, IodosoBenzoate, Proline_Endopept, Trypsin_R,
$_{315}$**DIGGSVSAGFSNSNS**$_{329}$	0.8265	1398.41	3.80	Trypsin, Clostripain, Cyanogen_Bromide, IodosoBenzoate, Proline_Endopept, Staph_Protease, Trypsin_K, Trypsin_R, AspN
$_{385}$**LGKNQTLATIKAKENQ**$_{400}$	0.7593	1757.02	9.70	Chymotrypsin, Clostripain, Cyanogen_Bromide, IodosoBenzoate, IodosoBenzoate, Proline_Endopept, Trypsin_R, AspN
$_{563}$NIATYNFENGRVRV$_{576}$	0.8575	1652.83	8.75	Cyanogen_Bromide, IodosoBenzoate, Proline_Endopept, Trypsin_K, AspN,
$_{637}$KILSGYIVEI$_{647}$	0.7649	1134.38	6.00	Clostripain, Cyanogen_Bromide, IodosoBenzoate, Proline_Endopept, Trypsin_R, AspN
$_{644}$VEIEDTEGLKEVINDRYD$_{684}$	0.8462	2137.28	4.08	Cyanogen_Bromide, IodosoBenzoate, Proline_Endopept
$_{687}$YISNPNYKVNVYAV$_{700}$	1.0997	1643.86	8.43	Clostripain, Cyanogen_Bromide, IodosoBenzoate, Staph_Protease, Trypsin_R, AspN
$_{691}$NPNYKVNVYAVTKENT$_{706}$	0.9906	1854.05	8.43	Clostripain, Cyanogen_Bromide, IodosoBenzoate, Trypsin_R, AspN
$_{700}$AVTKENTIINPSENGDTSTNGI$_{721}$	0.9138	2275.41	4.14	Chymotrypsin, Clostripain, Cyanogen_Bromide, IodosoBenzoate, Trypsin_R, Chymotrypsin (modified)

Final T-cell predicted epitopes (MHCII class)				
Sequence	V.J score	Mass	pI	Undigested enzyme
$_{118}$QYQREDPTEKGLDFKLYWTDSQNKKEVISSDNLQ$_{152}$	1.0003	4104.46	4.66	Cyanogen_Bromide
$_{168}$TSAGPTVPDRDNDGIPDSLEVEGYT$_{193}$	0.7173	2605.71	3.66	Cyanogen_Bromide, IodosoBenzoate, Trypsin_K,
$_{413}$APIALNAQDDFSSTPITM$_{433}$	0.5277	1892.11	3.56	Trypsin, Clostripain, Cyanogen_Bromide, IodosoBenzoate, Proline_Endopept, Staph_Protease, Trypsin_K, Trypsin_R,
$_{455}$**YGNIATYNF**$_{464}$	1.1019	1062.15	5.52	Trypsin, Clostripain, Cyanogen_Bromide, IodosoBenzoate, Proline_Endopept, Staph_Protease, Trypsin_K, Trypsin_R, AspN,
$_{533}$NEPNGNLQYQGKDITEFDFNFDQQ$_{558}$	0.9275	2861.97	3.77	Clostripain, Cyanogen_Bromide, IodosoBenzoate, Trypsin_R,
$_{640}$GYIVEIEDTEGLKEV$_{654}$	0.8338	1693.87	3.91	Clostripain, Cyanogen_Bromide, IodosoBenzoate, Proline_Endopept, Trypsin_R,
$_{661}$**MLNISSLQQ**$_{669}$	0.54001	1033.21	5.28	Trypsin, Chymotrypsin, Clostripain, IodosoBenzoate, Proline_Endopept, Staph_Protease, Trypsin_K, Trypsin_R, AspN,
$_{686}$LYISNPNYKY$_{695}$	0.9188	1274.44	8.43	Clostripain, Cyanogen_Bromide, IodosoBenzoate, Staph_Protease, Trypsin_R, AspN,
$_{701}$TKENTIINPSENGDTSTNGIKRIL$_{725}$	0.6127	2615.88	5.85	Cyanogen_Bromide, IodosoBenzoate, Chymotrypsin (modified)

* Bold and highlighted sequences related to final selected epitopes based on higher score of antigencity given by Vaxyjen and more undigested enzyme.

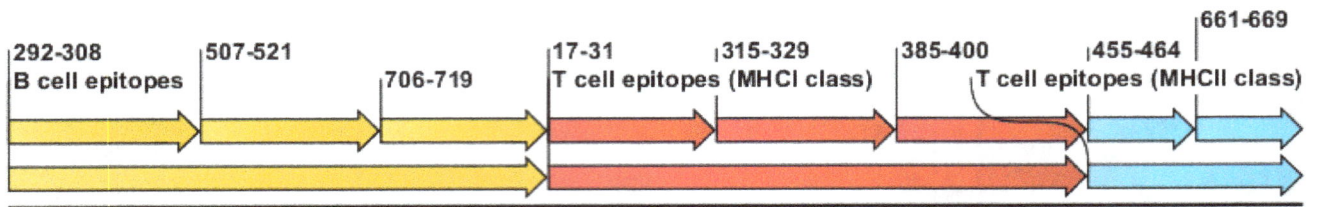

Figure 2 Final predicted epitopes
Yellow arrows represent B-cell epitopes: 292-308, 507-521 and 706-719 amino acids residual, red arrows represent T-cell (MHCI class) epitopes: 17-31, 315-329 and 385-400 amino acids residual, blue arrows represent T-cell (MHCII class) epitopes: 455-464 and 661-669 amino acids residual

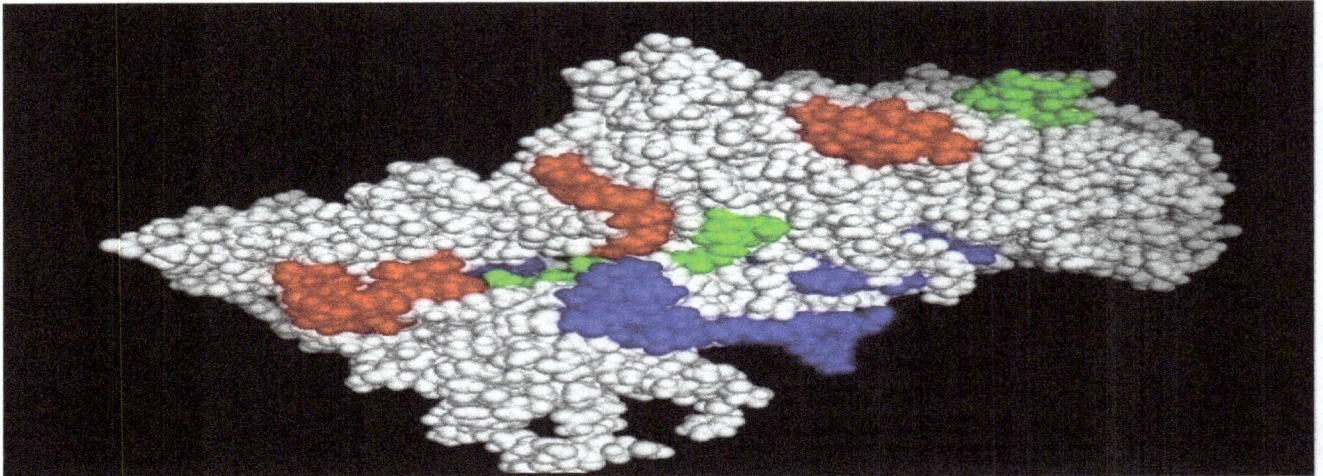

Figure 3 Tertiary structure of local situation of predicted epitopes in the PA protein
Red spheres are the B-cell epitopes, the green ones are the MHCI class of T-cell epitopes and the blue ones related to MHCII class of T-cell epitopes
White spheres are the remainder part of PA protein

Selecting a proper antigens, rational adjuvant design and a good delivery system are the most important factors in a successful vaccine designing approach (Yang *et al.* 2013). Using developed bioinformatics tools in epitopes prediction which are based on multi-parameter and -method analysis methods, designing a dominant epitope through epitope prediction process seems to be more accurate and significant (Li *et al.* 2013).

In order to examine the functionality of the software used in this study, a comparison of the experimental epitopes of four antigens submitted in *ideb* server (http://www.iedb.org) with the results of epitope prediction of the selected antigens, has been done using bioinformatical tools (Yousefi *et al.* 2015). As it is demonstrated in Table 2, the reported experimental results overlap those outputs predicted by Bioinformatical tools.

The empirical epitopes in subunit vaccines are too costly and need molecular biology and immunological technologies. In this respect, in a study (Forouharmehr and Nassiry, 2015) B and T-cells epitopes, secondary and tertiary structures, and antigenicity prediction of P40 protein of mycoplasma agalactiae bacteria were analyzed using alternative online softwares.

And possible antigenic epitopes and their immunogenicity of predicted peptides were determined. Another research on epitope prediction belongs to Yousefi *et al.* (2015) which concentrate on the most desirable epitopes of OMP25 antigen of *Brucella melitensis* bacteria. In that study they used a wide range of on-line epitope prediction software and reported the most probable epitopes with high antigenicity and less restriction site for enzyme digestibility. In the current study, B and T-cell epitope prediction of PA antigen of *B. anthracis* has been conducted using well-known online epitope prediction servers.

As it is shown in Figure 2, final B and T-cell epitope prediction suggested three epitopic region for B cells, three epitopic region for MHCI T-cells and two for MHCII T-cells. Kaur *et al.* (2009) have identified three main regions; ID-I: 604-622, ID-II: 626-676 and ID-III: 707-723 residues as B-cell epitopes through BCPred, BcePRED servers; these regions are not the same as the predicted B cell epitopes in this study. Random coil regions, which are located on the surface, are essential in binding ligands, since they are both exposed and hydrophilic.

The high rate of random coil structures implies to most protein-forming antigenic epitopes (Li *et al.* 2013).

Through secondary analysis of final predicted epitopes, our findings revealed that all of our final predicted B cell epitopes contain 100% random coil structure, the same as MHCI and MHCII T-cell predicted epitopes. Consequently, these recommended epitopes (due to the suitable random coil structure) could be exposed to protein surface, making them appropriate candidates to be used in recombinant subunit of vaccines epitope with stronger antigenicity. In order to prevent degradation and decomposition of epitopes during antigen processing, epitopes with less restriction site of proteosomal should be selected (Toes *et al.* 2001).

Subsequently, the predicted B and T-cell epitopes have been analyzed based on the presence of enzymatic restriction sites. The results have demonstrated that some enzymes such as Trypsin, Clostripain, CyanogenBromide, IodosoBenzoate, Proline_Endopept, Staph_Protease, Trypsin_K, Trypsin_R and AspN, which are the central enzymes responsible for protein degradation, have no restriction sites in the final selected epitopes. Hence it could be concluded that these epitopes can be used not only in injective vaccines but also in oral ones. Bioinformatic analysis revealed that these epitopes have more antigenic effect on the body via their highest persistence in gastrointestinal tract and by avoiding enzyme digestion.

CONCLUSION

Finally, it can be concluded that, since the using recombinant vaccines has many advantages over the use of killed or live attenuated bacteria, the identification of epitopic zones of pathogenic bacteria and their use in the sub-unit vaccines consist of multiplex epitopic, can be an important step in creating a safe immunity in animals and humans. In the case of anthrax bacteria, based on the results obtained in this study, the use of final predicted areas with the highest immunogenic scores through bioinformatics processes can be a good alternative to initiating experimental experiments against this bacterium.

ACKNOWLEDGEMENT

This Project was supported by a grant in aid of research from the Ferdowsi University of Mashhad.

REFERENCES

Abboud N. and Casadevall A. (2008). Immunogenicity of *Bacillus anthraci*s protective antigen domains and efficacy of elicited antibody responses depend on host genetic background. *Clin. Vaccine Immunol.* **15(7)**, 1115-1123.

Brey R.N. (2005). Molecular basis for improved anthrax vaccines. *Adv. Drug. Deliv. Rev.* **57**, 1266-1292.

Cassataro J., Estein S.M. and Pasquevich K.A. (2005). Vaccination with the recombinant *Brucella* outer membrane protein 31 or a derived 27-amino-acid synthetic peptide elicits a CD4+ T helper 1 response that protects against *Brucella melitensis* infection. *Infect. Immun.* **73**, 8079-8088.

Farchaus J.W., Ribot W.J., Jendrek S. and Little S.F. (1998). Fermentation, purification, and characterization of protective antigen from a recombinant, avirulent strain of *Bacillus anthracis*. *Appl. Environ. Microbiol.* **64**, 982-991.

Flick S.H.C., Walker N.J., Gibson P., Bullifent H., Hayward S., Miller J., Titball R.W. and Williamson E. D. (2002). A recombinant carboxy terminal domain of the protective antigen of *Bacillus anthracis* protects mice against anthrax infection. *Infect. Immun.* **70**, 1653-1656.

Forouharmehr A. and Nassiry M.R. (2015). B and T-cell epitopes prediction of the P40 antigen for developing mycoplasma agalactiae vaccine using Bioinformatic Tools. *Genet. Millennium.* **13(1)**, 3954-3961.

Geourjon C. and Deléage G. (1995). SOPMA significant improvements in protein secondary structure prediction by consensus prediction from multiple alignments. *Comput. Appl Biosci.* **11**, 681-684.

Gordon V.M., Klimpel K.R., Arora N., Henderson M.A. and Leppla S.H. (1995). Proteolytic activation of bacterial toxins by eukaryotic cells is performed by furin and by additional cellular proteases. *Infect. Immun.* **63**, 82-87.

Inglesby T.V., O'Toole T., Henderson D.A., Bartlett J.G., Ascher M.S., Eitzen E., Friedlander A.M., Gerberding J., Hauer J., Hughes J., McDade J., Osterholm M.T., Parker G., Perl T.M., Russell P.K. and Tonat K. (2002). Anthrax as a biological weapon, 2002: updated recommendations for management. *J. Am. Med. Assoc.* **287**, 2236-2252.

Kaur M., Chug H., Singh H., Chandra S., Mishra M., Sharma M. and Bhatnagar R. (2009). Identification and characterization of immunodominant B-cell epitope of the C-terminus of protective antigen of *Bacillus anthracis*. *Mol. Immunol.* **46(10)**, 2107-2115.

Leppla S.H. (1982). Anthrax toxin edema factor: a bacterial adenylate cyclase that increases cyclic AMP concentrations of eukaryotic cells. *Proc. Natl. Acad. Sci. USA.* **79**, 3162-3166.

Leppla S.H., Robbins J.B., Schneerson R. and Shellac J. (2002). Development of an improved vaccine for anthrax. *Clin. Invest.* **110**, 141-144.

Little S.F., Ivins B.E., Fellows P.F. and Friedlander A.M. (1997). Passive protection by polyclonal antibodies against *Bacillus anthracis* infection in guinea pigs. *Infect. Immun.* **65**, 5171-5175.

Little S.F. and Ivins B.E. (1999). Molecular pathogenesis of *Bacillus anthracis* infection. *Microbes Infect.* **2**, 131-139.

Li Y., Liu X. and Zhu Y. (2013). Bioinformatic prediction of epitopes in the Emy162 antigen of *Echinococcus multilocularis*. *Exp. Ther. Med.* **6**, 335-340.

Pitt M.L., Little S.F., Ivins B.E., Fellows P., Barth J., Hewetson J., Gibbs P., Dertzbaugh M. and Friedlander A.M. (2001). *In vitro* correlate of immunity in a rabbit model of inhalational anthrax. *Vaccine.* **19**, 4768-4773.

Ponomarenko J.V. and Van R. (2009). B-cell epitope prediction. *Struct. Bioinform.* **35**, 849-879.

Stanley J.L. and Smith H. (1961). Purification of factor I and recognition of a third factor of the anthrax toxin. *J. Gen. Microbiol.* **26**, 49-63.

Tabatabai L.B. and Pugh J. (1995). Modulation of immune responses in Balb/c mice vaccinated with *Brucella abortus* Cu-Zn superoxide dismutase synthetic peptide vaccine. *Vaccine.* **12**, 919-924.

Toes R.E., Nussbaum A.K. and Degermann S. (2001). Discrete cleavage motifs of constitutive and immuno proteasomes revealed by quantitative analysis of cleavage products. *J. Exp. Med.* **194(1)**, 1-12.

Vitale G., Pellizzari R., Recchi C., Napolitani G., Mock M. and Montecucco C. (1998). Anthrax lethal factor cleaves the N-terminus of MAPKKs and induces tyrosine/threonine phosphorylation of MAPKs in cultured macrophages. *Biochem. Biophys. Res. Commun.* **248**, 706-711.

Vizcaíno N., Zygmunt M.S., Verger J.M., Grayon M. and Cloeckaert A. (1997). Localization and characterization of a specific linear epitope of the Brucella DnaK protein. *FEMS Microbiol. Lett.* **154**, 117-122.

Wang W., Wu J. and Qiao J. (2014). Evaluation of humoral and cellular immune responses to BP26 and OMP31 epitopes in the attenuated *Brucella melitensis* vaccinated sheep. *Vaccine.* **32**, 825-833.

Wass M.N., Kelley L.A. and Sternberg M.J. (2010). 3DLigandSite: predicting ligand-binding sites using similar structures. *Nucleic Acids Res.* **38**, 469-473.

Yamaguchi H., Miura H. and Ohsumi K. (1996). Analysis of the epitopes recognized by mouse monoclonal antibodies directed to *Yersinia enterocolitica* heat-shock protein 60. *Microbiol. Immunol.* **40**, 77-80.

Yang X., Jerod A., Sky B., Ling C., Beata C., Theresa T. and David W.P. (2013). Progress in Brucella vaccine development. *Front. Biol.* **8(1)**, 60-77.

Yousefi S., Tahmoorespur M. and Sekhavati. M.H. (2015). B and T-cell epitope prediction of the OMP25 antigen for developing *Brucella melitensis* vaccines for sheep. *Iranian J. Appl. Anim. Sci.* **5(3)**, 629-638.

Zhang W., Liu J., Zhao M. and Li Q. (2012). Predicting linear B-cell epitopes by using sequence - derived structural and physico-chemical features. *Int. J. Data Mining. Bioin.* **6(5)**, 557-569.

20

Comparison of the Fatty Acid Composition of the Longissimus Dorsi Muscle of Kids, Lambs and Calves Produced under Iranian Transhumant Production System

A. Kiani[1*], M.H. Gharoni[2] and R. Shariati[3]

[1] Department of Animal Science, Faculty of Agricultural Science, Lorestan University, Khoramabad, Iran
[2] Department of Veterinary Science, Faculty of Veterinary Medicine, Lorestan University, Khoramabad, Iran
[3] Research and Development Section, Garrin Dam Simorg, Khorramabd, Iran

*Correspondence E-mail: kiani.a@lu.ac.ir

ABSTRACT

Transhumant production system (TPS) is a type of extensive livestock production practiced by transhumant pastoralists in which indigenous livestock breeds are mainly fed a pasture-based diet. The hypothesis tested in this work was whether differences existed between fat samples from the different species in respect to ratios of n-6/n-3, and polyunsaturated fatty acids to saturated fatty acids. To test our hypothesis, fatty acids (FAs) composition of the longissimus dorsi muscle of kids (n=10), fat-tailed lambs (n=10), and calves (n=10) produced in a pasture-based system were determined. All animals were indigenous intact male and randomly selected from nomads in the Zagros mountains (Noorabad, Lorestan province). The live body weight of kids, lambs and calves were 21 ± 6, 27 ± 4, and 158 ± 35 kg respectively. Meat samples were analyzed either without (lean meat) or mixed with 30% of sirloin subcutaneous fat (fat meat). Results showed that saturated FAs (as percentage) in kids meat was lower than those in lambs and calves (41.4 *vs.* 46.2 and 47.4% P=0.02). Kid meat had higher α-linolenic (C18:3 n-3), eicosapentaenoic acid (C20:5 n-3), docosapentaenoic acid (C22:5 n-3), but lower undesirable FAs (C16:0+C14:0) and n-6/n-3 ratio in comparison with lambs and calves. In conclusion, goat meat produced under TPS conditions, compared with lambs and calves, showed more promising healthy source of FAs for human nutrition.

KEY WORDS omega-3, red meat, ruminants, transhumant pastoralists.

INTRODUCTION

Transhumant production system (TPS) is practiced in Iran predominantly in the Zagros mountains where livestock are mainly raised in a pasture-based feeding (Badripour, 2004). The annual migration of nomads' takes place from mountainous cold rangelands towards the warmer plains at the beginning of autumn, with the reverse movement in the spring. In transhumant system animal husbandry, the stock comprise on average 48% sheep, 47% goats, 3% cattle and 2% draught animals (Badripour, 2004). Nomadic and trans-humant pastoralists own 13.6, 8.3 and 0.25 million heads sheep, goats and cattle, respectively and they produces about 20% of red meat production (e.g. 800 thousand metric ton) in the country (Organization for Nomadic People of Iran, 2008). Red meat production under extensive systems (such as TPS) is different from that of intensive production systems in many aspects including management, breeds, feeding regimes, animal activity, and environmental condition (Zervas and Tsiplakou, 2011). In TPS, ruminants are usually finished on diets containing high proportions of green forages which might produce meat with more desirable fatty acids compositions than that of intensive production systems (Fincham *et al.* 2009; Daley *et al.* 2010; Howes *et al.* 2015).

Red meat is a good dietary source of essential amino acids, minerals, vitamins, and fatty acids (FAs) (Williams, 2007), however, its consumption is under question mainly due to unbalanced ratio of n-6 to n-3 FA, and ratio of polyunsaturated FA (PUFA) to saturated FA (SFA) (Binnie *et al.* 2014). These unhealthy nutritional facts of red meat could be to some extent improved by dietary inclusion of green grass in animal daily ration (French *et al.* 2000; Nuernberg *et al.* 2005; Fincham *et al.* 2009). There are opportunities in extensive production systems in which ruminants are fed green forage in which enhancement of desirable FAs compositions might occur (Nuernberg *et al.* 2005; Talpur *et al.* 2008; Fincham *et al.* 2009; Ponnampalam *et al.* 2014; Howes *et al.* 2015; Kiani and Fallah, 2016). In TPS, ruminants are finished on diets containing high proportions of green forages, thus they might produce meat with more desirable FAs compositions. To our knowledge, information on fatty acid composition of red meat produced in Iranian transhumant production system practicing by nomads is lacking. The hypothesis tested in this work was whether differences existed between fat samples, with respect to n-6/n-3, and PUFA/SFA ratios, from the goats, sheep and cattle reared under TPS. To test our hypothesis, we chose the longissimus dorsi (LD) muscle to compare FAs composition and ratios of n-6/n-3 and PUFA/SFA in kids, lambs, and calves produced under TPS.

MATERIALS AND METHODS

Animals and meat samples

In total, 30 intact indigenous male animals including 10 Lori goat kids, 10 fat-tailed lambs, and 10 Lori calves; all reared in a pasture-based feeding system by nomads were randomly selected. All animals had free access to their dams' milk even when they were able to graze. The animals had grazed on the same natural ranges at least for two months prior to slaughter day. The ranges were located at Zagros Mountains, near Noorabad, Lorestan province, Iran (34 °02'34.8"N, 48 °17'53.0"E). Predominant plant species of the grazed ranges were ryegrass (*Lolium perenne)*, clover (*Trifolium* spp.), and other legumes (such as *Astragalus* spp.). Kids had free access to green leaves of oak trees (*Quercus brantii*). Slaughter age for lambs and kids were 5.5 ± 1 and for calves were 8 ± 2 months. Live body weight (Mean±SD) of the animals averaged 21 ± 6, 27 ± 4, and 158 ± 35 kg for kids, lambs and calves, respectively. All animals were slaughtered in a commercial processing plant (Gholshan Abbattoir, Khoramabad, Lorestan) according to the Halal procedure. Slaughter procedure was approved by

the Animal Ethics Committee of Lorestan University. Meat samples (about 30-50 g) were taken from the loin portion of *longissimus dorsi* (between ribs 12th and 13th) muscles on the left side of the carcasses. Samples were cut off within 2 h after slaughter and were chilled t +4 °C for 24 h. Meat samples were divided into two sub-samples; either without subcutaneous fat (lean meat) or mixed with 30% sirloin subcutaneous fat of each animal(fat meat). Samples were ground by means of a food processor (3×5 s), and stored at -80 °C pending analysis.

Lipid extraction and methylation

Fatty acid methyl esters (FAME) were determined using the procedures described by (Sukhija and Palmquist, 1988) with some modification. Briefly, 0.1 g of meat sample was weighed (0.0001 g, KERN, Germany) and then was freeze dried. Sample was then placed into screw cap Pyrex culture tube (16×125 mL, Scott glass tube). One mL of heptane was added and mixed. After that, 0.2 mL of sodium methylate (25%) was added, and the tube was put in a 50 °C water bath for 10 min. Sample was then cooled for about 5 °C and 3 mL of freshly made methanolic HCl 10% (prepared by adding 20 mL of acetyl chloride to 100 mL of anhydrous methanol) was added to the sample and the tube was shaken vigorously. The tube was put in steam bath (90 °C) for 30 min and then the tube sample was cooled with ice. Finally, one mL of heptane and three mL of potassium carbonate 10% was added and mixed for one minute using a shaker. The sample was centrifuged (Centrifuge 5415 R; Rotofix 32A, Germany) at 300 *g* for 5 min. Heptane phase (upper phase) was transferred to the GC vial (1.5 mL) using Pasteur pipette.

Determination of fatty acid composition

Fatty acid methyl esters were analyzed using a GC (GC-FID; HP 6890 chromatograph, Hewlett-Packard, Avondale, PA, USA) fitted with a flame ionization detector. The GC was equipped with a Chrompack CP-Sil 88 TM fused silica capillary column (100 m×0.25 mm i.d., 0.2 mm film thickness; Varian Inc., Walnut Creek, CA, USA). The injector and detector temperatures were maintained at 280 °C and 300 °C, respectively. Initially, the column temperature was held at 150 °C for 5 min and then increased at 5 °C min^{-1} to 180 °C (held for 30 min), then increased at 1 °C min^{-1} to 190 °C (held for 5 min) and finally increased at 1 °C min^{-1} to 200 °C (held for 35 min). Hydrogen was used as the carrier gas at a flow rate of 1.0 mL min^{-1}. Identification of common FAs was accomplished by comparison of sample peak retention times with those of FAME standard mixtures (Supelco^{TM}37 component FAME Mix, Supelco-47885-U, Sigma-Aldrich Chemie GmbH, Germany) and by using published chromatograms obtained under similar analytical conditions. Result for each FA was expressed as percentage of total FAME.

Statistical analysis

Data were analyzed using the MIXED procedures in SAS Software (SAS, 2008). The model considered the fixed effects of meat species (kids, lambs, and calves) and meat type (lean meat *vs.* fat meat) and the interaction effects. Non-significant interactions were excluded from the model. For all statistical analyses, significance was declared at P ≤ 0.05 and trends at P ≤ 0.10, unless otherwise stated. The Fisher's protected least significant difference (LSD) test was used for multiple treatment comparisons using the LSMEANS with letter grouping obtained using SAS pdmix800 macro (Saxton 1998). The residual analysis was carried out to test the model assumptions using the UNIVARIATE procedure with NORMAL and PLOT options.

RESULTS AND DISCUSSION

Fatty acid characterizations of red meat produced under TPS

The major FAs of red meat produced in TPS were oleic acid (C18:1), stearic acid (C18:0) and palmitic acid (C16:0) irrespective of the species (Table 1). Stearic and palmitic acid values were numerically similar to those reported in concentrate fed ruminants (Enser *et al.* 1996; Banskalieva *et al.* 2000; Wood *et al.* 2008).

Red meat produced in TPS had relatively less SFAs and more oleic acid (C18:1) than those reported in concentrate fed ruminants (Enser *et al.* 1996; Banskalieva *et al.* 2000; Wood *et al.* 2008). Under TPS condition, animals are generally fed with green forages. Linolenic acid (7 to 37 g), linoleic acid (2 to 10 g), and palmitic acid (3 to 8 g) contribute to the 90% of FAs in forages (Clapham *et al.* 2005). Fatty acids of dietary forages are hydrogenated in the rumen, causing a decrease in linolenic acid and an increase in monounsaturated FAs (MUFAs) (Jenkins *et al.* 2008; Fincham *et al.* 2009).

In the present study, n-6/n-3 ratio in meat produced under TPS was notably lower than those reported for forage-fed and concentrate-fed ruminants.

The low n-6/n-3 ratio in LD muscle found in animals under TPS was mainly due to higher percentage of n-3 fatty acids compared to other reports. For instance, Lori goat reared under TPS in comparison with Pateri goats (Pakistani breed) under traditional feeding system (Talpur *at al.* 2008) produced meat with almost two-fold higher ALA (C18:3 n-3), eicosapentaenoic acid (EPA, C20:5 n-3), and docosapentaenoic acid (DPA, C22:5 n-3). In general, inclusion of high quantity of green

forages in ruminant diet positively increased amounts of ALA, and EPA in meat (Nuernberg *et al.* 2005; Bas *et al.* 2005; Fincham *et al.* 2009). Pasture finishing animals produced higher quantities of n-3 fatty acids; consequently improved both the PUFAs/SFAs and n-6/n-3 ratios in comparison with concentrate-fed animals (reviewed by Howesb *et al.* 2015). The meat of cattle grazed on pasture had higher amount of ALA and lower linolenic acid (LA; C18:2 n-6) compared to meat from concentrate-fedcattle (Daley *et al.* 2010). French *et al.* (2000) reported that beef cattle on pasture had higher amount of omega-3 FAs, consequently better ratio of n-6/n-3 compared to concentrate-fed ones. Low n-6/n-3 ratio in meat is nutritionally important character because of its relationship with cardiovascular disease mortalities (Simopoulos, 2006). The n-6/n-3 ratio in a healthy diet should be kept below 4.0 (Simopoulos, 2006). Long chain fatty acids such as EPA and DHA have been given much attention due to their biological potency and that they are beneficial to human health. In present study, red meat produced under TPS showed relatively high concentrations of EPA, DHA, and DPA suggesting an alternative source for these important FAs. These findings were in agreement with other reports suggesting that finishing animals in green forage diets enhanced the long chain FAs composition in meat (Howes *et al.* 2015). Therefore, some fatty acid characteristics of red meat produced under TPS such as relatively high concentrations of EPA, DHA, and DPA deserve further investigations.

Lean meat versus fat meat

Irrespective of the species, lean meat had higher percentage of myristic acid (C14:0), LA, ALA, arachidonic acid (AA; C20:4 n-6), EPA, and DPA than fat meat (Table 1). Lean meat had higher PUFAs, n-6 FAs, n-3 FAs, and higher ratio of PUFAs/SFAs than fat meat (Table 2). However, no significant differences were found between lean and fat meat in respect to the unhealthy FAs, SFAs, and ratio of n-6/n-3 (Table 2). Furthermore, lean meat showed higher PUFAs/SFAs ratio than fat meat (0.35 *vs.* 0.18). The value of PUFAs/SFAs found in lean meat produced under TPS conditions was very close to that of desirable value (i.e. >0.4) in a healthy diet (Simopoulos, 2006). The higher PUFAs/SFAs ratio in lean red meat compared to fat meat might be explained by a low capacity of storing long chain FAs in adipose tissue of ruminants (Wood *et al.* 2008). In general, adipose tissue has less phospholipid, but more triglycerides than does muscle tissue. Phospholipids contain 10 times more LA and 8 times more ALA than triglycerides, whereas triglycerides have higher contents of palmitic acid, stearic acid, and oleic acid compared to phospholipids (Scollan *et al.* 2006; Wood *et al.* 2008).

Table 1 Fatty acids composition (percentage of total fatty acids) of longissimus dorsi muscle either without (lean meat) or with subcutaneous fat (fat meat) in indigenous kids, lambs and calves produced under the same transhumant production system

Items	Species			SEM	P-value	Type meat		SEM	P-value
	Kids	Lambs	Calves			Lean meat	Fat meat		
Lauric acid (C12:0)	0.32	0.35	0.17	0.11	0.50	0.29	0.27	0.10	88.0
Myristic acid (C14:0)	1.77[b]	2.62[ab]	3.22[a]	0.27	< 0.01	2.83	2.24	0.18	0.01
Palmitic acid (C16:0)	20.0[b]	23.5[a]	26.0[a]	0.78	< 0.001	23.0	23.2	0.66	0.84
Margaric acid (C17:0)	1.18	2.06	1.16	0.27	0.06	1.13	1.81	0.23	0.07
Stearic acid (C18:0)	17.7	16.9	16.1	1.22	0.66	16.2	17.6	1.5	0.19
Myristoleic acid (C14:1 n-9 cis)	0.11[b]	0.69[a]	0.05[b]	0.11	< 0.01	0.33	0.23	0.08	0.27
Palmitoleic acid (C16:1 n-9 cis)	2.30[ab]	1.51[b]	3.10[a]	0.40	0.05	1.77	2.84	0.26	< 0.01
Palmitoleic (C17:1 n-9 cis)	0.63	0.76	0.91	0.25	0.73	0.70	0.83	0.16	0.38
Oleic acid (C18:1 n-9 cis)	39.8[a]	39.8[a]	35.1[b]	1.38	0.01	36.1	40.4	1.06	0.05
Linoleic acid (C18:2 n-6 cis)	4.55	5.10	6.53	0.87	0.28	7.19	3.60	0.68	< 0.01
α-Linolenic acid (ALA; C18:3 n-3 cis)	1.81[a]	0.78[b]	0.69[b]	0.16	< 0.001	1.24	0.95	0.11	0.04
Arachidonic acid (AA; C20:4 n-6)	2.04	1.82	1.63	0.44	0.81	2.76	0.90	0.37	< 0.01
Eicosapentaenoic acid (EPA; C20:5 n-3)	1.37[a]	0.41[b]	0.25[b]	0.18	< 0.01	0.85	0.50	0.12	0.02
Docosapentaenoic acid (DPA; C22:5 n-3)	1.44[a]	0.49[b]	0.56[b]	0.19	< 0.01	1.12	0.54	0.14	< 0.01
Docosahexaenoic acid (DHA; C22:6 n-3)	0.26	0.24	0.10	0.08	0.27	0.28	0.11	0.06	0.07

The means within the same row with at least one common letter, do not have significant difference (P>0.05).
SEM: standard error of the means.

Table 2 Means of unhealthy, saturated, unsaturated and the ratio of n-6/n-3 fatty acids of longissimus dorsi muscle either without (lean meat) or with subcutaneous fat (fat meat) of indigenous kids, lambs, and calves produced under the same transhumant production system

Items	Species			SEM	P-value	Type meat		SEM	P-value
	Kids	Lambs	Calves			Lean meat	Fat meat		
Unhealthy fatty acids (C14:0 + C16:0)	24.0[b]	27.6[b]	32.4[a]	1.22	0.001	27.1	28.9	0.92	0.14
Saturated fatty acids (SFAs)	41.4[a]	46.2[b]	47.4[b]	1.35	0.02	43.6	46.4	1.08	0.08
Monounsaturated fatty acids (MUFAs)	45.1	44.1	41.6	1.67	0.36	41.5	45.7	1.17	< 0.01
Polyunsaturated fatty acids (PUFAs)	13.3	9.38	10.9	1.69	0.29	14.6	7.80	1.28	< 0.01
n-6 fatty acids	7.54	7.1	8.87	1.26	0.60	10.6	5.05	1.00	< 0.01
n-3 fatty acids	5.1[a]	1.96[b]	1.60[b]	0.49	< 0.001	3.52	2.23	0.33	< 0.01
PUSAs / SFAs ratio	0.33	0.21	0.24	0.04	0.20	0.35	0.18	0.04	< 0.01
n-6 / n-3 ratio	1.5[c]	3.5[b]	6.0[a]	0.42	< 0.001	3.46	3.93	0.31	0.23

PUSAs: polyunsaturated fatty acids and SFAs: saturated fatty acids.
The means within the same row with at least one common letter, do not have significant difference (P>0.05).
SEM: standard error of the means.

Comparisons among three species (kids vs. lambs and calves)

Red meat of kids showed several nutritional merits over the meat of lambs and calves. Sum of SFAs in kids' meat was lower than that in lambs and calves (Table 2). Lower SFAs in kids could be partly explained by higher bio-hydrogenation in goats compared to the other ruminants. Tora et al. (2016) reported an interspecies variation in rumen microbiota and rumen biohydrognation between dairy goats and cows. They reported more rumen biohydrogna-tion and higher SFAs concentration (in particular, 16:0 and 18:0) in rumen fluid of goats compared with cows. How-ever, in the present study, kid meat had significantly lower palmitic acid (C16:0) than both calves and lambs (Table 1).

Kids had the lowest unhealthy FAs (C16:0+C14:0) and the highest desirable FAs (C18:0, C18:1, C18:2, C18:3, C20:5 and C22:5) among the three species. The content of desirable FAs in kids' meat in the present study (76%) was in range of reported 61 to 81% (Banskalieva et al. 2000). Stearic acid contributed to 43%, 37% and 34% of sum of SFAs in kids, lambs, and calves meat, respectively. Stearic acid is known to have neutral effect on diet–related disease risk (Hunter et al. 2010; Katan et al. 1994). In fact, undesir-able FAs and SFAs increased plasma cholesterol concentra-tion (Grundy and Denke, 1990; Katan et al. 1994; Astrup et al. 2011). Therefore, high stearic acid and low undesirable FAs in kid meat are considered as healthy nutritional mer-its.

Kids had higher ALA, EPA and DPA than both calves and lambs (Table1). Consequently, the sum of n-3 FAs in kid's meat was two-fold more than those in calves and lambs (Table 2). The main source of variations in the FAs profile of red meat is animal nutrition. In the present study, animals grazed same range; however, due to goats feeding behavior, kids also used green leaves of oak trees (*Quercus brantii*). The leaves of oak trees are tanninferous feeds. Vasta *et al.* (2009) found the highest UFAs concentrations in ruminal fluid from lambs supplemented with tannins, following to an increase in PUFAs and a decrease in SFAs levels in intramuscular fat. Similarly, the highest levels of LA, ALA and the sum of PUFAs in lamb muscle were found in the diet with the largest amount of tannins (Willems *et al.* 2014). Grazing animals on diverse pastures in mountainous areas accumulated more biohydrogenation intermediates in the rumen and an elevated PUFAs content

in inter-muscular fat. This suggested that rumen biohydrogenation was partially inhibited and PUFAs were further desaturated and elongated when animals were grazed on the botanically diverse pasture (Lourenco *et al.* 2007).

Fat meat in kids contained a high percentage of PUFAs (Figure 1), and high ratio of PUFAs/SFAs (0.34) regardless to the type of meat. Fat meat of kids contained lower oleic acid than lean meat whereas fat meat of lambs and calves had higher values compared to lean meat (Figure 1).

Monounsaturated fatty acids in lean meat compared to fat meat was lower in kids, while in calves lean meat had less MUFAs compared with fat meat. Polyunsaturated fatty acids in lean meat and fat meat of kids did not differ, whereas in lambs and calves lean meat had higher PUFAs compared to fat meat (Figure 1).

Lean meat of lambs and calves contained two-fold more PUFAs compared to fat meat.

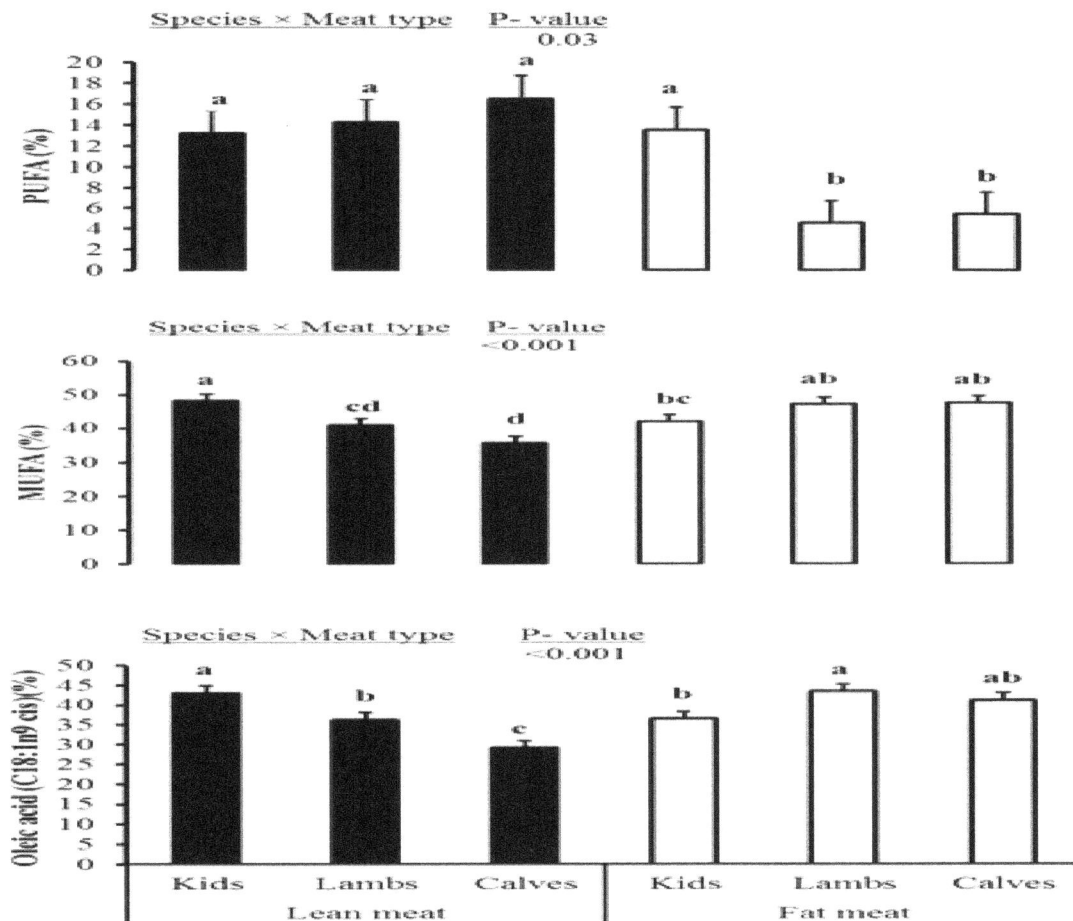

Figure 1 Significant interaction effects between species (kids, lambs and calves) and meat type (without or with subcutaneous fat) on oleic acid, monounsaturated fatty acids (MUFAs) and polyunsaturated fatty acids (PUFAs) of longissimus dorsi muscle presented as percentage of total fatty acid proportions.
[a, b and c] Bars in the same chart with a common superscript letter do not differ significantly (P<0.05).

The high percentage of PUFA in fat meats of kids in comparison with lambs and caves isinteresting since the PUFAs subcutaneous fat in ruminants is generally low (approximately 0.15). Unsaturated fatty acids of dietary lipids are mainly converted to SFAs through bacterial lipolysis and biohydrogenation in the rumen (Jenkins *et al.* 2008). In general, adipose tissue has less phospholipid, but more triglycerides than does muscle tissue.

Phospholipids contain more PUFAs (LA and ALA) than triglycerides, whereas triglycerides have higher contents of C16:0, C18:0 and C18:1 compared to phospholipids (Wood *et al.* 2008) presumably because the capacity for incorporation of PUFA into phospholipids is limited. Feeding ruminants a diet rich in PUFAs promotes fat deposition in the phospholipid fraction, and SFAs, when in excess, are deposited and stored in the triglyceride fraction. Compared to grass, mixed leguminous pastures produced significantly higher proportions of linoleic acid and α-linolenic acid in the abomasum and subcutaneous fats (Lourenco *et al.* 2007) suggesting greater duodenal flow of PUFAs. The feeding behavior of kids under TPS condition (eating tanniferous feeds) and lower rumen biohydrogenation might have led to more by-pass and intestine absorption of PUFAs consequently storing extra PUFAs in subcutaneous fat.

CONCLUSION

Red meat produced under TPS conditions showed desirable nutritional merits in terms of fatty acid profile. Transhumant production system enriches the meat quality characteristics in terms of PUFAs. Under TPS condition, kids showed several nutritional merits over the lambs and calves with higher ALA, EPA and DPA. Furthermore, lower unhealthy FAs (C16:0+C14:0) and n-6/n-3 ratio was found in kids compared to lambs and calves presumably due to kid different feeding behaviour and rumen biohydrogenation. This study suggests that under TPS, goats might produce more promising healthy red meat compared to sheep and cattle.

ACKNOWLEDGEMENT

This study was sponsored by Lorestan University of Medical Sciences (Grant number:200/56083).

REFERENCES

Astrup A., Dyerberg J., Elwood P., Hermansen K., Hu F.B. and Jakobsen M.U. (2011). The role of reducing intakes of saturated fat in the prevention of cardiovascular disease: where does the evidence stand in 2010? *Am. J. Clin. Nutr.* **93,** 684-688.

Badripour H. (2006). Country pasture/forage resources profile, range management expert, Technical Bureau of Rangeland, Forest, Range and Watershed Management Organization (FRWO), No. 131, Tehran, Iran. Available online at: http://www.fao.org/ag/AGP/AGPC/doc/pasture/forage.htm (accessed January 2008).

Banskalieva V., Sahlu T. and Goetsch A.L. (2000). Fatty acid composition of goat muscles and fat depots: a review. *Small Rumin. Res.* **37,** 255-268.

Bas P., Dahbi E., El A.A., Morand-Fehr P. and Araba A. (2005). Effect of feeding on fatty acid composition of muscles and adipose tissues in young goats raised in the Argan tree forest of Morocco. *Meat Sci.* **71,** 317-326.

Binnie M.A., Barlow K., Johnson V. and Harrison C. (2014). Red meats: Time for a paradigm shift in dietary advice. *Meat Sci.* **98,** 445-451.

Clapham W.M., Foster J.G., Neel J.P.S. and Fedders J.M. (2005). Fatty acid composition of traditional and novel forages. *J. Agric. Food Chem.* **53,** 10068-10073.

Daley C.A., Abbott A., Doyle P.S., Nader G.A. and Larson S. (2010). A review of fatty acid profiles and antioxidant content in grass-fed and grain-fed beef. *Nutr. J.* **9,** 10-22.

Enser M., Hallett K., Hewitt B., Fursey G.A.J. and Wood J.D. (1996). Fatty acid content and composition of english beef, lamb and pork at retail. *Meat Sci.* **42,** 443-456.

Fincham J.R., Fontenot J.P., Swecker W.S., Herbein J.H., Neel J.P.S. and Scaglia G. (2009). Fatty acid metabolism and deposition in subcutaneous adipose tissue of pasture- and feedlot-finished cattle. *J. Anim. Sci.* **87,** 3259-3277.

French P., Stanton C., Lawless F., O'Riordan E.G., Monahan F.J. and Caffrey P.J. (2000). Fatty acid composition, including conjugated linoleic acid, of intramuscular fat from steers offered grazed grass, grass silage, or concentrate-based diets. *J. Anim. Sci.* **78,** 2849-2855.

Grundy S.M. and Denke M.A. (1990). Dietary influences on serum lipids and lipoproteins. *J. Lipid. Res.* **31,** 1149-1172.

Howes N.L., Bekhit A.A., Burritt D.J. and Campbell A.W. (2015). Opportunities and implications of pasture-based lamb fattening to enhance the long-chain fatty acid composition in meat. *Comp. Rev. Food Sci. Food Safety.* **14,** 22-36.

Hunter J.E., Zhang J. and Kris-Etherton P.M. (2010). Cardiovascular disease risk of dietary stearic acid compared with trans, other saturated, and unsaturated fatty acids: a systematic review. *Am. J. Clin. Nutr.* **91,** 46-63.

Jenkins T.C., Wallace R.J., Moate P.J. and Mosley E.E. (2008). Recent advances in biohydrogenation of unsaturated fatty acids within the rumen microbial ecosystem. *J. Anim. Sci.* **86,** 397-412.

Katan M.B., Zock P.L. and Mensink R.P. (1994). Effects of fats and fatty acids on blood lipids in humans: an overview. *AM. J. Clin. Nutr.* **60,** 1017-1022.

Kiani A. and Fallah R. (2016). Effects of live weight at slaughter on fatty acid composition of longissimus dorsi and biceps femoris muscles of indigenous Lori goat. *Trop. Anim. Healt. Prod.* **48,** 67-73.

Lourenco M., Van Ranst G., De Smet S., Raes K. and Fievez V. (2007). Effect of grazing pastures with different botanical composition by lambs on rumen fatty acid metabolism and fatty acid pattern of longissimus muscle and subcutaneous fat. *Animal.* **1,** 537-545.

Nuernberg K., Dannenberger D., Nuernberg G., Ender K., Voigt J. and Scollan N.D. (2005). Effect of a grass-based and a concentrate feeding system on meat quality characteristics and fatty acid composition of longissimus muscle in different cattle breeds. *Livest. Prod. Sci.* **94,** 137-147.

Organization for Nomadic People of Iran. (2008). Census of Pastoralists, Organization for Nomadic People of Iran (ONPI), Sazmaan-e-Ashayar-e-Iran. Available at: http://ashayer.ir/index.aspx?siteid=1&pageid=237.

Ponnampalam E.N., Butler K.L., Jacob R.H., Pethick D.W., Ball A.J., Hocking Edwards J.E, Geesink G. and Hopkins D.L. (2014). Health beneficial long chain omega-3 fatty acid levels in Australian lamb managed under extensive finishing systems. *Meat Sci.* **96,** 1104-1110.

SAS Institute. (2008). SAS®/STAT Software, Release 9th. SAS Institute, Inc., Cary, NC. USA.

Scollan N., Hocquette J.F., Nuernberg K., Dannenberger D., Richardson I. and Moloney A. (2006). Innovations in beef production systems that enhance the nutritional and health value of beef lipids and their relationship with meat quality. *Meat Sci.* **74,** 17-33.

Simopoulos A.P. (2006). Evolutionary aspects of diet, the omega-6/omega-3 ratio and genetic variation: nutritional implications for chronic diseases. *Biomed. Pharm.* **60,** 502-507.

Sukhija P.S. and Palmquist D.L. (1988). Rapid method for determination of total fatty acid content and composition of feedstuffs and feces. *J. Agric. Food Chem.* **36,** 1202-1206.

Talpur F.N., Bhanger M.I. and Sherazi S.T.H. (2008). Intramuscular fatty acid profile of longissimus dorsi and semitendinosus muscle from Pateri goats fed under traditional feeding system of Sindh, Pakistan. *Meat Sci.* **80,** 819-822.

Tora P.G., Bernard L., Belenguer A., Rouel J., Hervas G., Chillard Y. and Frutos P. (2016). Comparison of ruminal lipid metabolism in dairy cows and goats fed diets supplemented with starch, plant oil, or fish oil. *J. Dairy Sci.* **99,** 301-316.

Vasta V., Mele M., Serra A., Scerra M., Luciano G., Lanza M. and Priolo A. (2009). Metabolic fate of fatty acids involved in ruminal biohydrogenation in sheep fed concentrate or herbage with or without tannins. *J. Anim. Sci.* **87,** 2674-2684.

Willems H., Kreuzer M. and Leiber F. (2014). Alpha-linolenic and linoleic acid in meat and adipose tissue of grazing lambs differ among alpine pasture types with contrasting plant species and phenolic compound composition. *Small Rumin. Res.* **116,** 153-164.

Williams P. (2007). Nutritional composition of red meat. *Nutr. Diet.* **64,** 111-119.

Wood J.D., Enser M., Fisher A.V., Nute G.R., Sheard P.R. and Richardson R.I. (2008). Fat deposition, fatty acid composition and meat quality: A review. *Meat Sci.* **78,** 343-358.

Zervas G. and Tsiplakou E. (2011). The effects of feeding systems on the charachteristics of producs from small ruminants. *Small Rumin. Res.* **101,** 140-149.

21

Effect of Adding Coenzyme Q10 and Ellagic Acid during Cryopreservation on Post-thaw Quality of Ram Semen

H. Daghigh Kia[1*], Z. Blooki[1], H. Vaseghi Dodran[1] and M. Mahdipour[1]

[1] Department of Animal Science, Faculty of Agriculture, University of Tabriz, Tabriz, Iran

*Correspondence E-mail: daghighkia@tabrizu.ac.ir

ABSTRACT

The aim of this study was to determine the effects of co-Q10 and ellagic acid on sperm parameters after the freeze–thawing of Ghezel ram sperm. Twenty ejaculates from five Ghezel rams were collected in this study. Semen samples, which were diluted with a soybean lecithin (SL) based extender containing 0.25 mM ellagic acid, 0.5 μM co-Q10, 0.25 mM ellagic acid + 0.5 μM co-Q10 and no antioxidant (control), were cooled to 4°C, frozen in 0.25 mL French straws and stored in liquid nitrogen. Sperm motility characteristics, membrane integrity, abnormal morphology, lipid peroxidation and antioxidant activities (glutathione peroxidase, superoxide dismutase and total antioxidant capacity) were evaluated following freeze-thawing. The results showed that 0.5 μM Co-Q10, improved viability, total motility parameters and decreased abnormal sperm and improved linearity (LIN), curvilinear velocity (VCL), straight-line velocity (VSL) and path velocity (VAP) parameters (P<0.05). Ellagic acid and the treatment with a combination (0.25 mM ellagic acid+0.5 μM co-Q10) improved viability and total motility parameters (P<0.05). The additives did not affect the maintenance of superoxide dismutase (SOD) and glutathione peroxidase (GPx), when compared to the control. It can be concluded that addition of 0.5 μM co-Q10 improved the post-thawing quality of ram semen.

KEY WORDS ellagic acid, Q10, ram, sperm.

INTRODUCTION

The potential benefits of artificial insemination are well known which are achieved by semen cryopreservation. Nonetheless the long-term storage of semen decreases sperm metabolic activity and thus fertility (Bailey *et al.* 2000). The highest damage to the sperm is caused by oxidative stress due to the production of reactive oxygen species (ROS). Oxidative stress results in the reduction of sperm activity and performance, that causes loss of motility, plasma membrane integrity and fertility (Najafi *et al.* 2014a). Semen cryopreservation causes some structural, biochemical and functional changes, which leads to various problems in sperm transport, survival and fertility rate in

domestic animals (Salamon and Maxwell, 2000). Polyunsaturated fatty acids (PUFA) are sensitive to oxidation and free radicals (Sheweita *et al.* 2005; Eskenazi *et al.* 2005). Ram sperm have a high ratio of unsaturated to saturated fatty acids and cholesterol to phospholipid ratio compared to other species. The large amounts of PUFA make the sperm membrane more vulnerable against oxidative damage especially by ROS. Sperm have multiple mechanisms of defense systems against ROS (Vaseghi-Dodaran *et al.* 2015). These mechanisms include catalase, uric acid, taurine, thiols, ascorbic acid and α-tocopherol but the most important ones are superoxide dismutase (SOD), glutathione peroxidase (GPx) and glutathione reductase (GSH) systems (Daghigh Kia *et al.* 2016a). Although these defense mechanisms are available, mature sperm do not have enough ability to fight against free radicals due to high concentrations of PUFA and failure in their plasma membrane enzyme activity, such as SOD. It has been reported that adding antioxidants to the freezing extender can reduce deteriorative effect of ROS and cold shock (Mata-Campuzano *et al.* 2015; Sariozkan *et al.* 2015) and improves sperm quality rams (Mehdipour *et al.* 2016), goats (Bucak *et al.* 2009), dogs (Funahashi and Sano, 2005) and human (Michael *et al.* 2007). The Co-Q10 is a part of the mitochondrial respiratory chain which has two important roles, first in metabolism and second acts as a fat-soluble antioxidant that protects the membrane and associated lipo-proteins (Ernster *et al.* 1993). The Co-Q10 is one of the most significant lipid antioxidants in mitochondrial respiratory chain. In addition, co-Q10 has fat like properties, so releases phospholipids in cell membrane and protects the sperm plasma membrane (Ernster and Dalner, 1995). Ellagic acid is a polyphenol such as Co-Q10 having a wide variety of biological activities, including strong antioxidant properties, anti-cancer, anti-proliferative, anti-mutation and anti-apoptotic (Hassoun *et al.* 2004). Co-Q10 has not been used so far in ram semen freezing and there are only two study on the effects of ellagic acid, one of them was used in freeze-thawing ram semen (Omur and Coyan, 2016) and the other one used it in cooling of ram semen (Omur *et al.* 2014). Therefore, this study investigated the effect of Co-Q10 and ellagic acid in the improvement of ram sperm quality after freeze-thawing process.

MATERIALS AND METHODS

Chemicals
All Chemicals were purchased from Sigma (St.Louis, MO, USA), Merck (Darmstadt, Germany) and Pars Tech Rokh (Mashhad, Iran), unless otherwise indicated.

Animal and semen collection
Semen samples were collected from 5 healthy mature Ghezel rams (3 and 4 years of age) using artificial vagina twice a week (totally twenty samples for experiment) then pooled to avoid an individual variations during experiment and transported at 34 °C to the laboratory for initial assessments. Samples were accepted for adding the antioxidant if the following parameters were observed: volume (0.75-2 mL), concentration (greater than 3×10^9 sperm/mL), motility (>70%) and abnormal sperm (<10%). A tris-based extender was used as the freezing extender (Tris 297.58 mM, citric acid 96.32 mM, fructose 82.66 mM, lecithin 1%, glycerol 7% (v/v), pH 6.8). Experimental treatments included four group with 1) ellagic acid 0.25 mM, 2) Co-Q10 0.5 µM, 3) ellagic acid 0.25 mM + Co-Q10 0.5 µM and 4) a group

without antioxidant considering as control group. The samples were placed in the refrigerator for two hours to reach at temperature of 4 °C and then loaded into 0.25 mL straws (IMV, L'Aigle, France). The samples were placed 4 cm above liquid nitrogen for 7 minutes and then plunged into liquid nitrogen and stored until thawing. For sperm evaluation, the straws were thawed individually at 37 °C for 30 seconds.

Sperm viability
For assessment of live sperm we used eosin-nigrosin staining. After freeze-thawing, 10 µL of the semen sample with 10 µL stain was mixed on a slide and for drying, the slides were held at 37 °C. Two hundred sperm were evaluated by microscope at magnification of × 1000. Sperm displaying unstained heads were considered viable, and sperm with stained or partially stained heads were counted as dead sperm (Najafi *et al.* 2014b).

Plasma membrane integrity
The hypo-osmotic swelling test (HOST) was used to assess integrity of plasma membrane. Fifty µL of the semen sample was mixed with 500 µL hypo-osmotic solution (100 mOsmol kg^{-1}) in water bath 37 °C for 30 minutes. Five µL of the sample was put on a slide and mounted with a coverslip. The phase contrast microscope slides were placed on a heating pad. Two hundred sperm were counted at 400 × magnification and the sperm with swollen tail and spun were considered as sperm with intact plasma membrane (Jeyendran *et al.* 1984).

Assessment of sperm motility
Computer assisted system (CASA; Video Test Sperm 3.1) was used to evaluate the parameters of total motility (TM, %), progressive motility (PM, %), path velocity (VAP, µm/s), straight-line velocity (VSL, µm/s), curvilinear velocity (VCL, µm/s), amplitude of lateral head displacement (ALH, µm), beat/cross frequency (BCF, Hz), straightness (STR, %) and linearity (LIN, %). To assess the parameters of motility, samples were incubated in warm bath for 5 minutes at 37 °C. Then, 5 µL of the sample was placed on a preheated slide, mounted with a cover slip, and then placed on a heating pad microscope. Each sample was randomly selected from at least 10 fields and 200 sperm were analyzed by CASA system. All parameters were photographed at 100 × magnification (Najafi *et al.* 2016).

Sperm morphology assessment
To evaluate sperm morphology, 10 µL of each semen sample was added to 150 µL of Hancock solution (Najafi *et al.* 2013). Then a drop of this mixture was placed on a slide and at least 200 sperm were counted under a microscope

phase contrast with a magnification of $\times$ 400 and the percentage of abnormal sperm and abnormal acrosome were estimated.

Lipid peroxidation

The concentrations of malondialdehyde (MDA) were measured by thiobarbituric acid reaction (TBARs). One mL of the diluted semen sample (250×106 sperm/mL) was mixed with 1 ml of cold 20% (w/v) tricholoroacetic acid to precipitate protein.

The precipitate was pelleted by centrifuging (960g for 15 min), and 1 mL of the supernatant was incubated with 1 mL of 0.67% (w/v) thiobarbituric acid in a boiling water bath at 100 °C for 10 min. After cooling, the absorbance was determined using a wave length of 532 nm spectrophotometer (Placer *et al.* 1996). All MDA concentrations were expressed as nmol/mL.

Total antioxidant capacity (TAC)

Randox kits were used to measure total antioxidant capacity of sperm according to TEAC. TEAC method is based on the inhibition of ABTS cation radical scavenging antioxidants. The ABTS incubated with peroxidase and H_2O_2 to produce a radical cation ABTS and stable blue-green color that has maximum absorbance at 600 nm, which can be measured by a spectrophotometer (Miller *et al.* 1993).

Glutathione peroxidase activity

The activity of GPx was determined using the Ransel Glutathione Peroxidase kit (Randox Laboratories, UK). In this assay, GPx catalyzes the oxidation of GSH with cumene hydroperoxide. In the presence of GR and NADPH GSSG is converted into GSH with concomitant oxidation of NADPH to NADP+. The decrease in absorbance was measured at 340 nm at 37 °C (pH 7.2). The GPx activity was normalized to g of protein and expressed in U/g of protein.

Superoxide dismutase activity

Activity SOD in the semen samples was measured using the method and using the Ransod Company Randox kit (RANDOX Laboratories Ltd, UK).

Statistical methods

The experiment was conducted in a completely randomized design with four treatments and four replications. Since the semen samples were pooled, these were not real biological replicates. These are rather technical replicates. The data obtained were analyzed by GLM procedure of SAS (2004). All data were checked for normal distribution by PROC UNIVARIATE and the Shapiro-Wilk test. Mean comparisons were performed using least square means and the sig-

nificant level was considered P < 0.05. Results are shown as Lsmean ± SEM.

RESULTS AND DISCUSSION

The diluents with the addition of Co-Q10 (0.5 μM), ellagic acid (0.25 mM) and the treatment with combination of the two compounds ellagic acid (0.25 mM) + Co-Q10 (0.5 μM) improved viability and total motility parameters and were significantly different with control group (P<0.05).

In addition, Co-Q10 (0.5 μM) decreased abnormal sperm and improved LIN, VCL, VSL and VAP parameters (P<0.05). Ellagic acid also improved VAP and VCL compared to the control group (Tables 1 and 2).

The treatment groups did not show any significant difference in GPx and SOD enzymes (Table 3).

Ellagic acid improved total antioxidant capacity compared to the control group (P<0.05). The combination of two antioxidants increased MDA levels compared to control groups (P<0.05; Figure 1 a, b).

Plasma membrane of ram semen is very sensitive to lipid peroxidation due to ROS because of being rich in unsaturated fatty acids. Ram sperm has high sensitivity against cold shock compared to other species such as cattle, rabbits and human (Mata-Campuzano *et al.* 2015).

Lipid peroxidation of the sperm membrane leads to loss of membrane fluidity and cell activity and infertility (Aitken and Sawyer, 2003). Therefore, additives with antioxidant properties are necessary in order to reduce the adverse effects of sperm cryopreservation. Ellagic acid and Co-Q10 have antioxidant and phenol properties that conserve sperm from adverse damages. Many studies have examined the effects of extracts and phenolic compounds after freeze-thawing rams and bull sperm (Daghigh Kia *et al.* 2015).

Accordingly, comparing the present study with the studies on compound phenols property in ram sperm shows that Co-Q10 improve the motility of ram sperm after freeze-thawing which are the same with the results of Daghigh Kia *et al.* (2016b) and the study on bull sperm by Daghigh Kia *et al.* (2015) and Daghigh Kia *et al.* (2016a) which improved motility of sperm by adding antioxidant to semen extender.

In our study, Co-Q10 (0.5 μM), as a phenolic compound reduced sperm abnormality. But it did not have any effect on membrane integrity, this outcome was contrary with result of Daghigh Kia *et al.* (2016b) on ram sperm freeze-thawing and Daghigh Kia *et al.* (2015) and Daghigh Kia *et al.* (2016a) on bull sperm freeze-thawing.

The results of this study indicate that there is a significant difference in total motility and viability of sperm in Co-Q10 at 0.5 μM level compared to the control group.

Table 1 Effects of different levels of ellagic acid and Co-Q10 on sperm motility parameters after freezing-thawing process

Sperm motility parameters	Antioxidants				SEM	P-value
	Control	EA (0.25)	Co-Q10 (0.5)	EA + Q10 (0.25+0.5)		
TM (%)	48.2[b]	64.6[a]	66.3[a]	63.2[a]	2.24	0.0001
PM (%)	18.6	27.0	24.6	25.8	3.30	0.0066
VAP (μm/s)	20.5[c]	34.6[b]	68.1[a]	31.8[bc]	2.83	0.0001
VSL (μm/s)	15.5[b]	23.9[b]	46.7[a]	21.7[b]	2.20	0.0001
VCL (μm/s)	50.1[c]	73.8[b]	91.7[a]	75.0[b]	4.38	0.0001
STR (%)	76.8	71.6	68.8	68.8	4.68	0.0001
LIN (%)	30.9[b]	32.7[b]	50.9[a]	29.0[b]	1.92	0.0001

EA: ellagic acid; TM: total motility; PM: progressive motility; VAP: path velocity; VSL: progressive velocity; VCL: track velocity; STR: straightness and LIN: linearity.
The means within the same row with at least one common letter, do not have significant difference (P>0.05).
SEM: standard error of the means.

Table 2 The effect of different levels of ellagic acid and Co-Q10 on sperm viability, host and hancock parameters after freezing-thawing process

Sperm assessment parameters	Antioxidant				SEM	P-value
	Control	EA (0.25)	Co-Q10 (0.5)	EA+Q10 (0.25+0.5)		
Viability (%)	55.6[b]	74.45[a]	81.31[a]	77.23[a]	2.25	0.0001
Membrane integrity (%)	37.61[b]	54.41[a]	56.68[a]	54.76[a]	4.88	0.0001
Abnormal sperm (%)	25.52[a]	26.13[a]	19.53[b]	25.49[a]	0.99	0.0001

EA: ellagic acid.
The means within the same row with at least one common letter, do not have significant difference (P>0.05).
SEM: standard error of the means.

Table 3 The effect of different levels of ellagic acid and Co-Q10 on sperm viability, host and hancock parameters after freezing-thawing process

Sperm assessment parameters	Antioxidant				SEM	P-value
	Control	EA (0.25)	Co-Q10 (0.5)	EA + Q10 (0.25+0.5)		
Superoxide dismutase u/mL	101	118.3	105	109.3	6.15	0.0001
Glutathioneperoxydaseu/g protein	10.03	9.37	9.27	8.93	0.26	0.0001

EA: ellagic acid.
SEM: standard error of the means.

These results may be due to the bioenergy role of Co-Q10 in the respiratory chain and the production of ATP (Almeida and Ball, 2005). The important role of recycling of vitamin E by Co-Q10 in the prevention of lipid peroxidation and also in energy production in semen is well-known (Lewin and Lavon, 1997). Co-Q10 is a part of the respiratory chain in mitochondria which plays a role in sperm cell metabolism and fat-soluble antioxidants (Ernstr *et al.* 1993). It is noteworthy that the only lipid-soluble antioxidant producing in our body is Co-Q10, preventing oxidation of proteins, fats and DNA (Ernster, 1993). Co-Q10 is effective on sperm quality and motility, especially in infertile men (Mancini *et al.* 1998; Alleva *et al.* 1997; Lee *et al.* 2006) which are in accordance with the results of our study, due to the improvement of sperm motility. Generally, the better effect of Co-Q10 on sperm motility in comparison ellagic acid may be due to the bioenergetics role of Co-Q10 in the respiratory chain and the production of ATP (Turunen *et al.* 2004; Almeida and Ball, 2005).

Membrane lipid peroxidation leads to disruption of membrane activity, decrease in membrane fluidity and inactivation of receptors and enzymes bands, which ultimately increases non-specific permeability to ions (Esterbauer *et al.* 1990). In the present experiment, TAC was improved with the use of ellagic acid and the combination of two antioxidants.

The MDA level was also higher in the combination treatment group, the difference between results in total antioxidant capacity and MDA justify other results which are based on a mixture of two antioxidants; due to the improved antioxidant capacity and increased levels of MDA, that are likely related to adverse effects or being ineffective on other traits which cause an increase in free radicals and lipid peroxidation.

The two treatment groups were not significantly different from the control group and combination treatments in MDA. Similarly, two antioxidants used in the study had no significant effect on GPx and SOD.

There are few studies on the effects of ellagic acid on ram semen and we compare our results with the studies used extracts that have active ingredient is ellagic acid. In the present study, ellagic acid showed significant effect on viability and total motility that agreed with Reda *et al.* (2016), who observed an improvement in the motility and viability of bull semen using pomegranate juice (PJ) compared to control group and Mansour *et al.* (2013) that used PJ in human semen. Ellagic acid is a polyphenol having variety of biological activities, such as antioxidant (Hassoun *et al.* 2004), anti-mutagenic, anti-apoptosis properties. Although the mechanisms of its action are unknown.

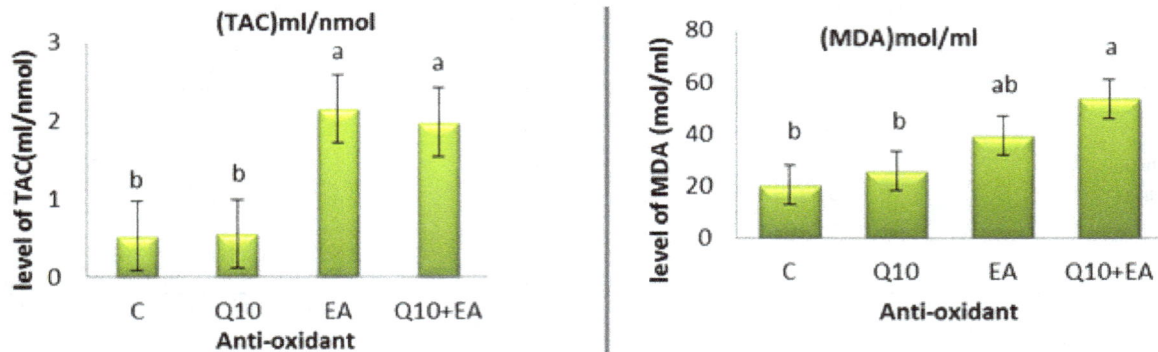

Figure 1 Effects of different levels of ellagic acid and Co-Q10 on TAC and MDA levels after freezing-thawing process

The discussed properties may explain the effect of ellagic acid for increasing sperm motility, viability and capacity to scavenge free radicals (Priyadarsini *et al*. 2002; Seeram *et al*. 2005).

Cisplatin-induced abnormal growth of rat sperm was blocked using ellagic acid (Turk *et al*. 2008). They used different levels of ellagic acid in rat that observed a reduction in MDA level and a significant increase in activity of GSH, GSH-Px and CAT. This result is contrary with our result that ellagic acid did not have significant effect on sperm parameters. They also showed an increase in sperm concentration, sperm motility, sperm cell density, diameter and thickness of the seminiferous tubules of germ cells and decrease in abnormal sperm compared to control group and study on sperm motile and abnormal sperm showed the same result with ours. Similarly, in another study, cyclophosphamide induced lipid per oxidation damage to the structure of the sperm and testicular tissue in rabbits showed the protective effect of eEllagic acid (2 mg/kg) (Ceribasi *et al*. 2010) but in our study ellagic acid did not have any effect on sperm. Ellagic acid has four phenol OH groups with a fused benzofuran structure, therefore has the property of scavenging ROS (reactive oxygen species) and RNS (reactive nitrogen species) which can prevent lipid peroxidation (Bondet *et al*. 1997).

The results of the present study are in contrast to those of Omur *et al*. (2014), who observed an improvement in sperm membrane integrity in the process of cooling ram semen with by using ellagic acid at level of 2 mM. However, they showed in another study Omur and Coyan (2016) that all levels of ellagic acid improved acrosome integrity in the process of freezing ram semen.

In a study of Yousefian *et al*. (2014), which was carried out on the Stallion, Co-Q10 at 1 μM improved sperm motility and parameters of sperm membrane. In present study, Co-Q10 at the level of 0.5 μM improved total motility, survival, abnormal sperm, LIN, VSL and VAP parameters, as well as the improvement of sperm membrane integrity.

CONCLUSION

The results showed that Co-Q10 at level of 0.5 μM improved the sperm quality parameters and reduced the oxidative parameters compared with all treatments. We recommend use of Co-Q10 at level of 0.5 μM in order to raise the quality of frozen ram semen.

ACKNOWLEDGEMENT

The authors wish to thank Mr. Abouzar Najafi for his kind cooperation during the experiment.

REFERENCES

Aitken R.J. and Sawyer D. (2003). The human spermatozoon-not waving but drowning. *Adv. Exp. Med. Biol.* **518**, 85-98.

Alleva R., Scararmucci A., Mantero F., Bompadre S., Leoni L. and Littarru G. (1997). The protective role of ubiquinol-10 against formation of lipid hydroperoxides in human seminal fluid. *Mol. Aspects. Med.* **18**, 221-228.

Almeida J. and Ball B.A. (2005). Effect of α-tocopherol and tocopherol succinate on lipid peroxidation in equine spermatozoa. *Anim. Reprod. Sci.* **87**, 321-337.

Bailey J.L., Bilodeau J. and Cormier N. (2000). Semen cryopreservation in domestic animals: a damaging and capacitating phenomenon. *J. Androl.* **21**, 1-7.

Bondet V., Brand-Williams W. and Berset C. (1997). Kinetics and mechanisms of antioxidant activity using the DPPH. Free radical method. *J. Food Sci. Technol.* **30**, 609-615.

Bucak M.N., Sarıözkan S., Tuncer P.B., Ulutaş P.A. and Akçadağ H.İ. (2009). Effect of antioxidants on microscopic semen parameters, lipid peroxidation and antioxidant activities in Angora goat semen following cryopreservation. *Small Rumin. Res.* **81**, 90-95.

Çeribaşi A.O., Türk G., Sönmez M., Sakin F. and Ateşşahin A. (2010). Toxic effect of cyclophosphamide on sperm morphology, testicular histology and blood oxidant antioxidant balance and protective roles of lycopene and ellagic acid. *Basic. Clin. Pharmacol. Toxicol.* **107**, 730-736.

Daghigh Kia H., Farhadi R., Ashrafi I. and Mehdipour M. (2016a). Anti oxidative effects of ethanol extract of Origanum vulgare on kinetics, microscopic and oxidative parameters of cryopreserved Holstein bull spermatozoa. *Iranian J. Appl. Anim. Sci.* **6,** 901-907.

Daghigh Kia H., Razavian S., Dodran H.V. and Najafi A. (2016b). Effect of adding *Rosa canina* extract and ascorbic acid as natural and synthetic antioxidants on freeze-thawing process of ram semen. *Anim. Rep.* **13,** 567-567.

Daghigh Kia H., Shahbaz Zadeh R. and Ashrafi I. (2015). Effect of antioxidant *Satureja makranta* the microscopic and biochemical parameters of bull sperm after freezing - thawing process. *J. Anim. Sci.* **108,** 101-112.

Ernster L. (1993). Lipid peroxidation in biological membranes: Mechanisms and implications. Pp. 1-38 in Active Oxygens, Lipid Peroxides, and Antioxidants. K. Yagi, Ed. CRC Press, Boca Raton, Florida. USA.

Ernster L. and Dallner G. (1995). Biochemical, physiological and medical aspects of ubiquinone function. *Biochim. Biophys. Acta (BBA)-Mol. Basis Dis.* **1271,** 195-204.

Eskenazi B., Kidd S., Marks A., Sloter E., Block G. and Wyrobek A. (2005). Antioxidant intake is associated with semen quality in healthy men. *Hum. Reprod.* **20,** 1006-1012.

Esterbauer H., Dieber-Rotheneder M., Waeg G., Striegl G. and Juergens G. (1990). Biochemical structural and functional properties of oxidized low-density lipoprotein. *Chem. Res. Toxicol.* **3,** 77-92.

Funahashi H. and Sano T. (2005). Select antioxidants improve the function of extended boar semen stored at 10 C. *Theriogenology.* **63,** 1605-1616.

Hassoun E.A., Vodhanel J. and Abushaban A. (2004). The modulatory effects of ellagic acid and vitamin E succinate on TCDD induced oxidative stress in different brain regions of rats after subchronic exposure. *J. Biochem. Mol. Toxicol.* **18,** 196-203.

Jeyendran R.S., Van der Ven H.H., Perez-Pelaez M., Crabo B.G. and Zaneveld L.J. (1984). Development of an assay to assess the functional integrity of the human sperm membrane and its relationship to the other sperm characteristics. *J. Reprod. Fertil.* **70,** 219-28.

Lewin A. and Lavon H. (1997). The effect of coenzyme Q10 on sperm motility and function. *Mol. Aspects. Med.* **18,** 213-219.

Lee W., Li K. and Huang Y. (2006). Biological function of CoQ10 and its effect on the quality of spermatozoa. *Zhonghua Nan. Ke. Xue.* **12,** 1119-1122.

Mancini A., Conte G., Milardi D., De Marinis L. and Littarru G. (1998). Relationshipbetween sperm cell iquinone and seminal parameters in subjects with and without varicocele. *Andrologia.* **30,** 1-4.

Mansour S.W., Sangi S., Harsha S., Khaleel M.A. and Ibrahim A. (2013). Sensibility of male rats fertility against olive oil, Nigella sativa oil and pomegranate extract. *Asian Pac. J. Trop. Biomed.* **3,** 563-568.

Mata-Campuzano M., Alvarez-Rodriguez M., Alvarez M., Tamayo-Canul J., Anel L., de Paz P. and Martinez-Pastor F. (2015). Post-thawing quality and incubation resilience of cryopreserved ram spermatozoa are affected by antioxidant supplementation and choice of extender. *Theriogenology.* **83,** 520-528.

Mehdipour M., Daghigh K.I.A.H., Najafi A., Vaseghi-Dodaran H. and García-Álvarez O. (2016). Effect of green tea (*Camellia sinensis*) extract and pre-freezing equilibration time on the post-thawing quality of ram semen cryopreserved in a soybean lecithin-based extender. Cryobiology. **73,** 297-303.

Michael A., Alexopoulos C., Pontiki E., Hadjipavlou-Litina D., Saratsis P. and Boscos C. (2007). Effect of antioxidant supplementation on semen quality and reactive oxygen species of frozen-thawed canine spermatozoa. *Theriogenology.* **68,** 204-212.

Miller N.J., Rice-Evans C., Davies M.J., Gopinathan V. and Milner A. (1993). A novel method for measuring antioxidant capacity and its application to monitoring the antioxidant status in premature neonates. *Clin. Sci.* **84(4),** 407-412.

Najafi A., Daghigh Kia H., Mohammadi H., Najafi M.H., Zanganeh Z., Sharafi M., Martinez-Pastor F. and Adeldust H. (2014a). Different concentrations of cysteamine and ergothioneine improve microscopic and oxidative parameters in ram semen frozen with a soybean lecithin extender. *Cryobiology.* **69,** 68-73.

Najafi A., Najafi M.H., Zanganeh Z., Sharafi M., Martinez-Pastor F. and Adeldust H. (2014b). Cryopreservation of ram semen in extenders containing soybean lecithin as cryoprotectant and hyaluronic acid as antioxidant. *Reprod. Domest. Anim.* **49,** 934-940.

Najafi A., Zhandi M., Towhidi A., Sharafi M., Akbari Sharif A., Khodaei Motlagh M. and Martinez-Pastor F. (2013). Trehalose and glycerol have a dose-dependent synergistic effect on the post-thawing quality of ram semen cryopreserved in a soybean lecithin-based extender. *Cryobiology.* **66,** 275-282.

Najafi A., Daghigh-Kiaa H., Vaseghi Dodarana H., Mehdipoor M. and Alvarez-Rodriguez A. (2016). Ethylene glycol, but not DMSO, could replace glycerol inclusion in soybean lecithin-based extenders in ram sperm cryopreservation. *Anim. Reprod. Sci.* **177,** 35-41.

Omur A. and Coyan K. (2016). Protective effects of the antioxidants curcumin, ellagic acid and methionine on motility, mitochondrial transmembrane potential, plasma membrane and acrosome integrity in freeze-thawed Merino ram sperm. *Vet. Med.* **61,** 10-16.

Omur A., Coyan K., Ozturk C., Gungor S. and Bucak M. (2014). The effects of curcumin, ellagic acid and methionine on postthawed Merino rams sperm parameters (759.1). *FASEB J.* **28,** 751-759.

Placer Z.A., Cushman L.L. and Johnson B.C. (1966). Estimation of product of lipid peroxidation (malonyl dialdehyde) in biochemical systems. *Anal. Biochem.* **16,** 359-364.

Priyadarsini K.I., Khopde S.M., Kumar S.S. and Mohan H. (2002). Free radical studies of ellagic acid, a natural phenolic antioxidant. *J. Agric. Food Chem.* **50,** 2200-2206.

Reda El-Sheshtawy I., Gamal El-Sisy A. and Walid El-Nattat S. (2016). Effects of pomegranate juice in Tris-based extender on cattle semen quality after chilling and cryopreservation. *Asian Pac. J. Reprod.* **4(1),** 26-31.

Salamon S. and Maxwell W. (2000). Storage of ram semen. *Anim. Reprod. Sci.* **62,** 77-111.

Sariozkan S., Tuncer P.B., Buyukleblebici S., Bucak M.N., Canturk F. and Eken A. (2015). Antioxidative effects of cysteamine, hyaluronan and fetuin on post-thaw semen quality, DNA integrity and oxidative stress parameters in the Brown Swiss bull. *Andrologia.* **47,** 138-147.

SAS Institute. (2004). SAS®/STAT Software, Release 9.1. SAS Institute, Inc., Cary, NC. USA.

Seeram N.P., Adams L.S., Henning S.M., Niu Y., Zhang Y., Nair M.G. and Heber D. (2005). *In vitro* anti-proliferative, apoptotic and antioxidant activities of punicalagin, ellagic acid and a total pomegranate tannin extract are enhanced in combination with other polyphenols as found in pomegranate juice. *J. Nutr. Biochem.* **16,** 360-367.

Sheweita S.A., Tilmisany A.M. and Al-Sawaf H. (2005). Mechanisms of male infertility: role of antioxidants. *Curr. Drug Metab.* **6,** 495-501.

Turunen M., Olsson J. and Dallner G. (2004). Metabolism and function of coenzyme Q. *Biochim. Biophys. Acta.* **1660(2),** 171-199.

Türk G., Ateşşahin A., Sönmez M., Çeribaşi A.O. and Yüce A. (2008). Improvement of cisplatin-induced injuries to sperm quality, the oxidant-antioxidant system, and the histologic structure of the rat testis by ellagic acid. *Fertil. Steril.* **89,** 1474-1481.

Vaseghi Dodaran H., Zhandi M., Sharafi M., Nejati-Amiri E., Nejati-Javaremi A., Mohammadi-Sangcheshme A., Shehab-El-Deen M.A.M. and Shakeri M. (2015). Effect of ethanol induced mild stress on post-thawed bull sperm quality. *Cryobiology.* **71,** 12-17.

Yousefian I., Zare-Shahneh A. and Zhandi M. (2014). The effect of coenzyme Q10 and α-tocopherol in skim milk-based extender for preservation of Caspian stallion semen in cool condition. *J. Equine Vet. Sci.* **34,** 949-954.

22

Association between Melatonin Receptor 1A (*MTNR1A*) Gene Polymorphism at the *MnlI* Site and Production Traits in Shal and Crossbreeding between Shal and Romanov

I. Afrooznia[1], M. Zandi[1*], M.R. Sanjabi[1] and M.H. Hadi Tavatori[2]

[1] Department of Agriculture, Iranian Research Organization for Science and Technology (IROST), Tehran, Iran
[2] Agricultural and Natural Resources Research Center of Qazvin Province, Qazvin, Iran

*Correspondence E-mail: mz1075@yahoo.com

ABSTRACT

The objective of this study was to evaluate the relationship between polymorphism at the *MnlI* site of *MTNR1A* gene and production traits of Shal and Romanov crossbreeding. The crossbreeding between Shal and Romanov was done on 600 Shal ewes. For this reason, Shal ewes were laparoscopically and artificially inseminated with the semen of Romanov breed. For the preparation of genomic DNA, a total of 90 ewes, either Shal (n=50) or Shal × Romanov (n=40), were used and an 824 bp fragment of exon 2 of the *MTNR1A* gene was amplified. In this research, two genotypes MM and Mm were detected in the study Shal population with frequencies of 0.78 and 0.22, respectively, while MM and mm genotypes of the study crossbreeds were detected with 0.125 and 0.0875 frequencies, respectively. However, the mm and Mm genotypes were not detected in Shal and Shal × Romanov crossbreeds, respectively. The *MnlI* marker site in Shal × Romanov breeds had a significant effect (P>0.05) on body weight after 3 months (BW3) as the ewes with MM genotypes had significantly higher BW3 than mm individuals. However, no significant difference was observed between MM and mm marker sites for birth weight (BW) and 6-month body weight (BW6). The same results were observed for Shal breeds as when MM and Mm genotypes were detected for BW, BW3 and BW6 traits. In conclusion, in this study, mm genotype as a measurement for litter size in sheep breeds was successfully introduced into Shal × Romanov crossbreeds, while their production traits were not affected by *MnlI* genotypes.

KEY WORDS crossbreed, *MTNR1A*, Romanov, Shal.

INTRODUCTION

Remarkable seasonal variations are observed in the reproductive activities of small ruminants living in temperate latitudes (Luridiana *et al.* 2015). Thus, a great alteration is induced by lamb reproductive seasonality and its consequent meat market price. Alternatively, out-of-season reproduction is induced through a wide use of hormonal treatments in some countries. However, there is a need for a search for some other methods due to the increasing demand for free-hormone products (Martinez-Royo *et al.* 2012). Nonetheless, identification of informative genetic markers has been suggested to improve selection efficiency for reducing breeding seasonality in sheep (Moradi *et al.* 2014). In a response to photoperiodism, animals are engaged in the secretion of differential melatonin to regulate their reproductive activities (Saxena *et al.* 2014). The light signals relayed to the pineal gland by several neurons from the retina, known as photoperiodic information, are translated into the daily cycle of melatonin secretion, the level of which is high at night and low during day time (Shahroudi *et al.* 2006). Accordingly, the secretion of gonadotrophin-releasing hormone (GnRH) is controlled by the seasonal rhythms of melatonin secretion that serves as a seasonal messenger in endocrine tissues at the hypothalamic level (Luridiana *et al.* 2014). The alternating presence or absence of female ovulation and male sperm production is dependent on luteinizing hormone secretion, the levels of which are induced by the corresponding changes in GnRH release (Shahroudi *et al.* 2006).

Melatonin reproductive effects occur in the premammillary hypothalamus (Luridiana *et al.* 2015) and the hypothalamic suprachiasmatic nucleus is the site of the circadian clock where melatonin circadian effects are mediated by the pharmacologically specific G-protein-coupled melatonin receptors with high affinity (Jia *et al.* 2012), through which melatonin exerts its effects when binding to them. However, it seems that the reproductive activity is only regulated by MT1 from among the receptors involved (Luridiana *et al.* 2015).

In an investigation for sheep sexual activity, Notter *et al.* (2003) showed that the genotype of this gene might become as a particular marker based on the association between different allelic forms at MT1 locus and sheep reproductive activity.

Reppert *et al.* (1994) were the first to clone melatonin receptor 1A gene (*MTNR1A*) as a high-affinity subtype in mammals. It should be noted that from among any other melatonin receptor subtypes like *MTNR1B* and *MTNR1C*, only *MTNR1A* appears to be engaged in the seasonal reproductive activity regulation (Saxena *et al.* 2014; Trecherel *et al.* 2010). An association between *MTNR1A* gene and seasonal reproductive activity has been evidenced by several studies on different animal species (Luridiana *et al.* 2014).

Circadian variations in melatonin concentrations have been repeatedly proposed to be caused by *MTNR1A* as a candidate gene via photoperiodic control of seasonality (Mura *et al.* 2014). *MTNR1A* gene, consisting of two exons divided by a large intron, has been mapped in the ovine chromosome 26. The exon one shows low degree of polymorphism. It has been shown that exon 2 of *MTNR1A* gene, which is involved in the coding process of ovine MT1 receptors is highly polymorphic, while changes in the second exon would make the differences in the structures of the receptors. The 2 restriction fragment length polymorphism (RFLP) sites for *MnlI* and *RsaI* enzymes are derived from exon 2 of the gene encoding MT1 receptor in sheep. The characteristic pattern of digestion by this enzyme is characterized by a mutation at *MTNR1A/MnlI* site, which leads to the absence (–or M) of *MnlI* specific cleavage site at position 605 of the coding sequence (Hatami *et al.* 2014). The 3 genotypes of "M/M", "M/m" and "m/m", which were pre-

viously described as +/+, +/– and –/– (Trecherel *et al.* 2010), are discriminated by the presence or absence of this mutation.

Sheep, goat and buffalo were selected to carry out the studies of relationship between this gene polymorphism and reproductive activity. In this regard, a reproductive association with the polymorphic sites identified in *MTNR1A* gene was shown by some sheep breeds like Merinos d'Arles, small tailed Han sheep, Awasi and Sarda (Saxena *et al.* 2014).

A local sheep breed is Shal, which constitutes a population of more than 600000 heads in Iran. The breed is large-sized, fat-tailed, and predominantly black or brown with white spots on the front head, while being well-adapted to harsh climates. It is mainly raised for its meat as the most important protein source in Iran. Following a random exposure of Ewes, which usually lamb 3 times every 2 years, to 18-month rams, lambing occurred within one season, i.e. from mid-January to mid-March (Hossein-Zadeh, 2015).

One of the breeds renowned for its long breeding season, early sexual maturity and high prolificacy is Romanov. However, poor carcass quality and relatively low growth rate are of the typical features of this breed in comparison to the traditional meat species. Application of a commercial crossing approach with the meat-type breeds is a simplest and fastest way of improving growth and carcass quality in Romanov lambs (Kuchtík *et al.* 2012). Crossbreeding between Romanov and Shal were used to improvement of prolificacy in Shal breed. In the current study, we considered the polymorphism of exon 2 of *MTNR1A* at *MnlI* site as a marker for prolificacy and investigated its relationship with the production traits of Shal and Romanov crossbreeding.

MATERIALS AND METHODS

Experimental animals and their management:
This experiment was carried out at a sheep farm in Ismaeel Abad (longitude: 49.66° E and latitude: 36.46° N) located in Qazvin province, Iran. Shal and Romanov crossbreeding were done on 600 Shal ewes. To this purpose, Shal ewes (1-2 years old) were laparoscopically and artificially inseminated with the semen of Romanov breed. An integration of natural pasture and concentrate feed under natural photoperiod was followed the ewes since birth. All the lambs were born indoors and a similar nutrition status was considered for all sampled individuals within each breed.

Preparation, amplification, and digestion of genomic DNA
A total of 90 ewes, either Shal (n=50) or Shal × Romanov (n=40), were used. Five mL of blood was collected from

the jugular vein in ethylenediaminetetraacetic acid (EDTA)-coated tubes. Ovine genomic DNA was extracted using phenol-chloroform method as previously described by Psifidi *et al.* (2015) and maintained at -20 °C until use. Approximately, 200 ng of genomic DNA was subjected to polymerase chain reaction (PCR) by using specific primers synthesized by Sina colon, Iran. The primers were those used by Messer *et al.* (1997) (sense primer 5'-TGTGTTTGTGGTGAGCCTGG-3' and antisense primer: 5'-ATGGAGAGGGTTTGCGTTTA-3') from the sequence of Exon II of the ovine MTNR1A gene (GeneBank U14109). Polymerase chain reaction (PCR) was carried out in a volume of 25 μl containing 200 ng of DNA, 1XPCR buffer, 2.0 mmol of $MgCl_2$, 0.2 mmol of each dNTP, 10 pmol of each primer, and 2 U of TaqDNA polymerase (Fermentas, Canada). The PCR conditions were as follows: initial denaturation at 94 °C for 5 min followed by 35 repeated cycles of denaturation at 94 °C for 1 min, annealing at 62 °C for 1 min, extension at 72 °C for 1 min, and final extension at 72 °C for 10 min on a thermal cycler (Bio-Rad, T100). The PCR products were resolved by electrophoresis on 2% agarose gel in parallel with 250 bp of DNA marker ladder (Fermentas). Then, they were separated by electrophoresis on 2% agarose gel. After amplification, 7 μL of the PCR product was digested with 2 units of *MnlI* endonuclease at 37 °C for 4 hours following a deactivation process at 65 °C for 20 minutes.

Statistical analysis
For the genotyping of the study samples, the digested fragments were electrophoresed on 3% agarose gel and stained with ethidium bromide. Allelic frequencies were determined by direct counting of the observed genotypes. The X^2 test was used to determine Hardy-Weinberg (HW) equilibrium of the mutation by using following equation (GenAllex Software).

$$X^2 = \sum_{i=1}^{111} \sum_{j=1}^{i} \left[\frac{(O_{ij} - E_{ij})^2}{E_{ij}} \right]$$

The significance of differences between the treatments was evaluated using the t-test. The results were expressed as mean ± SEM and the statistical significance was accepted at P < 0.05. The data were analyzed using a statistical software (SPSS, 2011).

RESULTS AND DISCUSSION

PCR-RFLP analysis of exon 2 of *MTNR1A* gene
In current study 824 bp fragment of exon 2 of Shal and Shal × Romanov *MTNR1A* gene were analyzed by RFLP (Figure 1).

The presence and absence of 303 bp fragment were referred to as alleles m and M, resulting in 236 and 67 bp products. Thus, the 3 genotypes of MM, Mm, and mm were detected based on 236, 236 and 303, and 303 fragments, respectively. In our study, the 2 genotypes of MM and Mm were detected in the study population of Shal with the frequencies of 0.78 and 0.22, respectively (Figure 2 and Table 1). However, MM and mm genotypes were detected in the crossbreeds under study with 0.125 and 0.0875 frequencies, respectively. Mm and mm genotypes were not detected in Shal × Romanov and Shal breeds, respectively (Figure 3 and Table 1). MM and Mm frequencies were significantly higher in Shal compared to Shal × Romanov crossbreeds, and that of mm was significantly higher in the study crossbreeds (P<0.05). The allelic frequencies of 0.89 and 0.125 were observed for M allele and 0.875 and 0.11 for m allele at *MTNR1A* locus in Shal and Shal × Romanov crossbreeds, respectively (Table 1).

Contrary to our results, Chu *et al.* (2006) demonstrated an association between MM genotype and non-seasonal estrus in ewes, as well as a relationship between mm genotype and seasonal estrus in ewes when investigating non-seasonal estrous breeds (Small Tail Han and Hu ewes) and seasonal estrous breeds (Dorset, Suffolk, and German Mutton Merino ewes).

Many studies indicated an association between the homozygous genotype for the absence of a polymorphic *MnlI* site at position 605 of exon 2 of *MTNR1A* gene and seasonal anovulation in ewes (Chu *et al.* 2006). In our research, the X^2 test confirmed that only the study Shal sheep population was in HW equilibrium. Perhaps, this was due to the limited number of crossbred population and non-random mating.

Nevertheless, congruent with our results, Carcangiu *et al.* (2009) reported that the ewes of MM genotypes lambed in autumn following an oestrus in spring. Having a very high frequency (0.81%) among their study ewes, M allele surely played a very important role in this process. Though Notter *et al.* (2003) demonstrated that the various sheep breeds reared in North America were sufficiently influenced by a single M allele leading to a reproductive seasonality, Pelletier *et al.* (2000) found that MM homozygous ewes of Merino d'Arles sheep breed showed oestrus in spring. In their study on native Iranian breeds of sheep, Ghiasi *et al.* (2006) demonstrated that the sheep, which genetically had a potential to show estrus during short and long days, indicated 2 genotypes at *MTNR1A* locus with the frequencies of 0.7 (MM) and 0.3 (Mm) in the Karakul breed, and 0.58 (MM) and 0.42 (Mm) in the Shal breed. Also, in another study on native Iranian breeds, i.e. Zel and Naeini, Moradi *et al.* (2014) reported that M allele was predominant in either Zel or Naeini lambs.

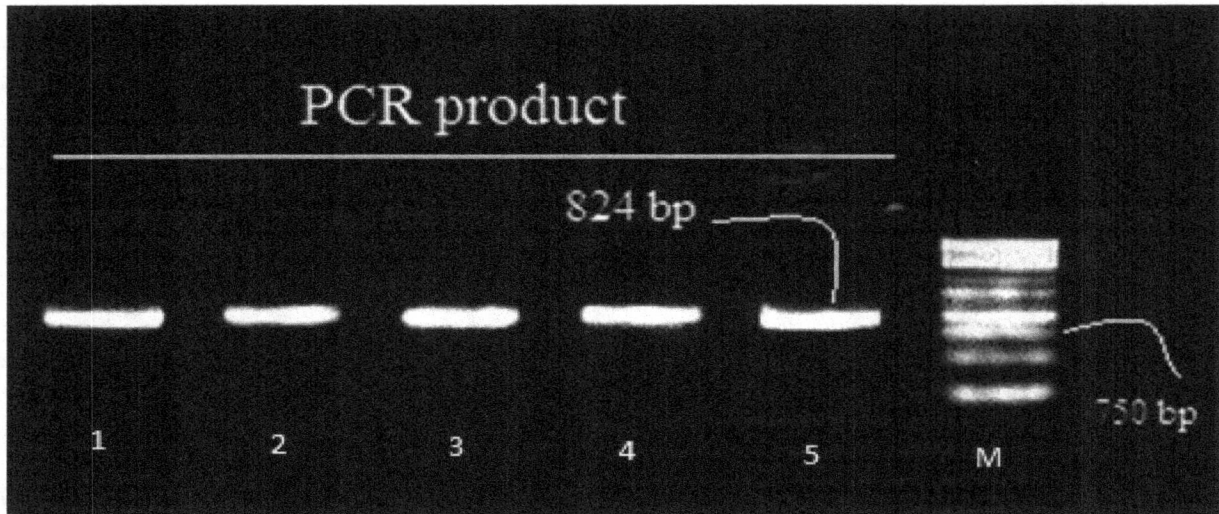

Figure 1 Electrophoresis pattern of PCR product of exon 2 of *MTNR1A* gene (824 bp) in Shal and Shal × Romanov sheep (2% agarose gel) Lanes 1-3 and 4-5, PCR amplification product for Shal and Shal × Romanov, respectively, and M: DNA molecular marker (250 bp)

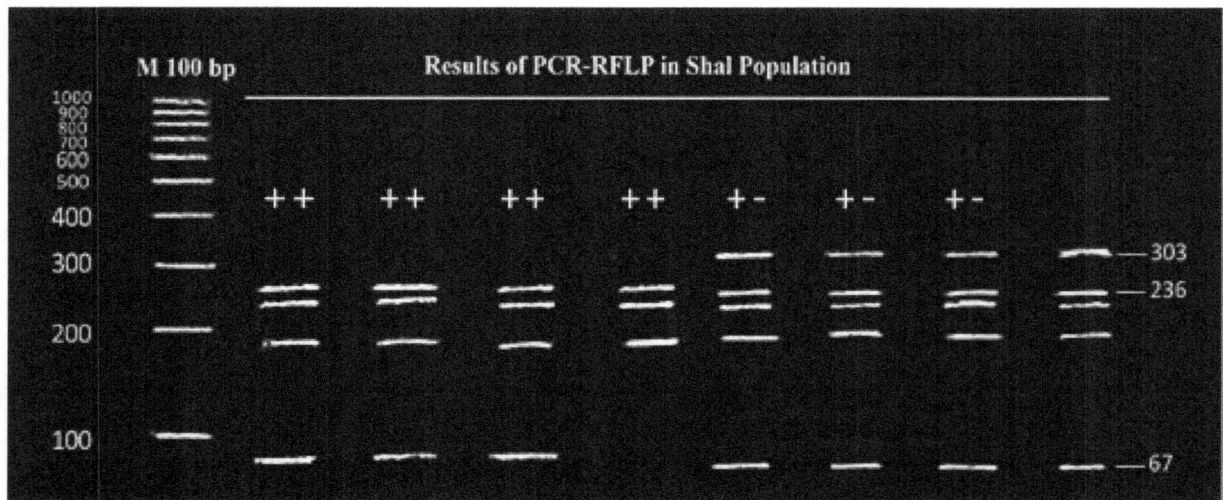

Figure 2 RFLP Genotyping of *MTNR1A* gene by *MnlI* in Shal breed

Table 1 Allele and genotype frequencies of exon 2 of *MTNR1A* gene for *MnlI* site in Shal and Shal × Romanov breeds

Breed	Allele frequency		Genotype frequeny		
	M	m	MM	Mm	mm
Shal	0.89	0.11	0.78	0.22	0
Shal × Romanov	0.125	0.875	0.125	0	0.875
P-value	P < 0.001	P < 0.001	P < 0.001	P < 0.001	P < 0.001

The genotypic frequencies for MM, Mm, and mm genotypes were 0.52, 0.25 and 0.23 in Zel breed and 0.60, 0.22 and 0.18 in Naeini breed. Furthermore, they showed that MM genotype had the lowest mean for litter size significantly. In general, the highest mean measurements for litter size were obtained from mm followed by Mm, whereas the minimum mean measurements were achieved from MM in both assessed breeds.

Even in Sarda sheep breed exhibiting an anoestrous period in late-winter/spring, polymorphisms within *MTNR1A* gene lead to some advances in reproductive activity resumption (Carcangiu *et al.* 2009).

Additionally, Luridiana *et al.* (2015) evidenced an relationship between *MTNR1A* gene polymorphism and a resumption of reproductive activity in adult Sarda sheep breed in spring.

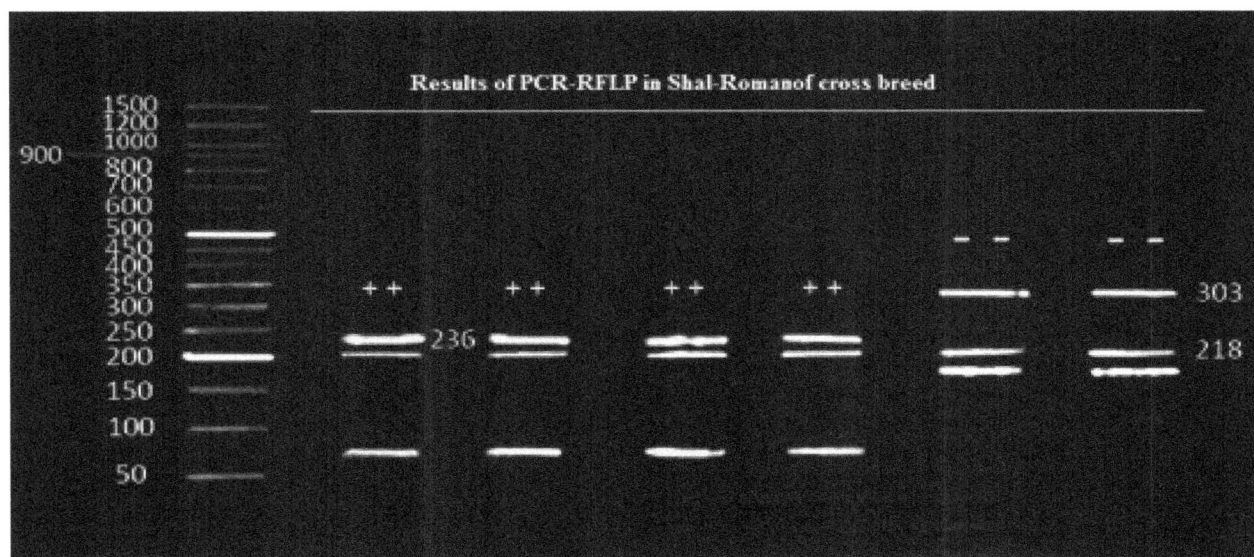

Figure 3 RFLP Genotyping of *MTNR1A* gene by *MnlI* in Shal × Romanov breed

However, no association between *MNTR1A* polymorphisms and seasonality of reproduction in Ile-de-France ewes was found by Hernandez *et al.* (2005). Moreover, some European breeds have not proven any impacts of *MTNR1A* gene polymorphisms on their reproductive activities.

The difference might be well attributed to breed traits or body and environmental conditions. Thus, there is a need to expand the knowledge of the effects of *MTNR1A* gene polymorphism for the clarification of the mechanisms regulating the different reproductive responses of the different sheep breeds (Mura *et al.* 2014).

Association between *MTNR1A* gene polymorphism and birth and body weights

There are restricted opportunities for a selection within the breeds due to the fact that less heritability (nearly 0.05-0.15) occurs for most reproductive traits compared to many other traits.

Nonetheless, to optimize reproductive potentials, a within-breed selection must be done after crossing the divergent breeds to the genetic potentials rapidly reset for those traits, through which basic changes in litter size or seasonal breeding patterns are best achieved.

Another opportunity can be gained by quickly adjusting genetic potentials via various mutations that affect the ovulation rate and litter size of sheep though a careful breeding management is required, particularly for production traits (Notter, 2012).

Our results revealed that *MnlI* marker site in Shal × Romanov breeds had a significant effect ($P>0.05$) on a 3-month body weight (BW3) as ewes with MM genotypes had significantly higher BW3 than mm individuals. However, no significant differences were observed between MM and mm marker sites for 6-month birth and body weights (BW6) (Table 2). The same results were observed for Shal breed as MM and Mm genotypes were detected for BW, BW3, and BW6 traits (Table 3).

Table 2 Production performance (Mean±SEM) of Shal × Romanov breed

Traits	P-value	Genotype	
		MM	mm
BW	0.185[ns]	4.38±0.13	4.75±0.23
BW3	0.034[*]	20.09±0.47	17.88±0.97
BW6	0.197[ns]	28.06±0.73	26.02±1.45

BW: birth weight; BW3 and 6: body weight at 3 and 6 months of age.
* (P<0.05).
SEM: standard error of the means.
NS: non significant.

Table 3 Production performance (Mean±SEM) of Shal breed

Traits	P-value	Genotype	
		MM	**Mm**
BW	0.106^{ns}	4.02±0.25	3.51±0.11
BW3	0.013^{*}	25.3±0.77	22.77±0.35
BW6	0.552^{ns}	37.3±1.18	36.61±0.40

BW: birth weight; BW3 and 6: body weight at 3 and 6 months of age.
* ($P<0.05$).
SEM: standard error of the means.
NS: non significant.

Although in our study, BW3 was affected by *MnlI* genotypes in both Shal and Shal × Romanov breeds, the production traits of Shal and Shal × Romanov breeds were not affected by *MnlI* genotypes based on BW and BW6 traits. Also, they were not influenced by mm or Mm genotypes in the current studies. In their study on the seasonal reproduction of Zandi breed, Hatami *et al.* (2014) reported that *MTNR1A/MnlI* marker site had a significant effect on BW1 as the ewes with Mm genotypes were found to have higher BW1 compared to MM individuals. No significant association was observed between *MTNR1A/MnlI* marker site and the other study traits, including BW, BW3, and BW6.

CONCLUSION

In the current research, the 2 genotypes of MM and Mm were detected in Shal population involved in seasonal reproduction. However, MM and mm genotypes were detected in Shal with Romanov crossbreeding. Thus, mm genotype observed as a measurement for litter size in sheep breeds successfully introduced into the crossbreeds. Therefore, based on BW and BW6 traits, the production traits of Shal and Shal × Romanov breeds were not affected by *MnlI* genotypes. Nevertheless, further studies are recommended to reveal the relationship between *MTNR1A* gene polymorphism at *MnlI* site and reproductive performance of Shal × Romanov ewes.

ACKNOWLEDGEMENT

We would like to thank Amir Hossein Ahadi for his helpful comments during the research. This project supported by Iran national science foundation.

REFERENCES

Carcangiu V., Mura M.C., Vacca G.M., Pazzola M., Dettori M.L., Luridiana S. and Bini P.P. (2009). Polymorphism of the melatonin receptor *MT1* gene and its relationship with seasonal reproductive activity in the Sarda sheep breed. *Anim. Reprod. Sci.* **116,** 65-72.

Chu M.X., Cheng D.X., Liu W.Z., Fang L. and Ye S.C. (2006). Association between melatonin receptor 1A Gene and expression of reproductive seasonality in sheep. *Asian-Australas J. Anim. Sci.* **19,** 1079-1084.

Ghiasi H., Nasiry M.R., Heravi-Mousavi A.R., Mousavizadeh A.Z. and Manesh A.J. (2006). Genetic polymorphism of the melatonin receptor 1A locus in Iranian Shal and Karakul sheep. *Iranian J. Biotechnol.* **4,** 201-203.

Hatami M., Mianji G.R. and Farha A. (2014). Association of melatonin receptor 1A gene polymorphisms with production and reproduction traits in Zandi sheep. *Iranian J. Appl. Anim. Sci.* **4,** 75-78.

Hernandez X., Bodin L., Chesneau D., Guillaume D., Chemineau P., Mal-paux B. and Migaud M. (2005). Relationship between *MT1* melatonin receptor gene polymorphism and seasonal physiological responses in Ile-de-Franceewes. *Reprod. Nutr. Dev.* **45,** 151-162.

Hossein-Zadeh N.G. (2015). Modeling the growth curve of Iranian Shal sheep using non-linear growth models. *Small Rumin. Res.* **130,** 60-66.

Jia L., Feng T., Chu M.X., Chen H., Di R., Sun J., Cao G. and Fang L. (2012). Polymorphism and structure of exon 2 of caprine melatonin receptor 1b gene and its relations to fertility and seasonal estrus. *Anim. Sci. Pap. Rep.* **30,** 169-179.

Kuchtík J., Zapletal D. and Sustova K. (2012). Chemical and physical characteristics of lamb meat related to crossbreeding of Romanov ewes with Suffolk and Charollais sires. *Meat Sci.* **90,** 426-430.

Luridiana S., Mura M.C., Daga C., Cosso G., Bodano S., Zidda F., Carcangiu V. and Farci F. (2014). Influences of melatonin treatment, melatonin receptor 1A (*MTNR1A*) and kisspeptin (*KiSS-1*) gene polymorphisms on first conception in Sarda ewe lambs. *Reprod. Fertil. Dev.* **28,** 750-756.

Luridiana S., Mura M.C., Daga C., Diaz M.L., Bini P.P., Cosso G. and Carcangiu V. (2015). The relationship between melatonin receptor 1A gene (*MTNR1A*) polymorphism and reproductive performance in Sarda breed sheep. *Livest. Sci.* **171,** 78-83.

Martínez-Royo A., Lahoz B., Alabart J.L., Folch J. and Calvo J.H. (2012). Characterisation of the melatonin receptor 1A (*MTNR1A*) gene in the Rasa Aragonesa sheep breed: association with reproductive seasonality. *Anim. Reprod. Sci.* **133,** 169-175.

Messer A.L., Wang L., Tuggle C.K. and Rothschild M.F. (1997). Mapping of the melatonin receptor 1A (*MTNR1A*) gene in pigs, sheep and cattle. *Mamm. Gen.* **8,** 368-370.

Moradi N., Rahimi-Mianji G., Nazifi N. and Nourbakhsh A. (2014). Polymorphism of the melatonin receptor 1A gene and its association with litter size in Zel and Naeini sheep breeds. *Iranian J. Appl. Anim. Sci.* **4,** 79-87.

Mura M.C., Luridiana S., Bodano S., Daga C., Cosso G., Diaz M.L., Bini P.P. and Carcangiu V. (2014). Influence of melatonin receptor 1A gene polymorphisms on seasonal reproduction in Sarda ewes with different body condition scores and ages. *Anim. Reprod. Sci.* **149,** 173-177.

Notter D.R. (2012). Genetic improvement of reproductive efficiency of sheep and goats. *Anim. Reprod. Sci.* **130,** 147-151.

Notter D.R., Cockett N.E. and Hadfield T.S. (2003). Evaluation of melatonin receptor 1A as a candidate gene influencing reproduction in an autumn lambing sheep flock. *J. Anim. Sci.* **81,** 912-917.

Pelletier J., Bodin L., Hanocq E., Malpaux B., Teyssier J., Thimonier J. and Chemineau P. (2000). Association between expression of reproductive seasonality and alleles of the gene mel1a receptor in the ewe. *Biol. Reprod.* **62,** 1096-1101.

Psifidi A., Dovas C.I., Bramis G., Lazou T., Russel C.L., Arsenos G. and Banos G. (2015). Comparison of eleven methods for genomic DNA extraction suitable for large-scale whole-genome genotyping and long-term DNA banking using blood samples. *PLoS One.* **10,** e0115960.

Reppert S.M., Weaver D.R. and Ebisawa T. (1994). Cloning and characterization of a mammalian melatonin receptor that mediates reproductive and circadian responses. *Neuron.* **13,** 1177-1185.

Saxena V.K., Jha B.K., Meena A.S. and Naqvi S.M.K. (2014). Sequence analysis and identification of new variations in the coding sequence of melatonin receptor gene (*MTNR1A*) of Indian Chokla sheep breed. *Meta Gene.* **2,** 450-458.

Shahroudi F.E., Nassiry M.R., Valizadh R., Heravi A., Mojtaba M., Pour T. and Ghiasi H. (2006). Genetic polymorphism at *MTNR1A*, CAST and CAPN loci in Iranian Karakul sheep. *Iranian J. Biotechnol.* **4,** 117-122.

SPSS Inc. (2011). Statistical Package for Social Sciences Study. SPSS for Windows, Version 20. Chicago SPSS Inc.

Trecherel E., Batailler M., Chesneau D., Delagrange P., Malpaux B., Chemineau P. and Migaud M. (2010). Functional characterization of polymorphic variants for ovine *MT1* melatonin receptors: possible implication for seasonal reproduction in sheep. *Anim. Reprod. Sci.* **122,** 328-334.

23

Calpastatin Gene Polymorphism in Raini and Tali Goat in the Kerman Province

V. Bahrampour[1*] and A. Mohammadi[2]

[1] Department of Animal Science, Technical and Vocational University, Kerman, Iran
[2] Department of Animal Science, Faculty of Agriculture, University of Zabol, Zabol, Iran

*Correspondence E-mail: v_bahrampur@yahoo.com

ABSTRACT

Restriction fragment length polymorphisms (RFLP) have been identified at the goat. The objective of the present study was to determine polymorphisms of calpastatin locus in Raini and Tali goats of the goats in Kerman Province, improving meat quality traits superior meat quality by selection. Calpastatin gene effect on quality and tenderness meat, to identify different genotype of this gene was randomly selected 150 Tali and 150 Raini goats, and blood samples were collected from total of animals using ethylenediamine-tetraacetic acid (EDTA) tubes. DNA was extracted from blood according to DNA preparation kit to determine the genetic relationship among studied goat animals. Amplification was performed using polymerase chain reaction (PCR). Two alleles (A and B) and three genotypes, (AA, BB and AB) were observed. The frequencies of the observed genotypes for Tali and raini were 38.4, 9.6, 52 and 17, 71, 12 for AA, BB and AB, respectively. And allele frequencies were 0.47, 0.53 and 0.12, 0.88 for A and B, in Tali and Raini goats respectively.

KEY WORDS calpastatin gene, genotype, PCR RFLP, Raeini goat, Tali goat.

INTRODUCTION

Goat (*Capra aegagrus*), is one of the livestock, first domesticated in Asia. Goats live in different areas. Goat is one of the most adaptable animals with different environmental conditions. Calpastatin, initially identified in skeletal muscle (Busch *et al.* 1972), is the endogenous inhibitor of the calpains (Goll *et al.* 2003). Calpastatins are rich in proline and glutamate, but poor in aromatic amino acids (Murachi, 1983). The rate of protein degradation post mortem affects meat quality (Koohmaraie *et al.* 2002). The calpastatin gene has also been found to have a large effect on pork quality (Rohrer *et al.* 2012). Genetic variation in calpastatin gene and the effect on meat quality traits both in cattle and sheep has been reported by many researchers (Zhou and Hickford, 2008). At the molecular level, calpastatin protein is com-

posed of five-domains having molecular weight of 76 kDa which bind and inhibit the calpains system. The function of N terminal domain is to enhance the targeting of the inhibitory region of calpain system by other inhibitory domain but itself don't play any inhibitory role (Averna *et al.* 2001). Marker assisted selection is one of the new DNA based methods that improves accuracy and progress in animal selection programs (Bastos *et al.* 2001). Genetic markers are important for the determination of allelic polymorphism at any specific locus. A two-allele system of polymorphic variants (M and N) of the ovine calpastatin gene by a PCR-RFLP method has been described (Palmer *et al.* 1998). Tenderness is one of the attributes of quality of beef that is most appreciated by consumers (Alfnes *et al.* 2008). PCR-RFLP is one of the most commonly used methods for polymorphism genotyping due to its simplicity. Initially, the RFLP analysis required a radio-actively labeled probe for detection, and now the method is coupled with PCR and simple agarose gel electrophoresis. In a study two allels (M and N) determined in region of the ovin calpastatin by PCR-RFLP method (Palmer *et al*. 1999).

MATERIALS AND METHODS

In this study, 150 Tali and 150 Raeni goats was sampled in the Kerman province. Blood samples were taken mainly from the jugular vein. In tubes containing EDTA and immediately transferred to ice, and the samples were stored at -20 °C until DNA extraction. DNA was extracted from 0.3 mL blood using the DNA purification kit (Fermentas, EU) according to manufacturer's instructions. Quality and quantity of DNA was measured by agarose gel (2%). The DNA amplification of the calpastatin gene was achieved by PCR. PCR-RFLP genotyping was used to detect the polymorphism in the region between exons 6 and 7 of the calpastatin gene. One pair of primers was used as follows.

Forward primer: 5'-AGCAGCCACCATCAGAGAAA-3'
Reversed primer: 5'-TCAGCTGGTTCGGCAGAT-3'

The final volume was 25 μL reactions, thermal program included: first denaturation temperature of 33 stages with 94 to 30 seconds, 56 °C connection temperature for 30 seconds, temperature 72 °C for 30 seconds and develop the final amplificon temperature of 72 °C for 4 minutes. The correctness of the resulting part of product on 1.5% agarose gel and noticeable with ethidium bromide voltage 82 and was established for 25 minutes.

The PCR products by restriction endonucleases XmnI were digested. Cut pieces again on 2% agarose gel and stained with ethidium bromide voltage 72, after 2 hours determined genotypes. Gel documentation system was used for Photography.

Statistical analysis

For determine gene, genotypic frequency and Hardy-Weinberg equilibrium use Pop-Gene32 program, version 1.31, Canada (Yeh *et al*. 1997).

RESULTS AND DISCUSSION

The samples of PCR showed on the gel agarose that was area between 1500 and 1600 bp. PCR product without any nonspecific band (Figure 1).

The allelic variation in the calpastatin gene was tested by PCR-RFLP. The PCR-RFLP method revealed two alleles; allele B was the not digested and had 1552 bp and allele A was the PCR product with the restriction site for XmnI

which upon digestion produced two fragments of roughly 960 and 592 bp PCR product (Figure 2).

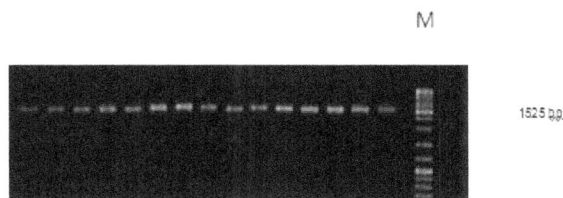

Figure 1 PCR products for calpastatin gene in Raini and Tali goats
Line M is the 100 bp molecular weight marker

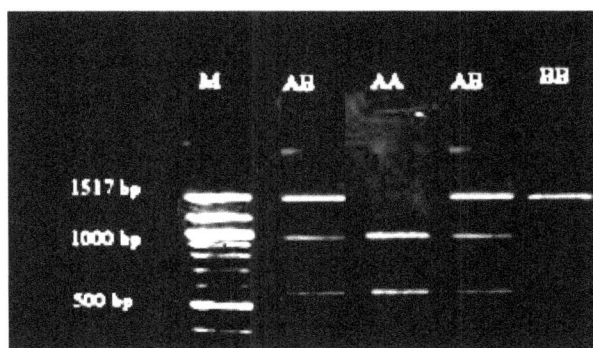

Figure 2 RFLP polymorphism of caprine CAST gene
M shows the 100 bp molecular weight marker
Three different PCR-RFLP patterns (genotype) were identified (AA, AB and BB)

In this study, a total of three genotypes were observed in the test population. The frequencies of the observed genotypes to Tali goat were 38.4, 9.6 and 52.0 for AA, BB and AB, respectively. Allele frequencies were 0.47 and 0.53 for A and B, and for Raini goat also genotypes were 0.17, 0.71 and 0.11 for AA, BB and AB respectively. Allele frequencies were 0.12 and 0.88 for A and B (Table 1).

The chi-square test showed significant (P<0.01) deviation from the Hardy-Weinberg equilibrium for this locus in the investigated these two populations. In the present study two alleles (A and B) and three genotypes (AA, BB and AB) were observed for calpastatin gene in Tali and Raini goat in kerman province, Iran.

The most frequent allele and genotype in the Tali goat were 53.0 and 67.0% for allele A and allele B, respectively (Table 2) and to Raini goat were 12.0 and 88.0 respectively. The results obtained from this study revealed the polymorphism pattern of the calpastatin gene in Tali and Raini goats.

Several methods including PCR-RFLP a good method to determine gene polymorphism is Calpastatin gene in domestic animals (Palmer *et al*. 1998).

Study on calpastatin gene in goat is very limited (Zhou and Hickford, 2008), this result shows that the polymorphism was detected in *CASTI* segment, as previously observed (Palmer *et al.* 1998) and (Casas *et al.* 2006).

Table 1 Observed alleles and genotypic frequencies for *CAST* gene in Tali and Raeini goats

Items	A	B	AA	BB	AB
Tali	0.47	0.53	38.4	9.6	52.0
Raeini	0.12	0.88	17.0	71.0	12.0

Table 2 Estimated statistically parameters for *CAST* gene in Tali and Raini gots

Items	Exp-Het	Exp-Hom	Het (Nei)	Ave-Het	Obs-Hom	Obs-Het
Tali	0.533	0.467	0.51	0.51	0.48	0.52
Raeini	0.139	0.871	0.70	0.70	0.88	0.12

Small ruminants calpastatin gene mRNA transcript variants 2 and 4 have also been reported (Zhang *et al.* 2012). Who observed frequencies of 61, 36 and 3% of the MM, MN, and NN genotypes respectively in Iranian karakul sheep and Iranian kurdi sheep, the frequencies of 76, 24 and 0% were observed for MM, MN and NN genotype respectively, showing the existence of M allele more frequent then N allele (Nassiry *et al.* 2006).

Similar data were found by (Mohammadi *et al.* 2008). There are several studies on the association of *CAST* gene polymorphism with meat quality by PCR-RFLP analysis in animals (Schenkel *et al.* 2006). The polymorphism in the *CAST* in goat was also reported by other researchers using PCR-RFLP technique (Zhou and Hickford, 2008). Higher frequencies of *CAST* gene A allele compared to the B allele have been reported in Brangus (0.78) and Pardo Suico (0.80) cattle (Asadi and Khederzadeh, 2015). Reported a significant association between A allele of bovine *CAST* gene and meat tenderness (Kuryl *et al.* 2003).

Association between the D and F alleles of porcine *CAST* gene and meat quality traits was also reported (Kapelanski *et al.* 2004). The polymorphism in the exon 1 of the CAST in sheep was also reported by other researchers using PCR-RFLP technique (Mohammadi *et al.* 2008). In goats and bovine the exon 6 of *CAST* gene were investigated for polymorphisms and a number of allelic variants were identified in these species (Zhou and Hickford, 2008).

Two allelic systems of polymorphic variants (M and N) in the region of ovine *CAST* locus have been described by PCR-RFLP method (Shahroodi *et al.* 2005). The present study was the first attempt for identification of *CAST* gene variation in Tali and Raini goats. Further studies are required to investigate the relationship between *CAST* gene polymorphisms and performance traits in Tali and Raini goat.

CONCLUSION

The PCR-RFLP analysis of calpastatin gene in Tali and Raini goats revealed high level of polymorphism. This information could be utilized in future breeding plan to exploit the genetic potential of tali and raeni goats. The Tali and Raini goats breed showed genetic diversity for the calpastatin gene, but the polymorphism found in the *CAST* gene may be helpful in selection programs for genetic improvement of meat traits. Though, previous to submission in the genetic development of the indigenous goat breeds, the relationship of these polymorphisms with meat traits needs to be recognized.

REFERENCES

Alfnes F., Rickertsen K. and Ueland Ø. (2008). Experimental evidenceof risk aversion in consumer markets: the case of beef. *Appl. Econom.* **40,** 3039-3049.

Asadi N. and Khederzadeh S. (2015). Polymorphism of candi-date genes for meat quality in sheep. *Middle-East J. Sci. Res.* **23,** 2001-2004.

Averna M., Tullio R.D., Passalacqua M., Salamini F., Ponteremoli S. and Mellon E. (2001). Changes in intracellular calpastatin localization are mediated by reversible phosphorylation. *Biochem. J.* **354,** 25-30.

Bastos E., Cravador A., Azevedo J. and Guedes-Pinto H. (2001). Single strand conformation polymorphism (SSCP) detection in six genes in the portuguese indigenous sheep breed "Churra da Terra Quente". *Biotechnol. Agron. Soc. Environ.* **5,** 7-15.

Busch W.A., Stromer M.H., Goll D.E. and Suzuki A. (1972). Ca2+-specific removal of Z lines from rabbit skeletal muscle. *J. Cell. Biol.* **52(2),** 367-381.

Casas E., White S.N., Wheeler T.L., Shackelford S.D., Koohmaraie M., Riley D.G., Chase C.C.J., Johnson D.D. and Smith T.P. (2006). Effects of calpastatin and micro-calpain markers in beef cattle on tenderness traits. *J. Anim. Sci.* **84,** 520-525.

Goll D.E., Thompson V.F., Li H., Wei W. and Cong J. (2003). The calpain system. *Physiol. Rev.* **83,** 731-801.

Kapelanski W., Grajewska S., Kuryl J., Bocian M., Jan-kowiak H. and Wisniewska J. (2004). Calpastatin (*CAST*) gene polymorphism and selected meat quality traits in pigs. *Anim. Sci.* **22,** 435-411.

Koohmaraie M., Kent M.P., Shackelford S.D., Veiseth E. and Wheeler T.L. (2002). Meat tenderness and muscle growth: is there any relationship? *Meat Sci.* **62,** 345-352.

Kuryl J., Kapelanski W., Pierzchała M., Grajewska S. and Bocian M. (2003). Preliminary observations on the effect of calpastatin gene (*CAST*) polymorphism on carcass traits in pigs. *Anim. Sci. Pap. Rep.* **21,** 87-95.

Mohammadi M., Beigi Nasiri M.T., Alami-Saeid K., Fayazi J., Mamoee M. and Sadr A.S. (2008). Polymor-phism of calpastatin gene in Arabic sheep using PCR-RFLP. *African J. Biotechnol.* **7,** 2682-2684.

Murachi T. (1983). Calpain and calpastatin. *Trends Biochem. Sci.*

8, 167-169.

Nassiry M.R., Tahmoorespour M., Javadmanesh A., Soltani M. and Foroutani Far S. (2006). Calpastatin polymorphism and its association with daily gain in Kurdi sheep. *Iranian J. Biotechnol.* **4,** 188-192.

Palmer B.R., Robert N., Hickford J.G.H. and Bickerstaffe G. (1998). Rapid Comunication: PCR-RFLP for *MSPI* and *NCOI* in the ovine calpastatin gene. *J. Anim. Sci.* **76,** 1499-1500.

Palmer B.R., Robert N. and Kent M.P. (1999). A candidate gene approach to animal quality traits. *Proc. Soc. Anim. Prod.* **57,** 294-296.

Rohrer G.A., Nonneman D.J., Miller R.K., Zerby H. and Moeller S.J. (2012). Association of single nucleotide polymorphism (SNP) markers in candidate genes and QTL regions with pork quality traits in commercial pigs. *Meat Sci.* **92,** 511-518.

Schenkel F.S., Miller S.P., Jiang Z., Mandell I.B., Ye X., Li H. and Wilton J.W. (2006). Association of a single nu-cleotide polymorphism in the calpastatin gene with carcass and meat quality traits of beef cattle. *J. Anim. Sci.* **84,** 291-299.

Shahroodi F.E., Nassiry M.R., Valizadeh R., Nosrati M., Javadmanesh A. and Tahmourespour M. (2005). The genetic polymorphism of calpastatin gene in Karakul sheep. *J. Agric. Sci. Natal. Res.* **2,** 1-10.

Yeh F.C., Yang R.C., Timothy B.J., Ye Z. and Judy M. (1997). Popgene, the User Friendly Shareware for Population Genetic Analysis. Molecular Biology and Biotechnology Center. Alberta, Cnada.

Zhang Z.R., Jiang X.S., Du H., Zhu Q., Li X.C., Yang C.W. and Liu Y.P. (2012). Characterization of the expression profiles of calpastatin (*CAST*) gene in chicken. *Mol. Biol. Rep.* **39,** 1839-1843.

Zhou H. and Hickford J.G. (2008). Allelic variation of the bovine calpastatin (*CAST*) gene. *Mol. Cell. Probes.* **22,** 129-130.

24

Correlations among Certain Growth and Production Traits in Different Breeds of Goats

R. Khandaker[1], M.K.I. Khan[1*] and M.M. Momin[1]

[1] Department of Genetics and Animal Breeding, Chittagong Veterinary and Animal Science University Khulshi, Chittagong-4225, Bangladesh

*Correspondence E-mail: kik1775@yahoo.co.uk

ABSTRACT

The study was conducted to estimate the genetic and phenotypic correlations among body measurements, body weight, kidding interval, kidding rate and milk yield in the different breeds of goat. Data was collected on 95 goats of three breeds (Jamunapari, Black Bengal and their crosses (Jamunapari×Black Bengal) through direct observation. The average body weight of Jamunapari goat was 34.75 ± 3.33 kg higher than Black Bengal (27.54 ± 5.332 kg) goats. It was observed that Jamunapari goat (0.65 ± 0.187 L/day) produces higher daily milk during lactation length than Black Bengal (0.50 ± 0.094 L/day), leading to higher lactation milk production of this goat. The kidding rate and kidding interval were 1.33 ± 0.466 and 223.27 ± 11.19 days in Jamunapari, 1.475 ± 0.555 and 247.94 ± 17.23 days in crossbred, 1.75 and 160.83 ± 16.57 days in Black Bengal goats, respectively. The heritability estimates of body weight, milk yield and lactation milk yield were 0.29 to 0.32, 0.25 to 0.20 and 0.30 to 0.32 for all breeds of goat. The genetic correlation (r_g) and phenotypic correlation (r_p) of body weight were positive with all traits except the r_g with lactation yield. The body length had negative r_p with lactation production and negative r_g with daily milk yield and lactation production. The body height has negative r_p with kidding rate, kidding interval, milk yield and lactation production and positive r_g only with kidding interval in all breeds. The high and positive correlations (either genetic or phenotypic) are important predictors of traits for improvement. Hence, these can be valuable tools for making selection/culling decisions for improved productivity of goats.

KEY WORDS correlations, (co)variances, different traits, goats, heritability.

INTRODUCTION

Goat is an important species for the poor and landless people, especially in South East Asia including Bangladesh, due to its smaller body size, quick return and less capital investment. The population of goats in Bangladesh is 25.21 million (DLS, 2013). Among the total goat population 90% of the goat are Black Bengal and remaining 10% are Jamunapari and others (Faruque and Khandoker, 2007). Generally, goat plays an important role in generating employment for landless and destitute women, income and improving household nutrition. Goat farming can be used as effective tool for poverty reduction as it require less investment and less feeding and management costs than large animal farming (Khan and Naznin, 2013). Goats of Bangladesh are grazes on barren and road-side land grass and feed with available least cost roughages like neem tree (*Azadirachta indica*) leaves, mango (*Mangifera indica*) leaves, jackfruit (*Artocarpus heterophyllus*) leaves, and concentrates: rice gruel, rice polish, cooked rice and sometime various brans. The genetic erosion in goat's resources is occurring due to unsystematic breeding and to overcome it, a structured genetic improvement program should be undertaken with specific breeding goals. The main aim of breeding-selection is to produce new generations which would exceed in performance over previous generation in production of milk and meat (Pantelic *et al.* 2008; Kapell *et al.* 2009). For construction of selection index the knowledge of the breeding values along with estimates of genetic and phenotypic correlation and heritability of different traits is essential. Several studies (Khan and Naznin, 2013; Khan and Khatun, 2013; Talukder *et al.* 2010) were conducted on the productive and reproductive parameters of goats of Bangladesh. However, studies on the breeding objectives and genetic correlations between economic traits are very limited in these goats. Genetic and phenotypic correlations have great importance in intermediary or indirect selection when changes in one trait are induced through selection for other trait where a genetic correlation exists. Therefore, the present study was designed to focus on the association among traits with the aim of identifying relevant correlations that can be used in further selection process.

MATERIALS AND METHODS

Sources of data

The farm houses and goat rearing farmers were selected from Podia Upazila at Chittagong district of Bangladesh by visiting the area frequently. On the basis of the breed characteristics (phenotypic) a total of 95 goats from three different genotypes (35 for Black Bengal, 25 for Jamunapari and 35 crossbred (Jamunapari×Black Benal)) were chosen based on age, body shape and size and live weight. The selected goats were categorized based on their age and lactation numbers under a breed. The goats were allowed for normal feeding and management under subsistence production system.

Productive and reproductive traits of different goat breeds

The data recording was started from the date of conception up to parturition to know the kidding rate and from kidding to next batch of kidding to calculate the kidding interval. The mature body weight of the doe was recorded using a top loading balance. The body height and body length of the goats were taken by measuring tape. Milk production per goat per lactation was collected on a test day basis at a weekly interval from individual goat. For milk yield data, kids were separated from their dams overnight (12 hours) preceding the day of milk recording then the kids were allowed for milk consumption and the amount of milk production was estimated as the difference between the body weight before and after sucking of kid. The lactation length of each goat was calculated by counting from the date of kidding up to the date of milk withdrawal. The milk production was calculated by multiplying average daily milk yield with lactation length.

Genetic and phenotypic correlations between traits

Phenotypic correlations among different traits were calculated by Pearson (1905) formula for estimation of genetic correlations, the heritability values of the traits were first estimated by a univariate animal model using AIREML (average information restricted maximum likelihood) (Johnson and Thompson, 1995). The model of the study was as:

$$Y = Xb + Zu + e$$

Where:
Y: vector of observations for all traits.
b: vector of common fixed effects due to farm, test group and breed.
u: vector of random genetic effects.
e: vector of residuals.
X and Z: incidence matrices relating observations to the fixed and random (animal) effects.

After obtaining the heritability of each trait, the mean value for the traits was calculated and the breeding value of each trait was estimated as:

$$BV = h^2(X - \bar{X})$$

Where:
BV: breeding value of each trait.
h^2: heritability.
X: individual trait.
$\bar{X}$: average of the trait.

Then the genetic correlation between breeding values of two successive traits was calculated.

Statistical analysis

Least squares means were estimated for fixed effects: breed, age and lactation numbers using PROC GLM and PROC MIXED of SAS (SAS, 2008). The statistical model for analysis was:

$$Y_{ijkm} = \mu + B_i + A_j + L_k + e_{ijkm}$$

Where:
Y_{ijkm}: measurement on a particular trait.
μ: population mean.
B_i: random effect of ith breed (i=3 breed i.e. Jamunapari, crossbred, Black Bengal).
A_j: fixed effect of age (j=1, 2, 2.5, 3 and 3.5 years).
L_k: fixed effect of lactation number (k=1, 2 and 3).
e_{ijkm}: random error, associated with each record, distributed as N $(0, \sigma^2)$.

The mean value was compared using the least significant difference (LSD) test at P= 0.05 (Steel *et al.* 1997).

RESULTS AND DISCUSSION

Productive and reproductive performance of three different goat breeds:

The Jamunapari goat produced maximum average daily milk yield (0.65 kg/day) and a longer lactation length (142.02 days) than Black Bengal (0.5 kg/day) goats and its crosses produced intermediate which leads to higher lactation production of Jamunapari goat (Table 1). Similar amount of milk yield for Jamunapari and Black Bengal goat was recorded by Hassan *et al.* (2010) and Paul *et al.* (2014). However, Jamunapari goat produces more milk and heavier than Black Bengal goat this might be the breed characteristics of this goat these finding are supported with Bhowmik *et al.* (2013).

However, within age and lactation number groups, the Jamunapari goat have shown significant differences for lactation length, but for other traits, no variation was observed.

For crossbred, no difference was observed for lactation length for age and lactation numbers but other two traits were varied. In case of Black Bengal goat, the age group within breed shown difference in lactation length, but remaining two traits showed no variation. It was observed that two years aged Black Bengal goat produces higher milk than one and three years old. Similar trends were also observed in Jamunapari goats also. Jamunapari and Black Bengal goats produced more milk during two to three lactations.

Olechnowicz and Sobek (2008) also reported that goats produces more milk during two to three lactations.

The body weight of the three studied breeds was not differed statistically; however, numerically Jamunapari (34.75 kg) was higher than crossbred (32.03 kg) and Black Bengal (28.79 kg) goats.

Higher body weight of Jamunapari goat might be due to breed characteristics. However, Bhowmik *et al.* (2013) observed the higher body weight for Jamunapari (45.47±3.78 kg) and crossbred goats (35.72±2.97 kg), than present study however Rahman (2007) and Paul *et al.* (2011) reported the similar body weights of goats in their study.

Table 1 Production and reproductive performance of three different breeds of goat in respect of age and lactation numbers

Breed	Age	LN	MY (kg/day)	LL (days)	LP (kg)	Lwt (kg)	KR	KI (days)
Jamunapari	1	1	0.5 ±0.360(3)	140[b] ±10(3)	73.6 ±47.692(3)	29.33[b] ±2.081(3)	1.66 ±0.577(3)	226.66 ±20.816(3)
	2	1	1.13[a] ±0.351(3)	133.33[c] ±7.637(3)	152.26 ±62.151(3)	33.33[c] ±2.516(3)	2[a] (3)	223.13 ±5.773
	2.5	2	0.43[b] ±0.17(7)	139.28[ac] ±5.763(8)	53.73(13)	35.07[c] ±1.427(8)	1.86 ±0.243(8)	228.57 ±6.90(8)
	3	2	0.56 ±.057(3)	177.25[a] ±13.301(4)	79.46 ±11.582(3)	38[a] ±7.788(4)	1.5 ±0.577(4)	212.5 ±18.929(4)
	3.5	3	0.65 (2)	139.5[b](3)	85.35(2)	38[a]±2.8(3)	1.5[b] (3)	225.5±3.5(3)
Breed average			0.65[a] ±0.187	144.02[a] ±6.09	88.48 ±22.88	34.75 ±3.33	1.33 ±0.466	223.27[b] ±11.19
Crossbred	2	1	0.7[b] (2)	101.75 ±6.968(6)	162.95[a](2)	33.35 ±0.774(5)	2 (5)	239.38[b] ±4.269(5)
	2.5	2	0.9[a](1)	115 ±8.027(3)	114.4 (1)	32.16 ±2.753(3)	1.33 ±.577(3)	246.66 ±25.166(3)
	3	2	0.48[ab] ±0.327(5)	110 ±16.329(7)	52.66[b] ±34.583(5)	30.42 ±3.952(3)	1.57 ±0.534(7)	245.71 ±22.253(7)
	3.5	3	0.3[c](1)	110 (1)	33.7 (1)	32.2 (1)	1(1)	260[a](1)
Breed average			0.57[ab] ±0.082	107.75[b] ±7.62	61.42 ±8.65	32.03 ±1.87	1.475 ±0.555	247.94[a] ±17.23
Black Bengal	1	1	0.5(1)	88.33[abc] ±7.637(3)	80.6 (1)	25.66[b] ±2.886(3)	2 (3)	158.33 ±16.072(3)
	2	1	0.4 ±0.282(2)	85[b](2)	55.05 ±0.282(2)	23.5[ab] ±7.778(2)	2 (2)	165 ±17.07(2)
	2.5	2	-	80[c](1)	-	31[a] (1)	2 (1)	150 (1)
	3	3	0.6 (1)	90[a](1)	80.6 (1)	30[a](1)	1 (1)	170 (1)
Breed average			0.50[b] ±0.094	85.83[c] ±1.909	72.08 ±0.094	27.54 ±3.554	1.75	160.83[ab] ±11.04

LN: lactation number; MY: milk yield; LL: lactation length; LP: lactation production; Lwt: live weight; KR: kidding rate and KI: kidding interval.
The significant test showed between breed, between age within breed and between lactation within age and breed.
Parenthesis indicates the number of goat studied.
The means within the same row with at least one common letter, do not have significant difference (P>0.05).

The body weight of goats was changed with the changes of age and / or lactation numbers within breeds. For Jamunapari and Black Bengal goats higher body weight was obtained at the year 3 to 3.5 of age but no differences was found in case of crossbred.

For kidding rate among three different breeds, no significant difference was observed within breed between age and but differed in lactation numbers (Table 1). The kidding rate of Black Bengal, Jamunapari and crossbred goats was 1.75, 1.33 and 1.48, respectively. Numerically higher kidding rate of Black Bengal goat proves its reputation of high fecundity. The variations were due to the differences of breed, age of goats, feeding and managements of estrus. Khan and Khatun (2013) observed kidding rate for Black Bengal goat was 1.50 to 1.17. However, Rout et al. (1999) observed the kidding rate for Jamunapari was 1.60. The factors for variation of kidding rate were similar like milk yield. The kidding interval differed within breed between age and lactation numbers for crossbred only. Kidding interval of Black Bengal goat was lower (161 days) in other word the reproductive performance was better in Black Bengal goat than Jamunapari (223 days) and crossbred (248 days). Khan and Khatun (2013) and Talukder et al. (2010) observed higher kidding interval than the current study for Black Bengal goat they found kidding interval for Black Bengal goat was 172 to 188 days. The kidding interval was also dependent upon the age of the goats and also the length of post partum heat period. Season of previous kidding and period of kidding had significant effects on the intervals between successive kidding. In second and third lactation the kidding interval was higher due to ages of the goats. Similar factors were responsible to increase Kidding interval reported by Bhowmik et al. (2013).

Body length and body height of different breeds of goat in respect of age and lactation number

The crossbred goat (57.02 cm) had the highest body length than Jamunapari and Black Bengal goat (Table 2). For the age group of crossbred goat there is no significant differences were observed in body length and body height but the age groups of Black Bengal goat within breed showed significantly different body length and body height. Usually the body length and height is used to estimate the body weight of the animals. Luo et al. (1997) and Janssens and Vandepitte (2004) also reported difference of these measurements within breed and between age group. Body measurements are the key indicator of milk production in goats stated by Khan (2010). The result of present study was higher than Mandal et al. (2008) and similar with Bingol et al. (2012). The variation of body length might be due to breed and environmental differences.

Heritability, genetic and phenotypic correlations between different traits of different goats

Heritability, genetic and phenotypic correlations between different traits of Jamunapari, crossbred and Black Bengal goats are presented in Table 3.

The heritability of all the traits was positive (Table 3). In Jamunapari, crossbreds and Black Bengal goat the heritability, phenotypic correlations and genotypic correlations of milk yield and lactation production was ranged from 0.20 to 0.23 and 0.29 to 0.31, 0.96 to 0.98 and 0.994 to 0.96, respectively.

This indicated that milk yield had the positive correlation with lactation production. Bagnicka et al. (2004) reported the heritability values of milk yield from 0.19 to 0.324 and phenotypic correlation 0.247 to 0.355.

Table 2 Body length and body height of three different breeds of goat in respect of age and lactation numbers

Breed	Age	Lactation numbers	Body length (cm)	Body height (cm)
Jamunapari	1	1	43.18±4.399(3)	50.02±5.865(3)
	2	1	50.8±9.169(3)	50.02±6.391(3)
	2.5	2	50.62±3.315(8)	60.58±3.327(8)
	3	2	55.55±8.057(4)	61.59±8.89(4)
	3.5	3	57.79±0.899(3)	65.41±0.899(3)
Breed average	-	-	51.58^b±5.169	60.02±5.072
Crossbred	2	1	52.71±1.27(5)	60.33±0.734(5)
	2.5	2	54.18±2.931	61.80±3.918(3)
	3	2	57.68±2.824(7)	64.95±4.831(7)
	3.5	3	63.5(1)	66.04 (1)
Breed average	-	-	57.02^a±1.496	63.27±3.149
Black Bengal	1	1	43.18^c±2.54(3)	54.18^a±1.466(3)
	2	1	49.53^a±1.796(2)	58.42±3.592(2)
	2.5	2	43.18^b(1)	58.42^b(1)
	3	3	50.8^{ac}(1)	63.5^c(1)
Breed average	-	-	46.69^b±2.167	58.62±2.529

The significant test showed between breed, between age within breed and between lactation within age and breed.
Parenthesis indicates the number of goat studied.
The means within the same row with at least one common letter, do not have significant difference (P>0.05).

Table 3 Heritability, genetic and phenotypic correlations among traits of different genotypes of goat. Genetic correlation is shown below diagonal, heritability on the diagonal (bold) and phenotypic correlation above diagonal

	LWT	BLN	BHT	KI	KR	MY	LP
Jamunapari							
LWT	**0.312**	0.591	0.506	0.416	0.378	0.243	0.166
BLN	0.591	**0.162**	0.827	0.2113	0.150	0.0071	-0.0317
BHT	0.506	0.87	**0.05**	-0.0388	-0.069	-0.124	-0.176
KI	0.364	0.273	0.055	**0.013**	0.882	0.763	0.725
KR	0.280	0.147	-0.0087	0.665	**0.02**	0.733	0.738
MY	0.092	-0.076	-0.092	0.494	0.459	**0.25**	0.980
LP	-0.0022	-0.103	-0.160	0.441	0.496	0.966	**0.32**
Crossbred							
LWT	**0.302**	0.802	0.761	0.643	0.225	0.441	0.442
BLN	0.802	**0.154**	0.903	0.599	0.204	0.139	0.132
BHT	0.761	0.903	**0.04**	0.755	0.445	0.274	0.269
KI	0.555	0.433	0.462	**0.012**	0.767	0.652	0.643
KR	-0.218	-0.223	0.0043	0.043	**0.02**	0.534	0.521
MY	0.248	-0.081	-0.033	0.282	0.022	**0.23**	0.971
LP	0.251	-0.084	-0.029	0.289	0.018	0.942	**0.31**
Black Bengal							
LWT	**0.292**	0.767	0.839	0.509	0.497	0.372	0.385
BLN	0.767	**0.142**	0.919	0.467	0.526	0.016	-0.013
BHT	0.839	0.919	**0.03**	0.354	0.422	-0.018	-0.045
KI	0.386	0.386	0.324	**0.01**	0.892	0.705	0.699
KR	0.296	0.432	0.394	0.437	**0.018**	0.609	0.564
MY	0.131	-0.313	-0.268	0.254	0.047	**0.20**	0.977
LP	0.152	-0.347	-0.3004	0.256	-0.062	0.958	**0.30**

LWT: live weight; BLN: body length; BHT: body height; KI: kidding interval; KR: kidding rate; MY: milk yield and LP: lactation production.

The heritability value of body weight ranged from 0.29 to 0.32 for Black Bengal, crossbred and Jamunapari goats respectively (Table 3). Estimated heritability value for body weight was lower compared with Otuma and Osakwe (2008) and Khan (2010), who found heritability value of 0.42 for Sahelian goats and Beetal goats, respectively. The high and positive correlations (either genetic or phenotypic) among various traits are important predictor for attempting animal improvement (productive and reproductive performance). More specially, in case of milk yield positive and higher correlation with other productive and reproductive traits must be valuable tools for making selection/culling decision. Besides this, decision on other management factor/tools should take in account for improvement of animal.

CONCLUSION

It can be concluded that the Jamunapari goat produced higher daily average milk and higher lactation yield than Black Bengal and crossbred produced moderate. The kidding rate and kidding interval of Black Bengal goats were also better than Black Bengal breeds. The heritability value of body weight, milk yield and lactation production were moderate for all traits in all three breeds. The genotypic and phenotypic correlations of body weight and milk yield were positive with all traits except genetic correlation of body weight and lactation production which was negative.

The body length and height had negative phenotypic correlation with lactation production and negative genetic correlation with milk yield and lactation production in all three breeds. However, these two traits had negative phenotypic correlation with kidding rate, kidding interval, milk yield and lactation production and positive genetic correlation only with kidding interval in all breeds.

ACKNOWLEDGEMENT

We acknowledge the University Grant Commission of Bangladesh for providing funds for this research. We thank the authority of Chittagong Veterinary and Animal Science University, Bangladesh for giving permission to pursue this study. We acknowledge the staffs who were involved in this project.

REFERENCES

Bagnicka E., Distl O., Hamann H. and Lukaszewicz M. (2004). Heritabilities and genetic correlations between the dairy traits in goats estimated in first vs later lactations. *Anim. Sci. Pap. Rep.* **2,** 205-213.

Bhowmik N., Mia M.M., Rahman M.M. and Islam S. (2013). Preliminary study on productive and reproductive performances of Jamunapari, Black Bengal and crossbred goats at Chittagong region of Bangladesh. *Iranian J. Appl. Anim. Sci.* **4,** 89-93.

Bingöl M., Gökdal O., Aygün T. and Daşkiran I. (2012). Some productive characteristics and body measurements of Norduz goats of Turkey. *Trop. Anim. Health. Prod.* **44**, 545-452.

DLS. (2013). Development and Work Plan. Department of Livestock Services, Ministry of Livestock and Fishery, Dhaka, Bangladesh.

Faruque M.O. and Khandoker M.A.M.Y. (2007). Recent advances of goat genotyping in Bangladesh. Pp. 28-40 in Proc. Workshop. Recent Adv. Livest. Genot. Bangladesh, Dhaka, Bangladesh.

Hassan M.R., Talukder M.A.I. and Sultana S. (2010). Evaluation of the production characteristics of the Jamunapari goat and its adaptability to farm conditions in Bangladesh. *Bangladesh Vet.* **27**, 26-35.

Janssens S. and Vandepittte W. (2004). Genetic parameters for body measurements and linear type traits in Belgian bleue de mairie, Suffolk and Texel sheep. *Small Rumin. Res.* **54**, 13-24.

Johnson D.L. and Thompson R. (1995). Restricted maximum likelihood estimation of variance components for univariate animal models using sparse matrix techniques and average information. *J. Dairy Sci.* **78**, 449-456.

Kapell D.N., Ashworth C.J., Walling G.A., Lawrence A.B., Edwards S.A. and Roehe R. (2009). Estimation of genetic associations between reproduction and production traits based on a sire and dam line with common ancestry. *Animal.* **3**, 1354-1362.

Khan M.K.I. and Naznin M. (2013). Study on the live weight and live weight gain of black Bengal and Jamunapari breeds by fitting the linear regression under semi-intensive conditions. *Pakistan J. Biol. Sci.* **16**, 998-1003.

Khan M.K.I. and Khatun M.J. (2013). Different traits of black Bengal goats under two feeding regime and fitting the Gompertz curve for prediction of weaning weight in the semi-scavenging system. *Indian J. Anim. Res.* **47**, 498-503.

Khan (2010). Body Measurements of Goats- Their Genetic Control and Association with Milk Yield. A Report From the Faculty of Veterinary and Animal Science, University of Faisalabad, Faisalabad, Pakistan.

Luo M.F., Wiggans G.R. and Hubbard S.M. (1997). Variance components estimation and multitrait genetic evaluation for type traits of dairy goats. *J. Dairy Sci.* **80**, 594-600.

Mandal A., Roy R. and Rout P. (2008). Direct and maternal effects for body measurements at birth and weaning in Muzaffarnagari sheep of India. *Small Rumin. Res.* **75**, 123-127.

Olechnowicz J. and Sobek Z. (2008). Factors of variation influencing production level, SCC and basic milk composition in dairy goats. *J. Anim. Feed Sci.* **17**, 41-49.

Otuma M.O. and Osakwe I.I. (2008). Estimation of genetic parameters of growth traits in Nigeria Sahelian goats. *J. Anim. Sci.* **2**, 83-86.

Pantelic V., Petrovic M.M., Aleksic S., Ostojic D., Sretenovic L. and Novakovic Z. (2008). Genetic correlations of productive and reproductive traits of Simmental cows in Republic of Serbia. *Arch. Zootech.* **11**, 73-78.

Paul S., Khandoker M.A.M.Y., Moinuddin M.A. and Paul R.C. (2011). Characterization of black Bengal goat. *J. Bangladesh Agric. Univ.* **9**, 61-66.

Paul R.C., Rahman A.N.M.I., Debnath S. and Khandoker M.A.M.Y. (2014). Evaluation of productive and reproductive performance of Black Bengal goat. *Bangladesh J. Anim. Sci.* **43**, 104-111.

Pearson K. (1905). On the General Theory of Skew Correlation and Non Linear Regression. Dulau and Company, London, United Kingdom.

Rahman A.H.M.S. (2007). Morphometric characterization of black Bengal Buck. MS Thesis. Bangladesh Agricultural Univ., Mymensingh, Bangladesh.

Rout P.K., Mandal A., Roy R. and Singh L.B. (1999). Improvement and Conservation of Jamunaparigoats in Their Home Tract. Ministry of Agriculture Report, New Delhi, India.

SAS Institute. (2008). SAS®/STAT Software, Release 8. SAS Institute, Inc., Cary, NC. USA.

Steel R.G.D., Torrie J.H. and Dickey D.A. (1997). Principles and Procedures of Statistics Biometrical Approach. Mc Graw-Hill Co., Inc., London, United Kingdom.

Talukder M.A.I., Hassan M.R., Mohanta U.K. and Amin M.R. (2010). Productive and reproductive performances of brown Bengal goat in hilly area at Naikhongchari. *Proc. Annu. Res. Rev. Work.* **197**, 86-90.

25

Cloning and Expression of Heat Shock Protein 60 kDa Gene from *Brucella melitensis* as Subunit Vaccine

T. Abbassi-Daloii[1], M. Tahmoorespur[1*] and M.H. Sekhavati[1]

[1] Department of Animal Science, Faculty of Agriculture, Ferdowsi University of Mashhad, Mashhad, Iran

*Correspondence E-mail: tahmoorespur@um.ac.ir

ABSTRACT

Brucellosis is caused by the bacterium *Brucella* and affects various domestic and wild species. GroEL (Heat Shock Protein 60kDa) as one of the major antigens that stimulate the immune system, increases *Brucella* survival. The aim of the current study was to clone and express GroEL in *Escherichia coli* in order to design subunit vaccine. Amplifying was performed using specific primers. The full-length open reading frame of this gene was cloned into the expression vector pET-32a(+) and expressed in BL21 (DE3). The expressed antigen was purified and the molecular weight of the recombinant protein was about 70 kDa. Sequencing results along with SDS-PAGE and Western analysis confirmed the expression of recombinant GroEL in the heterologous *Escherichia coli*. The results of colony polymerase chain reaction (PCR), enzyme digestion and sequencing showed that the GroEL antigen has been successfully cloned and subcloned into pET-32a(+). The results showed that *Escherichia coli* was able to express GroEL protein appropriately. This protein was expressed by induction with isopropyl β -D-thiogalactoside (IPTG) at concentration of 1 mM and it was confirmed by Ni-NTA column, sodium dodecyl sulfate polyacrylamide gel electrophoresis (SDS-PAGE) and Western-blotting electrophoresis. The results of this study showed that *Escherichia coli* can be used as an appropriate host to produce the recombinant GroEL protein. This recombinant protein may be useful to simulate immune system, to produce recombinant vaccine and diagnostic kit in future studies after it passes biological activity tests *in vivo* in animal model and or other suitable procedure.

KEY WORDS expression, GroEL, immunogenicity, recombinant vaccine.

INTRODUCTION

Brucellosis, a common zoonotic disease, concerns veterinarians as well as public health authorities in developing countries. Brucellosis is caused by *Brucella*, a gram negative, facultative intracellular, partially acid fast coccobacillus lacking capsule or flagellea (Pappas, 2010). This disease is characterized by abortion and reduced fertility in animals, and also by chronic infections with symptoms such as undulant fever, arthritis, and osteomyelitis in human (Pappas *et al.* 2006). The genus of *Brucella* consists of more than ten species which *B. abortus*, *B. melitensis* and *B. suis* cause most of the animal and human diseases. *B. melitensis* that mainly infects goats and sheep is considered as the most pathogenic species of *Brucella* to humans (Franco *et al.* 2007). Many researchers demonstrated that the correct identification of pathogens is critical for epidemiological studies and for the development of effective preventative measures, including vaccination (Ahsani *et al.* 2010a; Ahsani *et al.* 2010b; Ahsani *et al.* 2011; Zandi *et al.* 2014; Shahdadnejad *et al.* 2016). Vaccination against diseases is very important as the main prophylactic to reduce losses due to the disease or minimize its severity (Ahsani *et al.* 2010a; Ahsani *et al.* 2010b; Ahsani *et al.* 2011; Zandi *et al.* 2014). In animals, immunization against *Brucella* infections is usually performed by administration of the live attenuated smooth *Brucella* strains like *B. abortus* S19, *B. melitensis* Rev.1, and non-smooth strain *B. abortus* RB51 (Corbel, 2006). However, live attenuated vaccines have the limitations of causing abortion in immunized pregnant animals, being pathogenic for humans, inducing resistant to antibiotics and interfering with the lipopolysaccharides-based serological tests (Cassataro *et al.* 2005). Up to now, there is no commercially available vaccine against human brucellosis and the disease is prevented by immunization of uninfected animals and elimination of the infected ones (Seleem *et al.* 2010). To develop a human *Brucella* vaccine, those *Brucella* proteins that exist in *Brucella* strains pathogenic to humans but absent in *Brucella* strains and not pathogenic to humans would be ideal for vaccine development (Golshani *et al.* 2015). GroEL gene encodes an inner membrane protein of *Brucella* with about 60 kDa molecular weight. This protein belongs to the heat shock protein family (HSPs) and can function as a chaperon, so it has an important role in the structure and folding of other proteins. *Brucella* produces a GroEL antigen in response to macrophage phagocytosis in order to increase its survival. GroEL is one of the major antigens that stimulate the immune system. Furthermore, this antigen has an important role in the disease cycle in humans and animals (Al Dahouk *et al.* 2007). In a comparative study on two different strains of *Brucella*, the GroEL antigen has been introduced as a good candidate for vaccine production and also for development of diagnostic kits (Amirmozafari *et al.* 2008). Polymerase chain reaction (PCR) is the most modern practical technology in diagnosing infectious diseases and compared with classical techniques, it has been shown to be more rapid, with results obtained in a few hours, and also more reliable (Ahsani *et al.* 2010a; Mohammadabadi *et al.* 2004; Mohammadabadi *et al.* 2011). Moreover, PCR allows a faster bacterial identification directly from clinical samples (Ahsani *et al.* 2010b). Genotyping, which is based on a more stable marker, DNA, is not dependent on gene expression. Another advantage of genotyping methods is that the discriminatory power of DNA-based methods is generally superior to that of phenotypic methods. The ability to distinguish between genomes is important to several disciplines of microbiological research, for example in studies on population genetics and microbial epidemiology (Ahsani *et al.* 2011; Zandi *et al.* 2014; Shahdadnejad *et al.* 2016) of great importance when choosing a method for genotyping are the typing ability, reproducibility, discriminatory power and also the ease and cost of performing the analysis. With PCR, selected segments of any DNA molecule can be am-

plified exponentially. The aim of the present study was to clone and express GroEL antigen from the *B. melitensis* strain *Rev 1*. Finally, the recombinant GroEL was purified to use as recombinant vaccine to stimulate the immune system in further studies.

MATERIALS AND METHODS

Bacterial strains, growth conditions and isolation

B. melitensis strain *Rev1* was obtained from the *Brucella* culture collection (Razi Institute, Mashhad, Iran) and was cultured as previously described (Delpino *et al.* 2007). DNA was extracted using a DNA extraction kit (Bioneer, Korea). The quality and quantity of the extracted DNA were analyzed by agarose gel electrophoresis and Nano-Drop ND-100 spectrophotometer (Thermo, USA). T/A cloning vector pTZ57R/T (Thermo, USA) were used for cloning and sequencing of the amplified gene. *E. coli* strain TOP10F` was used as host for cloning, sequencing and maintenance. The prokaryotic expression vector pET-32a(+) (Novagene, USA) and *E. coli*, BL21 (DE3) (Stratagene, USA) were used for rGroEL protein production.

PCR amplification

B. melitensis Rev1 genomic DNA was used as template for amplification full length open reading frame of GroEL gene using *EX Taq* DNA polymerase (Takara, Japan). The specific primers with restriction sites at the 5′ end were designed using primer Premier 5, according to the available nucleotide sequences in the NCBI GenBank database (Table 1). Polymerase chain reaction (PCR) was carried out by the Personal Cycler™ thermocycler (Biometra, Germany) with the reaction mixture containing 2.5 µL of 10X PCR buffer, 2 µL MgCl$_2$ and 2 µL dNTPs, 0.3 µL of the DNA solution (50 to 100 ng/µL), 1.5 µL of mix primer (5 pmol/µL) and 0.125 U/µL of *EX Taq* DNA polymerase, and deionized water up to final volume of 25 µL. The PCR program was performed with an initial denaturation step at 94 °C for 6 min followed by 27 cycles of denaturation at 94 °C for 30 sec, annealing at 58 °C for 30 sec and extension at 72 °C for 45 sec, and a final extension at 72 °C for 10 min.

Cloning and nucleotide sequences analysis

The PCR products were purified from the agarose gel by Ron`s Agarose Gel Mini prep Kit (BioRon, Germany) according to the instructions provided by the manufacturer. The purified PCR products were ligated into pTZ57R/T cloning vector by T/A cloning strategy according to the manufacturer's instruction. Competent cell preparation and transformation steps were followed as described by Sambrook and Russell (2001).

Table 1 The specific primers with restriction sites

Primer sequences (5' → 3')	Restriction enzyme	Length (bp)
F:5`-A**CCATGG**ATGGCTGCAAAAGACGTAAAATTCG-3`	*NcoI*	1641
R:5`-A**GAGCTC**TTATTAGAAGTCCATGCCGCCCATGC-3`	*SacI*	

The recombinant vectors were transformed into competent *E. coli* TOP10F`. The bacterial clones harbouring recombinant plasmid DNA were screened based on their ampicillin resistance. The fidelity of *E. coli* TOP10 F` transformants was verified by PCR reaction using M13 universal primers. The plasmids were purified using the Ron`s Plasmid Mini Kit (BioRon, Germany) and were confirmed by restriction sites enzyme digestion and sequencing (Bioneer, South Korea).

In order to sub-clone the recombinant GroEL into the expression vector, recombinant plasmid was subjected to digestion with *NcoI* and *SacI* restriction enzymes (Thermo, USA). After digestion, the digestion product was electrophoresed on low melting agarose gel 0.8% and then purified. The purified digested product was cloned into the expression vector pET-32a(+) in which the recombinant protein is included a six-Histidine tag (His-tag) at the N-terminal end for easier purification. Standard techniques for these steps such as ligation, competent cell preparation and transformation were followed as previously described (Sambrook and Russell, 2001). Recombinant vectors were transformed into competent *E. coli* BL21 (DE3). The recombinant clones harbouring plasmid DNA with inserts were screened based on their ampicillin resistance. The fidelity of *E. coli* BL21 (DE3) transformants was verified by PCR reaction using T7 universal primers. Recombinant plasmids were purified using the Ron`s Plasmid Mini Kit (BioRon, Germany) and confirmed by restriction enzyme digestion. Purified plasmids were subjected to sequencing (Bioneer, South Korea).

Expression, confirmation, and purification

The positive recombinant construct was cultured in LB ampicillin medium. Protein synthesis was induced using 0.1 mM isopropyl β -D-thiogalactoside (IPTG) in recombinant bacterial culture with OD 600. Bacteria were incubated for 5 h at 37 °C then harvested by centrifugation (3000 g, 20 min, 4 °C) and stored at -80 °C. The pellet from a 100 mL bacterial culture was suspended in lysis buffer (Tris 50 mM, EDTA 5.0 mM, urea 8.0 M, pH=8.0) and was lysate with sonication. Lysed cells were centrifuged at 9000 g for 15 min at 4 °C to separate the supernatant containing soluble materials from the pellet. Both the supernatant and the pellet were evaluated on SDS-PAGE 10% to analysis the expression of rGroEL.

Expressed protein was purified by chromatography through Ni-agarose (Thermo, USA) from insoluble phase of lysate using guanidine hydrochloride 6 M to dissolve the pellet according to the instructions provided by the manufacturer.

The quality and identity of the purified rGroEL protein were analysed by SDS-PAGE (10%) and western blotting assay. For western blotting, the SDS-PAGE gels were electro blotted onto nitrocellulose. The blotted nitrocellulose was then blocked with skim milk for 3 h. The membranes were washed three times and then Anti Poly-Histidine-HRP (Sigma) (1:2000 diluted in BSA 1%) was added. After 1 h incubation at room temperature and washing, diaminobenzidine (DAB) as chromogen was employed for visualization. Finally, the quantity of the recombinant protein was estimated using Bradford assay. The purified rGroEL protein was stored at -20 °C for further evaluation of immunogenicity and protective efficacy in mice.

RESULTS AND DISCUSSION

PCR amplification and cloning

The GroEL gene from *B. melitensis Rev1* was amplified and the accuracy of this fragment was visualized on agarose gel electrophoresis (Figure 1).

Consequently, the amplified products were successfully ligated into cloning vector and transformed into competent *E. coli* TOP10F` cells. After screening positive colonies using colony-PCR, the integrity of the recombinant plasmids was confirmed by restriction digestion.

Expression and purification of recombinant protein

Expression of the rGroEL gene in *E. coli* cells bearing recombinant plasmids was analyzed by SDS-PAGE and Western blotting (Figure 2). The best induction time was reached by selecting the induced bacteria samples at time T1, T2 and T3 and the highest expressing colony was in T3 (Figure 2.A). Ni-agarose chromatography was used to achieve proteins purification and the quality of purified protein was confirmed by SDS-PAGE and Western blotting with the Anti Poly-Histidine-HRP.

Recently, many efforts have been made to identify new immunogens in *Brucella* using immune approaches. However, not all of these new targets showed *in vivo* protective efficacy (Yang *et al.* 2013). Resistance to facultative intracellular bacterial pathogens mainly depends on acquired cell-mediated resistance which activates specific T lymphocytes followed by macrophages activation for increasing killing of such organisms (Zhan *et al.* 1996).

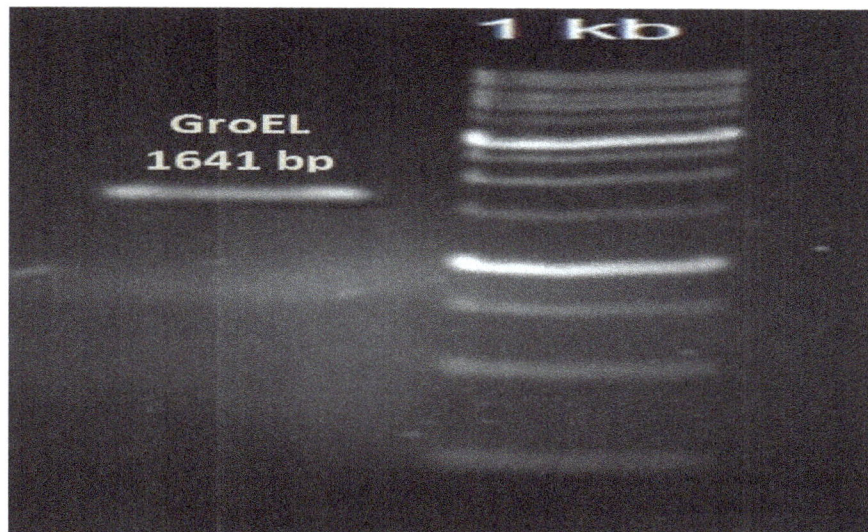

Figure 1 PCR products of GroEL gene with 1641 bp length

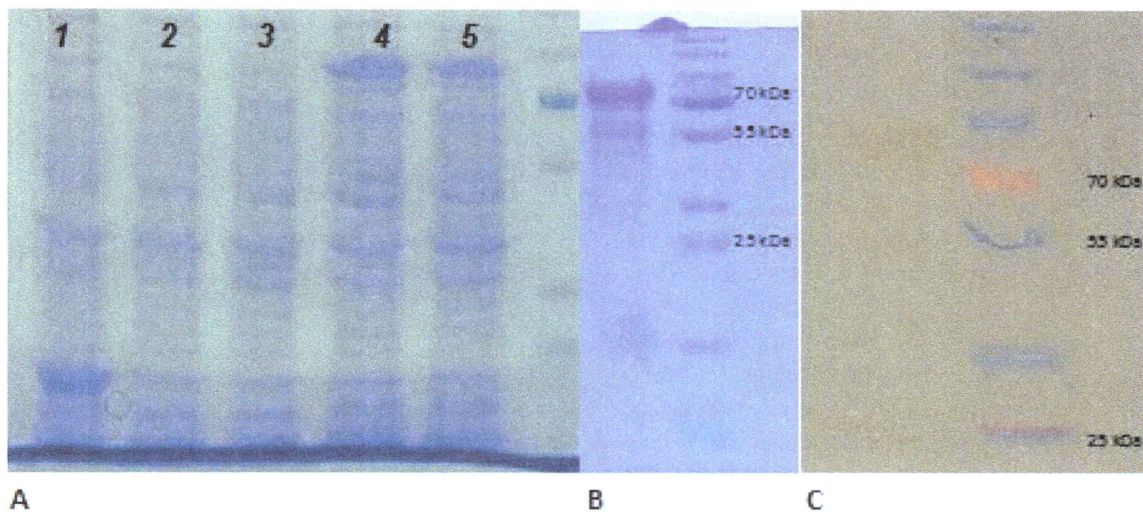

Figure 2 A: SDS-PAGE analysis of the recombinant protein
Line 1: pET-32a as negative control; 2: total cell lysate of *E. coli* BL21 (DE3) containing pET32a-GroEL showing the expression of before induction; Line 3, 4 and 5: total cell lysate of *E. coli* BL21(DE3) containing pET-32a- GroEL showing the expression in different time point (after 1, 2, and 4 hours) by IPTG
B: purified protein
C: Western blotting profile of the rGroEL protein
Pre-stained protein marker (Thermo, USA) with 9 bands

An important protection activity inducer in monocytes and macrophages is the interferon gamma (Ritchie *et al.* 2012). Th1 CD4 + T cells which producing interferon gamma are responsible for macrophage activation and attraction of inflammatory effector cells, and also play a role in acquired cellular resistance. Identification and characterization of specific antigens that induce preferentially a Th1 subset response would be important for the development of protective recombinant vaccines against *Brucella*. HSPs are major antigens which eliciting humoral and cell-mediated immune responses against different bacterial pathogens (Bae and Toth, 2000).

GroEL as a common heat-shock protein present in different type of pathogens induces humoral and cellular immune responses in host models (Silva, 1999). In this study, GroEL gene as a dominant *B. melitensis Rev 1* antigen was candidate for cloning and expression in order to use as efficient subunit vaccine.

Due to the importance of high-level production of recombinant protein in immunological studies, the fusion was cloned in pET-32a(+) expression vector. The pET is the most powerful system has so far been developed for the cloning and expression of the recombinant proteins in *E. coli*.

The pET32 series is designed for cloning and high-level expression of peptide sequences fused with the 109aa Trx•Tag™ thioredoxin protein. Cloning sites are available for producing fusion proteins also containing cleavable His•Tag® and S•Tag™ sequences for detection and purification. Target genes are cloned in pET plasmids under control of strong bacteriophage T7 transcription and translation signals where expression is induced by providing a source of T7 RNA polymerase in the host cells. T7 RNA polymerase is so active that when fully induced, almost all of the cell's resources are converted to the target gene expression (LaVallie *et al.* 1993). Sequencing of the cloning product confirmed the integrity of the cloning. The plasmid construct pET-32a-GroEL was transferred into *E. coli* BL21 (DE3) containing T7 RNA polymerase and the expression was induced by the addition of IPTG. Successfully induced expression by different concentration of IPTG and high level production of the fusion demonstrated the high efficiency of our fusion construct.

CONCLUSION

Brucellosis is a common zoonotic disease that can infect domestic animals and till today there is no recombinant vaccine for it. Protein recombinant vaccines are feasible than live or poor bacterial vaccine because it can be produced in large scale and also are safe for recipient host. Therefore, the aim of the present study was to clone and expression of one candidate *B. melitentis* antigen GroEL in order to designing suitable recombinant vaccine. In order to produce rGroEL the cloning and expression were done successfully and results of sequencing and also SDS-PAGE and Western blotting confirmed rGroEL production. The evaluation of humoral and cellular immune responses of this antigen against *Brucella melitensis* infection in mice has already initiated in our laboratory.

ACKNOWLEDGEMENT

The authors would like to express their gratitude to Ferdowsi University of Mashhad for financial support.

REFERENCES

Ahsani M.R., Bafti M.S., Esmailizadeh A.K. and Mohammadabadi M.R. (2011). Genotyping of isolates of Clostridium perfringens from vaccinated and unvaccinated sheep. *Small Rumin. Res.* **95,** 65-69.

Ahsani M.R., Mohammad Abadi M.R., Shamsodini Bafti M., Ezatkhah M., Derakhshan Hasani M. and Esmailzadeh A.K. (2010a). Application of triplex PCR technique in identification of *Clostridium perfringens* b, c and d types. *Iranian J. Anim. Sci. Res.* **2,** 185-190.

Ahsani M.R., Mohammadabadi M.R. and Shamsaddini M.B. (2010b). Clostridium perfringens isolate typing by multiplex PCR. *J. Venom Anim. Toxins. Incl. Trop. Dis.* **16,** 573-578.

Al Dahouk S., Fleche P.L., Nockler K., Jacques I., Grayon M., Scholz H.C., Tomaso H., Vergnaud G. and Neubauer H. (2007). Evaluation of *Brucella* MLVA typing for human brucellosis. *J. Microbiol. Method.* **69,** 137-145.

Amirmozafari N., Ghazi F., Mostafazadeh A., Mostafaie A. and Rajabnia R. (2008). Comparison of heat shock response in *Brucella abortus* and *Brucella melitensis. Pakistan J. Biol. Sci.* **11,** 188-194.

Bae J.E. and Toth T.E. (2000). Cloning and kinetics of expression of *Brucella abortus* heat shock proteins by baculovirus recombinants. *Vet. Microbiol.* **75,** 199-204.

Cassataro J., Estein S.M., Pasquevich K.A., Velikovsky C.A., de la Barrera S., Bowden R., Fossati C.A. and Giambartolomei G. H. (2005). Vaccination with the recombinant *Brucella* outer membrane protein 31 or a derived 27-amino-acid synthetic peptide elicits a CD4+ T helper 1 response that protects against *Brucella melitensis* infection. *Infect. Immun.* **73,** 8079-8088.

Corbel M.J. (2006). Brucellosis in Humans and Animals. World Health Organization. Publisher by World Health Organization, Geneva, Switzerland.

Delpino M.V., Estein S.M., Fossati C.A., Baldi P.C. and Cassataro J. (2007). Vaccination with *Brucella* recombinant DnaK and SurA proteins induces protection against *Brucella abortus* infection in BALB/c mice. *Vaccine.* **25,** 6721-6729.

Franco M.P., Mulder M., Gilman R.H. and Smits H.L. (2007). Human brucellosis. *Lancet Infect. Dis.* **7,** 775-786.

Golshani M., Rafati S., Jahanian-Najafabadi A., Nejati-Moheimani M., Siadat S.D., Shahcheraghi F. and Bouzari S. (2015). In silico design, cloning and high level expression of L7/L12-TOmp31 fusion protein of *Brucella* antigens. *Res. Pharm. Sci.* **10,** 436-445.

LaVallie E.R., DiBlasio E.A., Kovacic S., Grant K.L., Schendel P. F. and McCoy J.M. (1993). A thioredoxin gene fusion expression system that circumvents inclusion body formation in the E. coli cytoplasm. *Biotechnology (N Y).* **11,** 187-193.

Mohammadabadi M.R., Shaikhaev G.O., Sulimova G.E., Rahman O. and Mozafari M.R. (2004). Detection of bovine leukemia virus proviral DNA in Yaroslavsl, Mongolian and black pied cattle by PCR. *Cell. Mol. Biol. Lett.* **9,** 766-768.

Mohammadabadi M.R., Soflaei M., Mostafavi H. and Honarmand M. (2011). Using PCR for early diagnosis of bovine leukemia virus infection in some native cattle. *Genet. Mol. Res.* **10,** 2658-2663.

Pappas G. (2010). The changing *Brucella* ecology: novel reservoirs, new threats. *Int. J. Antimicrob. Agents.* **36(1),** 8-11.

Pappas G., Papadimitriou P., Christou L. and Akritidis N. (2006). Future trends in human brucellosis treatment. *Expert Opin Invest. Drugs.* **15,** 1141-1149.

Ritchie J.A., Rupper A., Cardelli J.A. and Bellaire B.H. (2012). Host interferon-gamma inducible protein contributes to *Brucella* survival. *Front. Cell. Infect. Microbiol.* **2,** 55-61.

Sambrook J. and Russell D.W. (2001). Molecular cloning: a laboratory manual. Cold Spring Harbor Laboratory Press,

Cold Spring Harbor, New York.

Seleem M.N., Boyle S.M. and Sriranganathan N. (2010). Brucellosis: a re-emerging zoonosis. *Vet. Microbiol.* **140,** 392-398.

Shahdadnejad N., Mohammadabadi M.R. and Shamsadini M. (2016). Typing of Clostridium Perfringens Isolated from Broiler Chickens Using Multiplex PCR. *Genet. 3rd Millennium.* **14(4),** 4368-4374.

Silva C.L. (1999). The potential use of heat-shock proteins to vaccinate against mycobacterial infections. *Microb. Infect.* **1,** 429-435.

Yang X., Skyberg J.A., Cao L., Clapp B., Thornburg T. and Pascual D.W. (2013). Progress in *Brucella* vaccine development. *Front. Biol.* **8,** 60-77.

Zandi E., Mohammadabadi M.R., Ezzatkhah M. and Esmailizadeh A.K. (2014). Typing of toxigenic isolates of *Clostridium perfringens* by multiplex PCR in ostrich. *Iran J. Appl. Anim. Sci.* **4,** 509-514.

Zhan Y., Liu Z. and Cheers C. (1996). Tumor necrosis factor alpha and interleukin-12 contribute to resistance to the intracellular bacterium *Brucella abortus* by different mechanisms. *Infect. Immun.* **64,** 2782-2786.

26

Effects of Utilizing Disposed Fish of White Nile River on Performance of Layers', Gezira, Sudan

G.E. Salih[1], M.S. Elnour[2], A.E. Abd-elgabar[1], M.A. Daffallah[1] and M.A.E. Elkhairey[1,3*]

[1] Department of Poultry Production, Faculty of Animal Production, University of Gezira, Sudan
[2] Department of Poultry Production, College of Animal Production, University of Bahri, Khartoum, Sudan
[3] College of Animal Science and Technology, Northwest A and F University, Yangling 712100, Shaanxi, China

*Correspondence E-mail: elkhairey11@yahoo.com

ABSTRACT

An experiment has been performed to utilize the wasted and disposed fish from White Nile River after simple heat treatments (Sun drying, roasting, direct boiling and indirect boiling) in layers' rations from 19 up to 40 weeks to replace the imported concentrate with the levels of 0, 1.5, 3.5 and 5% for all heat treatments. The crude protein of treated fish was 50.75, 52.50, 50.55 and 50.05% for sun drying, roasting, direct boiling and indirect boiling, respectively and for super concentrate 31.50%. Rations had been formulated according to NRC, 1994 recommendations. A total of 390 birds of Hy-line W-98 at 19 week of age were randomly distributed for groups, 5 group (6 replicate/13 birds). The performance of layers during the productive period (22 up to 40 weeks) showed significant differences (P<0.05) for feed intake, body weight gain, egg production and egg quality characteristics among treatment groups. The sun dried, roasted and supper concentrate respectively had the highest feed intake and body weight, while indirect boiling and boiling fish showed the lowest one. Sun dried 1.5%, roasted 1.5 and 3.5% levels and supper concentrates were the best egg production percentages and weights while the indirect boiling and direct boiling treatments were the lowest ones. There were no any significant differences among treatments for egg shell, weight, shell thickness and panel test. The study recommended the utilization of local wasted and disposal fish of White Nile River in substitution of imported concentrates in layers rations.

KEY WORDS disposed fish, egg characteristics, indirect boiling, layers, roasting.

INTRODUCTION

Poultry has seen the greatest increase in production in recent decades and this trend will likely continue. Both poultry eggs and meat are well positioned to meet demands for increased supply from our growing world population. The importance of poultry as a source of high quality protein in the form of meat and eggs had been established for a long time. The advance technology in management of poultry farms with specialized breeds and feed companies made eggs and poultry meat more widely available throughout the world than before. Recently, the wide fluctuation in ingredient costs has led to greater emphasis on poultry production economics. If wastes and neglected by-products from agriculture and food industry could be transferred into animal protein, this would solve a great problem, helping people of developing countries to avoid hunger and spare cereals and legumes for human consumption (Al-Harthi *et al.* 2009). The eggs production was increased as consumers become more educated about the nutritive value of the egg. Eggs are relatively inexpensive per unit of protein and energy contained in yolk and albumen, so eggs consumption continues to increase in developing countries. The crude protein for fish meal in Sudan is 43.5% with 9.6% content of ether extract (Omer, 2000). The sun-drying time for (5 and 7 days) were similar in crude protein (50.77%) and ether extract (17.13%) (Elobied, 2003). Based on these observations, some studies have shown that poultry by-product meal cannot replace more than 50% of fish meal in fish diets (Fowlerm, 1991), but other studies have shown that with the recent improvement of the quality of poultry by-product meal it could replace 75% or 100% of fish meal without significant decrease in fish growth (Alexis et al. 1985). Objectives of the study to make use of the disposed fishes of the White River Nile in Sudan after a certain simple local treatments, i.e. (sun drying, roasting, direct boiling and indirect boiling) and to replace the imported concentrate for layers diets from 19 weeks (pre-production period) through the production period up to 40 weeks of age.

MATERIALS AND METHODS

Site of the study

The experiment had been carried out at Rural Development and Extension Center, Faculty of Animal Production, University of Gezira, El-Managil town Gezira State, Sudan, (14.25 N-32.99 E, 76 km. western Wad-Medani town) during the period from May 2013 to August 2013. The temperature at the Gezira State ranged between 20 °C to 47 °C; with relative humidity 45-80% while the raining level during autumn season (July, August, September and October) ranges between 110-120 mL (ARC, 2008).

Local disposed fish collection and preparations

More than 500 kg of disposed fish was brought from the White Nile River, Khor Abugassaba site 15 km north Eldueim town, central Sudan in April 2012. Usually after the flood season some fish are trapped at cultivated rice and irrigation canals. When the water run out, trapped fish are naturally died with some spoilage symptoms. Different fish types have been collected and identified which include: *Tilapia* spp. (Bulti), *Mormyrius niloticus* (Khshmelbanat), *Shilbe mystus* (Shellbaya), *Labeo nilotcus* (Debsa), *Synodontis nilotcus* (Gurgor) and *Protopterus aethiopicus* (Um-koru).

The whole quantity has been grinded by commercial mill then the raw grinded fish stored in a plastic bag to avoid moisture, microbial contamination and parasites (Tables 1 and 2). Then the raw ground fish was treated by heat as the flowing treatments:

Sun drying treatment (SDT)

A quantity of fish were treated by sun drying in which 125 kg of raw ground fish were exposed to direct sun radiation.

A screen-net was used to protect processing fish meal from all direct contact with animals and insects. The sun-treated fish were exposed to direct sun radiation for 72 hours. The sun-dried fish was weighed and stored in plastic bags to avoid moisture, microbial contamination and parasites and put in a clean and aerated room.

Roasting treatment (RT)

Approximately 125 kg of raw ground fish was treated by roasting in special local designed metal drum. The metal drum was exposed to moderate stove gas fire for 15 minutes with manual turned drum. The roasted product was weighed and stored in a plastic bags and put in a clean and aerated room.

Direct boiling treatment (DBT)

The direct boiling treatment depend on cooking about 125 kg of raw grinding fish on an alumium pots and gas stove. Tap water was boiled, after water start boiling gradually raw fish added, then the mixture boiled for 20 minutes on gas stove, then the mixture dried by air in the cage which been used for the first treatment then the direct boiling fish meal was weighed and stored in a plastic bags and put in a clean and aerated room.

Indirect boiling treatment (IBT)

Locally constructed pots were made of double wall aluminum, similar to waterpass aim. The heats from gas stove during the treatment was transmitted via water between wall of aluminum pots. The raw ground fish were placed inside the inner pots and tap water was poured in the space between the two bowls, that will transferred the heat indirectly to the raw ground fish, which subjected to the indirect boiling fish. The double bowl was covered. The treatments for ground raw fish depend on gas stove fire process till the water boiling point, then left for fifteen minutes. Then indirect boiling fish subjected to drying via screen-net cage. Then, the indirect boiling fish was weighed and stored in a plastic bag and put in a clean and aerated room.

Chemical analyses of the treated samples

Table 3 show the samples chemically analyzed according to AOAC (2005) at biochemistry lab, Faculty of Veterinary, University of Khartoum and Soba National Laboratory. Metabolizable energy (ME) value of the feed ingredients were calculated according to equation of Ellis (1981):

ME= 1.549 + 0.0102 CP + 0.0275 EE + 0.0148 NFE - 0.0034 CF

Where:

CP: crude protein.

EE: ether extracts.

NFE: nitrogen free extract.

CF: crude fibre.

Experimental birds

A total of 390 pullets of Hy-line W-98 at 19 weeks of age were used. The pullets were randomly distributed in 13 treatment groups, each replicated 3 times with 10 birds per replicate, to test the disposal White River Nile fish with supper concentrate.

Statistical analysis

Analysis of variance ANOVA, Steel and Torrie (1980) was performed on all data using the CRBD procedure of SPSS (2012). Differences between dietary treatments were tested using Duncan (1955).

Experimental rations

The diet was formulated according to Hy-line W-98 performance standards manual (Tables 1 and 2). The diets were used during the pre-productive period and the production period.

Table 1 Ingredients and calculated analysis of the experimental diets fed to laying hens during pre-production and production periods (19-40 weeks of age)[*]

Ingredients (%)	Sun dried DF[1]. 5%	Sun dried DF. 3.5%	Sun dried DF. 1.5%	Roasted DF. 5%	Roasted DF. 3.5%	Roasted DF. 1.5%	Supper concentrate 5% (control group)
Sorghum grain	54	54	54	54	54	54	54
Groundnut cake	12.01	12.01	12.01	12.01	12.01	12.01	12.01
Wheat bran	18	18	18	18	18	18	18
Local treated fish meal	5	3.5	1.5	5	3.5	1.5	0
Imported concentrates	0	1.5	3.5	0	1.5	3.5	5
Dicalcium phosphate	0.25	0.25	0.25	0.25	0.25	0.25	0.25
Limestone	10	10	10	10	10	10	10
Salt (NaCl)	0.3	0.3	0.3	0.3	0.3	0.3	0.3
Lysine-HCl	0.1	0.1	0.1	0.1	0.1	0.1	0.1
DL-methionine	0.04	0.04	0.04	0.04	0.04	0.04	0.04
Antitoxins	0.2	0.2	0.2	0.2	0.2	0.2	0.2
Premix[2]	0.1	0.1	0.1	0.1	0.1	0.1	0.1
Total	100	100	100	100	100	100	100
Calculated analysis							
ME (kcal/kg)	2823.26	2829.72	2834.57	2823.25	2829.72	2834.57	2818
Protein (%)	16.88	17.20	17.44	16.917	17.28	17.56	16.64

[*] Diets formulated according to Hy-line W-98 Performance Standards Manual.

[1] Disposed fish.

[2] Supplied per kilogram of diet: vitamin A (retinyl acetate+retinyl palmitate): 6050 µg; vitamin D_3: 55 µg; vitamin E (α-tocopheryl acetate): 22.05 µg; vitamin K_3: 2 mg; vitamin B_1: 5 mg; vitamin B_2: 6 mg; vitamin B_3: 60 mg; vitamin B_6: 4 mg; vitamin B_{12}: 0.02 mg; Pantothenic acid: 10.0 mg; Folic acid: 6 mg; Biotin: 0.15 mg and Ethoxyquin: 0.625 mg.

Table 2 Ingredients and calculated analysis of the experimental diets fed to laying hens during pre-production and production periods (19-40 weeks of age)[*]

Ingredients (%)	Direct boiling DF[1]. 5%	Direct boiling DF. 3.5%	Direct boiling DF. 1.5%	Indirect boiling DF. 5%	Indirect boiling DF. 3.5%	Indirect boiling DF. 1.5%
Sorghum grain	54	54	54	54	53.97	54
Groundnut cake	12.01	12.01	12.01	12.01	12.04	12.01
Wheat bran	18	18	18	18	18	18
Local fish meal	5	3.5	1.5	5	3.5	1.5
Imported concentrates	3.5	1.5	0	3.5	1.5	0
Dicalcium phosphate	0.25	0.25	0.25	0.25	0.25	0.25
Lime stone	10	10	10	10	10	10
Salt (NaCl)	0.3	0.3	0.3	0.3	0.3	0.3
Lysine-HCl	0.1	0.1	0.1	0.1	0.1	0.1
DL-Methionine	0.04	0.04	0.04	0.04	0.04	0.04
Antitoxins	0.2	0.2	0.2	0.2	0.2	0.2
Premix[2]	0.1	0.1	0.1	0.1	0.1	0.1
Total	100	100	100	100	100	100
Calculated analysis						
ME (kcal/kg)	2823.25	2829.72	2834.57	2823.26	2829.70	2830
Protein (%)	16.89	17.22	17.40	16.88	17.18	17.44

[*] Diets formulated according to Hy-line W-98 Performance Standards Manual.

[1] Disposed fish.

[2] Supplied per kilogram of diet: vitamin A (retinyl acetate+retinyl palmitate): 6050 µg; vitamin D_3: 55 µg; vitamin E (α-tocopheryl acetate): 22.05 µg; vitamin K_3: 2 mg; vitamin B_1: 5 mg; vitamin B_2: 6 mg; vitamin B_3: 60 mg; vitamin B_6: 4 mg; vitamin B_{12}: 0.02 mg; Pantothenic acid: 10.0 mg; Folic acid: 6 mg; Biotin: 0.15 mg and Ethoxyquin: 0.625 mg.

Table 3 Proximate analyses of the ingredients used in feed formulation

Ingredients / Proximate analyses	*ME (kcal/kg)	CP (%)	EE (%)	CF (%)	Ash (%)	NFE (%)	Ca (g/kg)	P (g/kg)
Sun-dried fish	2395.55	50.75	9.20	4.06	30.45	6.04	49.10	29.62
Roasted fish	2418.61	52.50	8.32	5.05	29.85	6.22	48.81	30.54
Direct boiled fish	2370.62	50.55	9.43	6.12	32.95	4.56	48.84	29.37
Indirect boiled fish	2360.24	50.05	8.60	4.65	30.05	5.61	53.59	33.79
Imported Concentrates	2088.66	31.50	3.30	13.88	22.60	6.50	14.25	7.75
Sorghum	2306.82	13.26	2.63	8.41	19.25	56.45	0.44	0.26
Groundnut cake	2516.45	43.75	4.57	15.03	11.88	24.77	0.55	0.64
Wheat bran	2023.92	14.00	4.83	23.98	5.05	52.14	0.32	7.75

ME: metabolizable energy; CP: crude protein; EE: ether extracts; CF: crude fibre and NFE: nitrogen free extract.
* Calculated metabolizable energy according to equation of Ellis (1981).

RESULTS AND DISCUSSION

Chemical analyses of treated disposed fish samples

Table 1 shows chemical analyses of the treated disposed fish samples and feed ingredients used in the experiments. Metabolizable energy was calculated using Ellis (1981) equation ME MJ/kg (0.004184 kcal/kg= 1.549 + 0.012 CP + 0.0275 EE + 0.0148 NFE - 0.0034 CF). The crude protein content of treated disposed fish for sun-dried, roasting, direct and indirect boiled samples of treated disposed fish were very high (50.75, 52.50, 50.55 and 50.05%, respectively) compared to super concentrate (34.41%). Ether extract content of treated disposed fish samples scored < 8% compared to imported super concentrate-control (3.3%). The crude fibre of treated disposed fish samples scored < 4.06% and super concentrate was 13.88%. The nitrogen free extract value was scored < 4.56%, which was less than value of super concentrate (25.81%). All treated disposed fish ash content, calcium and phosphorus was high than imported supper concentrated. Calculated ME values for testing treated disposed fish scored < 9.876 MJ/kg with low value for imported super concentrate (9.294 MJ/kg) compared to all treated disposed fish. The chemical composition of treated disposed fish and the feed stuffs used were shown in Table 1. All of the treated disposal fish samples were a good source of protein and its protein content was very close to many previous studies (Omer Dar-elgalal, 2012; Salih *et al.* 2012).

Feed intake (g/bird/week)

Tables 4a, 4b and 4c show a significant differences (P<0.05) in the feed intake of treated disposed fish and control during the pre-production period. Mutayoba *et al.* (2003) found out that, the feed intake during 19th to 21th weeks were less than the present study, also the present study results of feed intake was higher than Chowdhury *et al.* (2005). Additionally Shim *et al.* (2013) reported that feed intake were 87.96 to 90.96 g/bird/day during the period (19-21 week), which was higher than the present study results.

Body weight

Table 5 show the growth performance of pre-layers body weight during the pre-production period showed the same variation of progress in body weight during the growing period with a little increase in weight gain. The present results of body weight were less than Al-Harthi *et al.* (2009) who found the body weight at the 20th week were (1303.8 to 1342.83 g/bird).

Body weight (g/bird/2 weeks)

Table 5 showed the body weight of the treated disposed fish concentrate compared to the control, which shown a significant difference (P<0.05) among the different treatments when compared with each other. Generally, the treatments of sun-dried fish (1.5%) was showed high body weight, followed by roasted fish (1.5 and 3.5%), control and roasted fish (5%), sun-dried fish (3.5 and 5%), while the different level of direct boiled fish and indirect boiled fish were showed lower body weight compared to other treatments. The body weight was affected significantly (P>0.05) as treated disposed fish substituted with supper concentrate. However, the best body weights achieved with the experiment were less than standards manual performance. The results in the present study were similar to Rao *et al.* (2011), while were less than the results found by Sirirat *et al.* (2013).

Egg production (% egg/week)

Data for table egg production (%) were summarized in Tables 6a and 6b. The average of weekly egg production (%) was showed significant different (P<0.05) among the treatment groups of treated disposed fish concentrate and supper concentrate.

The sun-dried fish (1.5%) showed the best egg production (%) throughout the production period followed by roasting fish (1.5 and 3.5%) and the control, especially during the 27th to 40th weeks. The results in the present study results were less than egg production (%) for Dickey *et al.* (2012), while the present results were similar to Rao *et al.* (2011) and higher than Novak *et al.* (2006).

Table 4a Means of weekly feed intake and weight gain during pre-production period (19-21 weeks)

Treatments	Feed intake/week (g/b/w)			Body weight/week (g)		
	19th	20th	21th	19th	20th	21th
Sun-dried fish 1.5%	641.67^{ab}	672.33^{abc}	674.68^{de}	1292.12^a	1294.13^a	1295.09^a
Sun-dried fish 3.5%	574.16^{cde}	600.68^{cde}	671.00^{de}	1229.17^e	1229.35^e	1232.11^e
Sun-dried fish 5%	648.00^{ab}	671.35^{abc}	694.02^{bcde}	1206.02^f	1207.12^f	1211.35^f
Roasted fish 1.5%	644.17^{ab}	688.66^{ab}	660.18^e	1281.11^{ab}	1282.85^{ab}	1284.35^{ab}
Roasted fish 3.5%	654.00^a	715.00^a	681.50^{cde}	1277.39^b	1280.70^b	1282.89^b
Roasted fish 5%	641.83^{ab}	676.00^{ab}	623.82^e	1263.73^c	1262.81^c	1261.20^c
Direct boiled fish 1.5%	566.66^{cde}	627^{bcde}	684.31^{bcde}	1170.54^g	1181.95^g	1183.57^g
Direct boiled fish 3.5%	537.68^{de}	587.33^{de}	737.00^{bcd}	1179.1^g	1179.94^g	1182.71^g
Direct boiled fish 5%	530.33^e	572.34^e	686.98^{bcde}	1163.50^g	1164.95^g	1163.64^g
Indirect boiled fish 1.5%	600.00^{bc}	598.71^{cde}	758.32^b	1163.61^{gh}	1169.23^{gh}	1167.71^{gh}
Indirect boiled fish 3.5%	585.01^{cd}	643.66^{abcde}	753.00^{bc}	1150.16^i	1154.19^i	1158.49^i
Indirect boiled fish 5%	615.66^{abc}	695.01^{ab}	848.64^a	1165.92^{gh}	1171.7^{gh}	1176.83^{gh}
Control 5% concentrates	667.00^a	650.00^{abcd}	753.00^{bc}	1234.64^{cd}	1234.95^{cd}	1250.86^{cd}
SEM	4.58	6.30	6.25	4.6	4.7	5.17

The means within the same column with at least one common letter, do not have significant difference (P>0.05).
SEM: standard error of the means.

Table 4b Means of weekly feed intake (g/bird/day) during the production period from 22-30 weeks

Age in weeks / Treatments	22th	23th	24th	25th	26th	27th	28th	29th	30th
Sun-drying fish 1.5%	86.038^a	88.34^a	89.61^b	90.23^{bc}	90.70^{abc}	92.20^a	92.71^a	92.78^a	92.74^a
Sun-drying fish 3.5%	82.47^d	84.79^d	87.80^d	88.16^e	90.67^{abc}	92.24^a	92.67^a	92.75^a	92.71^a
Sun-drying fish 5%	81.02^e	83.32^e	86.32^e	86.71^f	89.22^{cd}	90.74^{bc}	91.22^{bc}	91.30^{bc}	91.26^{bc}
Roasting fish 1.5%	85.37^b	87.69^b	90.70^a	91.11^a	91.87^a	92.23^a	92.71^a	92.79^a	92.77^a
Roasting fish 3 5%	84.98^b	87.31^b	90.27^a	90.69^{ab}	91.55^{ab}	91.27^{abc}	91.76^{abc}	91.85^{abc}	91.81^{abc}
Roasting fish 5%	83.59^c	85.86^c	88.89^c	89.31^d	91.17^{ab}	90.63^c	91.14^c	91.21^c	91.17^c
Direct boiling fish 1.5%	79.95^f	82.21^f	84.60^g	85.02^{gh}	86.20^e	87.23^e	87.75^e	87.81^e	87.77^e
Direct boiling fish 3.5%	79.56^{fg}	81.82^{fg}	84.15^{gh}	84.60^{hi}	85.74^e	87.39^e	87.54^e	87.62^e	87.58^e
Direct boiling fish 5%	78.26^h	80.53^h	83.56^h	83.89^i	86.04^e	87.20^e	87.71^e	87.79^e	87.75^e
Indirect boiling fish 1.5%	78.15^h	80.46^h	84.70^g	85.28^{gh}	87.76^d	89.25^d	89.75^d	89.84^d	89.83^d
Indirect boiling fish 3.5%	79.07^g	81.31^h	84.81^g	85.27^{gh}	87.80^d	89.27^d	89.78^d	89.91^d	90.21^d
Indirect boiling fish 5%	79.18^g	81.50^g	85.56^f	85.60^g	88.05^d	89.61^d	90.05^d	90.13^d	90.91^d
Control 5% concentrates	83.95^c	86.19^c	89.22^{bc}	89.61^{cd}	90.13^{bc}	91.60^{ab}	92.16^{ab}	92.21^{ab}	92.17^{ab}
Manual standard	92	92	95	96	97	97	98	98	98
SEM	0.141	0.141	0.230	0.231	0.475	0.309	0.309	0.309	0.309

The means within the same column with at least one common letter, do not have significant difference (P>0.05).
SEM: standard error of the means.

Table 4c Means of weekly feed intake (g/bird/day) during the production period from 31-40 weeks

Age in weeks / Treatments	31th	32th	33th	34th	35th	36th	37th	38th	39th	40th
Sun-drying fish 1.5%	93.74^a	95.75^a	97.19^a	97.94^a	98.00^a	98.38^a	98.46^a	98.54^a	99.24^a	100.23^a
Sun-drying fish 3.5%	93.71^a	95.72^a	97.21^a	97.89^a	97.97^a	98.34^a	98.43^a	98.51^a	99.21^a	100.24^a
Sun-drying fish 5%	92.25^b	93.67^{cd}	95.18^{cd}	95.89^{cd}	95.96^{bc}	96.31^b	96.40^{bc}	96.48^{bc}	97.18^{bc}	98.16^{bc}
Roasting fish 1.5%	93.65^a	95.78^a	97.26^a	97.97^a	98.03^a	98.40^a	98.49^a	98.56^a	99.27^a	100.26^a
Roasting fish 3.5%	92.81^{abc}	94.84^{ab}	96.29^{ab}	97.00^{ab}	97.69^{ab}	97.44^a	97.53^{ab}	97.61^{ab}	98.31^{ab}	99.30^{ab}
Roasting fish 5%	92.17^b	94.17^{bc}	95.67^{bc}	96.79^{bc}	97.31^a	97.67^a	97.76^a	97.84^a	98.54^a	99.53^a
Direct boiling fish 1.5%	88.79^d	90.77^e	92.28^e	92.96^e	93.77^e	94.01^c	94.16^e	94.18^e	94.88^e	95.87^e
Direct boiling fish 3.5%	88.58^d	90.57^e	92.08^e	92.80^e	93.41^e	93.92^c	94.0^e	94.09^e	94.79^e	95.78^e
Direct boiling fish 5%	88.19^d	90.54^e	91.86^e	92.98^e	94.24^{de}	94.01^c	94.72^{de}	94.80^{de}	95.50^{de}	96.49^{de}
Indirect boiling fish 1.5%	90.65^c	92.84^d	94.31^d	95.05^d	95.07^{cd}	95.43^b	95.54^{cd}	95.60^{cd}	96.31^{cd}	97.29^{cd}
Indirect boiling fish 3.5%	90.81^c	91.81^d	93.34^d	94.01^d	94.97^{cd}	95.44^b	95.43^{cd}	95.64^{cd}	96.29^{cd}	97.59^{cd}
Indirect boiling fish 5%	91.09^c	93.10^d	94.83^{cd}	95.54^d	95.59^c	96.10^b	96.19^c	96.27^c	96.97^c	97.96^c
Control 5% concentrates	92.83^{ab}	94.83^{ab}	96.33^{ab}	97.02^{ab}	97.09^{ab}	97.46^a	97.55^{ab}	97.63^{ab}	98.32^{ab}	99.33^{ab}
Manual standard	98	98	98	99	99	99	99	99	99	99
SEM	0.306	0.323	0.316	0.334	0.409	0.355	0.384	0.384	0.384	0.384

The means within the same column with at least one common letter, do not have significant difference (P>0.05).
SEM: standard error of the means.

Table 5 Means of Body weight (g/bird/2weeks) during the whole production period

Treatments \ Age in weeks	22th	24th	26th	28th	30th	32th	34th	36th	38th	40th
Sun-drying fish 1.5%	1296.01^a	1301.91^a	1310.07^a	1321.22^a	1330.34^a	1339.11^a	1350.74^a	1361.18^a	1370.62^a	1382.02^a
Sun-drying fish 3.5%	1234.35^d	1240.41^d	1248.30^d	1257.54^d	1268.30^d	1278.58^d	1289.61^d	1297.57^d	1308.09^d	1316.37^d
Sun-drying fish 5%	1209.42^e	1215.62^e	1223.81^e	1234.28^e	1242.75^e	1253.33^e	1265.27^e	1275.13^e	1283.51^e	1294.79^e
Roasting fish 1.5%	1284.66^b	1290.04^b	1298.27^b	1305.00^b	1319.51^b	1328.49^b	1338.58^b	1349.04^b	1358.45^b	1367.91^b
Roasting fish 3.5%	1278.49^b	1284.78^b	1292.95^b	1303.76^b	1312.08^b	1321.83^b	1332.00^b	1342.94^b	1353.67^b	1362.00^b
Roasting fish 5%	1253.59^c	1259.67^c	1268.63^c	1275.63^c	1286.19^c	1295.95^c	1307.67^c	1317.01^c	1326.55^c	1337.55
Direct boiling fish 1.5%	1190.17^f	1196.09^f	1204.25^{fg}	1215.05^g	1224.09^g	1236.91^g	1245.01^g	1256.24^g	1268.68^g	1276.34^g
Direct boiling fish 3.5%	1183.84^{fg}	1189.84^f	1197.31^g	1208.85^g	1216.34^g	1225.74^g	1237.29^g	1248.89^g	1252.40^g	1267.33^g
Direct boiling fish 5%	1161.19^h	1167.25^i	1174.33^i	1188.27^h	1195.49^h	1206.88^h	1216.55^h	1225.29^h	1237.92^h	1246.24^h
Indirect boiling fish1.5%	1165.58^h	1176.98^g	1189.11^h	1210.42^g	1222.16^g	1231.66^g	1239.75^g	1245.67^g	1259.00^g	1267.76^g
Indirect boiling fish3.5%	1160.23^h	1175.07^g	1182.94^h	1207.94^g	1218.88^g	1229.43^g	1236.98^g	1248.73^g	1255.33^g	1271.52^g
Indirect boiling fish 5%	1177.45^g	1192.94^f	1206.38^f	1224.53^f	1235.42^f	1247.82^f	1255.69^f	1264.37^f	1276.54^f	1285.48^f
Control 5% concentrates	1259.41^c	1265.38^c	1270.54^c	1285.31^c	1292.87^c	1304.12^c	1315.65^c	1324.74^c	1337.78^c	1345.07^c
SEM	2.440	2.435	2.435	2.435	2.435	2.435	2.435	2.435	2.435	2.435

The means within the same column with at least one common letter, do not have significant difference (P>0.05).
SEM: standard error of the means.

Table 6a Means of daily egg production (%) during 22-30 weeks

Treatments	Daily egg production at different age (%)								
	22th	23th	24th	25th	26th	27th	28th	29th	30th
Sun-dried fish 1.5%	59.21^a	62.28^a	65.91^a	70.77^a	73.70^a	77.01^a	80.81^a	85.34^a	89.21^a
Sun-dried fish 3.5%	56.75^{de}	60.06^{cde}	63.69^{cd}	68.46^{cde}	70.75^{de}	74.54^{de}	78.68^{bc}	83.25^{cd}	86.75^{cd}
Sun-dried fish 5%	56.12^{ef}	59.43^{ef}	62.45^{de}	67.48^{def}	70.43^{de}	73.50^{fg}	77.74^{cd}	82.17^{de}	85.79^{de}
Roasting fish 1.5%	58.68^{ab}	62.08^a	65.68^{ab}	70.15^{ab}	72.69^{ab}	76.48^{ab}	80.54^a	85.15^{ab}	88.92^a
Roasting fish 3.5%	58.43^{abc}	61.49^{ab}	65.43^{ab}	69.90^{abc}	72.80^{ab}	76.25^{abc}	80.36^a	84.65^{ab}	88.43^{ab}
Roasting fish 5%	57.48^{cd}	60.55^{bcd}	64.51^{bc}	69.04^{bcd}	71.13^{cd}	75.28^{cd}	79.75^{ab}	83.98^{bc}	87.48^{bc}
Direct boiled fish 1.5%	55.48^{fg}	58.62^f	61.91^e	56.88^g	67.91^f	71.76^h	75.51^e	83.84^f	83.85^f
Direct boiled fish 3.5%	54.74^{gh}	58.45^f	59.75^f	63.67^h	66.11^g	69.60^i	73.73^f	78.33^g	81.80^g
Direct boiled fish 5%	54.29^h	56.89^g	58.58^f	62.51^h	64.63^h	68.36^j	72.73^f	77.36^g	80.89^g
Indirect boiled fish 1.5%	55.45^{fg}	58.51^f	62.28^e	66.25^{fg}	69.51^e	72.71^{gh}	77.05^d	81.71^e	85.24^e
Indirect boiled fish 3.5%	56.16^{ef}	59.53^{def}	63.10^{de}	66.93^{efg}	70.55^{de}	73.97^{ef}	78.06^{cd}	82.60^{de}	86..15^{cde}
Indirect boiled fish 5%	55.79^{efg}	59.12^{ef}	63.03^{de}	67.25^{efg}	70.03^{de}	73.46^{fg}	77.59^{cd}	82.25^{de}	85.70^{de}
Control 5%	57.66^{bcd}	60.73^{bc}	64.100^{ab}	68.13^{de}	72.03^{bc}	75.46^{bcd}	79.59^{ab}	84.46^{abc}	87.33^{bc}
SEM	0.336	0.336	0.404	0.500	0.399	0.335	0.409	0.437	0.437

The means within the same column with at least one common letter, do not have significant difference (P>0.05).
SEM: standard error of the means.

Table 6b Means of daily egg production during 31-40 weeks

Treatments	Daily egg production at different age (%)								
	31th	32th	33th	34th	35th	36th	37th	38th	39th	40th
Sun-dried fish 1.5%	91.41^a	91.41^a	90.95^a	90.48^a	90.54^a	89.21^a	88.28^a	88.45^a	86.46^a	86.19^a
Sun-dried fish 3.5%	89.28^{cd}	89.32^c	88.48^{cd}	88.52^{cd}	88.08^{cd}	86.75^{cd}	86.15^{cd}	86.13^{de}	84.35^{cd}	83.72^{cd}
Sun-dried fish 5%	88.32^{de}	87.72^c	87.85^{de}	86.51^{de}	86.79^{ef}	85.79^{de}	85.19^{de}	85.50^{ef}	83.37^{de}	83.10^{de}
Roasting fish 1.5%	90.80^{ab}	90.23^{ab}	90.13^{ab}	90.07^{ab}	89.60^{ab}	88.60^{ab}	87.97^a	87.97^{ab}	86.28^{ab}	85.59^{ab}
Roasting fish 3.5%	90.65^{ab}	90.57^{ab}	89.40^{ab}	90.10^{ab}	89.70^{ab}	88.45^{ab}	87.85^{ab}	87.50^{abc}	86.07^{ab}	85.43^{ab}
Roasting fish 5%	90.08^{bc}	89.68^b	90.95^a	89.18^{bc}	88.78^{bc}	87.78^{bc}	86.88^{bc}	86.84^{bc}	85.15^{bc}	84.76^{bc}
Direct boiled fish 1.5%	86.04^g	85.96^d	85.58^f	85.44^f	84.84^g	84.18^f	83.24^f	83.22^g	81.76^f	81.16^f
Direct boiled fish 3.5%	84.43^h	84.08^e	83.53^g	83.40^g	83.13^h	82.13^g	81.20^g	81.51^h	79.71^g	79.11^g
Direct boiled fish 5%	83.15^i	82.84^f	82.48^h	8230^h	81.96^i	80.89^h	79.96^h	79.96^i	78.26^h	77.86^h
Indirect boiled fish 1.5%	87.11^f	88.74^d	86.85^e	86.71^e	86.11^f	85.45^e	84.51^e	84.49^f	82.69^{ef}	82.43^e
Indirect boiled fish 3.5%	88.57^{de}	88.19^c	87.66^{de}	87.52^{de}	87.26^{de}	86.59^{de}	85.32^{de}	85.30^{ef}	83.84^{de}	83.24^{de}
Indirect boiled fish 5%	88.31^e	87.96^c	87.39^{de}	87.26^e	87.33^{de}	86.33^{de}	85.39^{de}	85.04^f	83.57^{de}	82.97^{de}
Control 5%	90.10^{bc}	89.93^b	89.73^b	89.59^{ab}	88.95^{bc}	87.87^b	87.33^{ab}	87.04^{abc}	85.58^{ab}	84.64^{bc}
SEM	0.327	0.315	0.343	0.343	0.374	0.384	0.342	0.342	0.375	0.363

The means within the same column with at least one common letter, do not have significant difference (P>0.05).
SEM: standard error of the means.

Feed conversion ratio (FCR)

The effects of the experimental treatments on FCR are illustrated in Tables 7a and 7b. The results showed a significant difference (P>0.05) among the treatments during the experimental periods. It appears that the feed conversion value was high at the begging of the production period then gradually decreased till it become relatively stable at 31th weeks of the production period up to week 40, that, the feed conversion ratio was affected significantly (P<0.05) by as treated disposed fish change and control. The feed conversion ratio (g feed/g egg) for all treatments that including control agree with the results of (Sittiya and Yamauchi, 2014), while, the ratio of feed conversion ratio was high than the results of (Bryant et al. 2007). The results of the present study were less than (Perez-Bonilla et al. 2012) who found the means of feed conversion ratio were (1.89 to 2.05 g/g) during 24 to 59 weeks of age, also the results found by (Bonekamp et al. 2010).

Egg weight

The results of egg weights were illustrated in Tables 8a and 8b, which showed a significant differences (P<0.05) among the treatments during the experimental periods. Egg weight of layers chickens fed roasted fish (5%) have the highest egg weight during all the experimental period followed by roasted fish (3.5 and 1.5%), while the lowest egg weight was found at treatment of direct boiled fish (5%).

The results of the present results were high than that reported by (Rao et al. 2011) who found the means of egg weight for layers chickens at age of (21, 24, 29 and 32 to 36weeks) were (46, 52, 53.2 and 53.5 g/egg/day) respectively, while the present results were low than (Neijat et al. 2011). Also Perez-Bonilla et al. (2012) who found the means of egg weight were (63.1 to 64.1 g) during 24 to 59 weeks of age, while the present study results of egg weight were similar to (Park and Ryu, 2011). Egg weight was significantly (P<0.05) affected due to the different treated disposed fish. Generally the treatments of sun-dried fish and roasted fish levels and control have the best performance production which indicates the positive ability of treatment especially sun-dried fish and roasted fish compared to control as a source of protein to support layer production.

Panel test

Table 9 shows the panel test for test acceptability, colour, and smell. The panel which runs only for (5%) replicate to showed the effects of concentrations for different treated disposed fish concentrate and control to avoid the interaction effect on the level at (1.5 and 3.5%).

The panel test runs for all treated disposed fish and control for only (5%) replicate for accurate test for treated disposed fish concentrate or supper concentrate without interaction between the mixture of treated disposed fish and control.

Table 7a Means of feed conversion ratio (g feed/g egg) during production period of 24-32 weeks

Treatments	Age in weeks							
	24th	25th	26th	27th	28th	29th	30th	31th
Sun-drying fish 1.5%	3.26	2.97	2.91	2.81	2.65	2.63	2.46	2.41
Sun-drying fish 3.5%	3.36	3.10	3.05	2.90	2.79	2.68	2.48	2.38
Sun-drying fish 5%	3.43	3.15	3.06	2.99	2.82	2.65	2.53	2.43
Roasting fish 1.5%	3.28	3.07	2.99	2.82	2.73	2.54	2.37	2.35
Roasting fish 3.5%	3.24	3.05	2.97	2.79	2.62	2.60	2.41	2.39
Roasting fish 5%	3.18	2.92	2.92	2.73	2.62	2.50	2.35	2.31
Direct boiling fish 1.5%	3.47	3.62	3.06	2.88	2.80	2.47	2.44	2.34
Direct boiling fish 3.5%	3.53	3.23	3.14	3.04	2.84	2.67	2.63	2.48
Direct boiling fish 5%	3.69	3.41	3.31	3.22	3.12	2.92	2.67	2.57
Indirect boiling fish 1.5%	3.28	3.08	2.96	2.90	2.74	2.57	2.41	2.37
Indirect boiling fish 3.5%	3.22	3.01	2.94	2.80	2.68	2.52	2.39	2.32
Indirect boiling fish 5%	3.33	3.03	3.01	2.90	2.71	2.53	2.43	2.33
Control 5% concentrates	3.48	3.18	2.98	2.84	2.75	2.61	2.48	2.37

The means within the same column with at least one common letter, do not have significant difference (P>0.05).

Table 7b Means of feed conversion ratio (g feed/g egg) during production period 32-40 weeks

Treatments	Age in weeks							
	32th	33th	34th	35th	36th	37th	38th	39th	40th
Sun-drying fish 1.5%	2.45	2.43	2.42	2.40	2.43	2.44	2.42	2.48	2.54
Sun-drying fish 3.5%	2.44	2.53	2.46	2.44	2.49	2.50	2.50	2.57	2.60
Sun-drying fish 5%	2.48	2.42	2.45	2.41	2.44	2.43	2.41	2.46	2.49
Roasting fish 1.5%	2.44	2.43	2.43	2.42	2.45	2.47	2.47	2.54	2.58
Roasting fish 3.5%	2.43	2.46	2.40	2.38	2.41	2.41	2.42	2.48	2.51
Roasting fish 5%	2.32	2.31	2.36	2.35	2.36	2.36	2.35	2.41	2.44
Direct boiling fish 1.5%	2.47	2.44	2.43	2.43	2.45	2.49	2.49	2.55	2.58
Direct boiling fish 3.5%	2.52	2.60	2.59	2.59	2.63	2.66	2.65	2.73	2.79
Direct boiling fish 5%	2.72	2.68	2.69	2.71	2.71	2.77	2.77	2.85	2.88
Indirect boiling fish 1.5%	2.42	2.50	2.50	2.49	2.52	2.55	2.55	2.62	2.65
Indirect boiling fish 3.5%	2.40	2.41	2.39	2.38	2.40	2.44	2.48	2.50	2.54
Indirect boiling fish 5%	2.42	2.45	2.43	2.39	2.43	2.44	2.45	2.51	2.54
Control 5% concentrates	2.42	2.44	2.43	2.40	2.43	2.43	2.43	2.48	2.51

The means within the same column with at least one common letter, do not have significant difference (P>0.05).

Table 8a Means of weekly egg weight (g) during production period from 24-31 weeks

Treatments	Age in weeks							
	24th	25th	26th	27th	28th	29th	30th	31th
Sun-drying fish 1.5%	52.50bcd	54.00ab	53.33b	53.77b	54.60ab	52.00a	53.17bcde	53.69abc
Sun-drying fish 3.5%	51.67cde	25.36b	53.00b	53.80b	53.10abc	52.33a	54.20abcd	55.53ab
Sun-drying fish 5%	50.80def	51.47bc	52.17bc	52.03c	52.47bc	52.77a	53.03cde	54.23abc
Roasting fish 1.5%	53.00abc	53.23b	53.33b	53.90ab	53.13abc	54.00a	55.53ab	55.28ab
Roasting fish 3.5%	53.67ab	53.66ab	53.33b	54.03ab	55.00a	52.66a	54.20abcd	54.04abc
Roasting fish 5%	54.57a	55.9a	55.33a	55.47a	54.90a	54.66a	55.87a	55.93a
Direct boiling fish 1.5%	49.66fj	52.00bc	52.33bc	53.20bc	52.33c	53.33a	54.16abcd	55.52ab
Direct boiling fish 3.5%	50.33efj	51.83bc	52.00bc	51.97c	52.61bc	52.73a	51.20e	53.27abc
Direct boiling fish 5%	48.67j	49.66c	50.66c	49.90d	48.70d	49.00b	51.20de	52.03c
Indirect boiling fish 1.5%	52.17bcde	52.67ab	53.66b	53.33bc	53.54abc	53.86a	55.10abc	55.40ab
Indirect boiling fish 3.5%	52.67bcd	53.33b	53.33b	54.33ab	54.17abc	54.46a	55.00abc	55.72ab
Indirect boiling fish 5%	51.33cdef	53.00b	52.66b	53.00bc	54.00abc	54.53a	55.00abc	55.87a
Control 5% concentrates	50.33efj	52.17bc	52.83b	53.87ab	53.10abc	52.67a	53.66abcd	54.87ab
SEM	0.17	0.22	0.16	0.14	0.18	0.26	0.20	0.21

The means within the same column with at least one common letter, do not have significant difference (P>0.05).
SEM: standard error of the means.

Table 8b Means of weekly egg weight (g) during production period from 32-40 weeks

Treatments	Age in weeks								
	32th	33th	34th	35th	36th	37th	38th	39th	40th
Sun-drying fish 1.5%	53.92b	55.30abc	56.33ab	56.83ab	57.09ab	57.55ab	58.03ab	58.24bc	57.71bcd
Sun-drying fish 3.5%	55.45ab	54.66bc	56.53ab	57.40ab	57.43ab	57.63ab	57.65ab	57.67cd	58.03bcd
Sun-drying fish 5%	54.31b	56.43ab	57.03ab	57.90a	58.08ab	58.76a	59.10a	59.77ab	59.66ab
Roasting fish 1.5%	54.83b	55.87ab	56.46ab	57.07ab	57.09ab	57.12ab	57.14ab	57.16cd	57.23cd
Roasting fish 3.5%	54.29b	55.10bc	56.50ab	57.56ab	57.59ab	58.06a	58.09a	58.11bc	58.46abc
Roasting fish 5%	57.00a	57.47a	58.06a	58.67a	59.52a	60.01a	60.40a	60.43a	60.63a
Direct boiling fish 1.5%	53.83b	55.73abc	56.46ab	57.30ab	57.33ab	57.35ab	57.37ab	57.39cd	57.72bcd
Direct boiling fish 3.5%	53.77b	53.52cd	54.13cd	54.73cd	54.76cd	54.78cd	54.80cd	54.82ef	54.59e
Direct boiling fish 5%	50.68c	52.35d	52.93d	53.53d	53.95d	53.97d	53.99d	54.01f	54.28e
Indirect boiling fish 1.5%	54.56b	54.70bc	55.30bc	55.90bc	55.93bc	55.95bc	55.97bc	55.99de	56.06de
Indirect boiling fish 3.5%	55.21b	56.34ab	57.13ab	57.73ab	57.76ab	57.78ab	57.08ab	57.82bcd	58.09bcd
Indirect boiling fish 5%	55.20b	55.71abc	56.73ab	57.66ab	57.69ab	58.27a	58.29a	58.31bc	58.54abc
Control 5% concentrates	54.80b	55.55abc	56.17ab	57.33ab	57.42ab	57.98ab	58.11a	58.44bc	58.81abc
SEM	0.15	0.19	0.17	0.16	0.16	0.16	0.16	0.17	0.21

The means within the same column with at least one common letter, do not have significant difference (P>0.05).
SEM: standard error of the means.

Table 9 Egg quality panel test (%)

Panel test degree	Sun-drying fish	Roasted fish	Direct boiled fish	Indirect boiled fish	Control
Acceptability					
4-Very good	42	36	44	38	40
3-Good	50	54	42	46	54
2-Acceptable	8	10	12	16	6
1-Not acceptable	-	-	2	0	0
Totals	100	100	100	100	100
Smell					
4-Good smell	50	48	54	42	56
3-Normal smell	46	42	42	48	42
2-Little fishy smell	4	10	4	8	2
1-Fishy smell	0	0	0	2	0
Totals	100	100	100	100	100
Egg yolk color					
4-Extremely desirable	0	0	0	0	0
3-Moderate desirable	8	6	10	6	8
2-Slightly undesirable	36	42	38	28	34
1-Extremely undesirable	56	52	52	66	58
Totals	100	100	100	100	100

The panel test declared that all treated disposed fish and control were similarity among all treatment level if the trace difference been ignored at acceptability, smell and colour.

CONCLUSION

The results of this study showed that the disposed fish in the Sudan can be converted to useful conventional fish meal for poultry rations by simple means of heat processing.

ACKNOWLEDGEMENT

The authors would like to acknowledge the Faculty of Animal Production, University of Gezira, Sudan for their collaboration and allowing us to use their farm poultry and facilities.

REFERENCES

Alexis M.N., Paparaskeva-Papoutsoglou E. and Theochri V. (1985). Formulation of practical diets for rainbow trout (*Salmo gairdneri*) made by partial or complete substitutes for fish meal by poultry by-product and certain plant by-products. *Aquaculture*. **50**, 61-73.

Al-Harthi A.M., El-Deek A.A., Yakout H.M. and AL-Refaee M. (2009). The nutritive value of date waste meal as feed stuff for Lohmann brown pullets and layers. *Japan Poult. Sci. Assoc.* **46**, 303-312.

AOAC. (2005). Official Methods of Analysis. Vol. I. 16th Ed. Association of Official Analytical Chemists, Arlington, VA, USA.

ARC. (2008). Agricultural Research Corporation. Metrological Unit. Available at:
http://www.arcsudan.sd.

Bonekamp R.P.R.T.,Lemme A., Wijtten P.J.A. and Sparla J.K.W.M. (2010). Effects of amino acids on egg number and egg mass of brown (heavy breed) and white (light breed) laying hens. *Poult. Sci.* **89**, 522-529.

Bryant G., Wu M.M., Gunawardana P. and Roland Sr D.A. (2007). Effect of nutrient density on performance, egg components, egg solids, egg quality, and profits in eight commercial Leghorn Strains during phase one. *Poult. Sci.* **86**, 691-697.

Chowdhury Sachchidananda D., Zeenat S., Musabbir A., Bishan L.C., Shubash C.D. and Bimol C.R. (2005). The nutritionale value Khesari for growing and laying pullets. *J. Poult. Sci.* **42**, 308-320.

Dickey E.R., Johnson A.K., Stalder K.J. and Bregendahl K., (2012). Effects of a premolt calcium and low-energy molt program on laying hen performance, egg quality and economics. *Poult. Sci.* **91**, 292-303.

Duncan D.B. (1955). Multiple Range and Multiple F-tests. *Biometrics*. **11**, 1-42.

Ellis N. (1981). The nutrient composition of Sudanese animal feeds. Bulletin 1: Northern Central Sudan. Central Animal Nutrition Research Laboratory, Kuku, Khartoum, Sudan.

Elobied A.O. (2003). Improved sun-drying of fish. M.S Thesis. University of Khartoum, Sudan.

Fowlerm L.G. (1991). Poultry by-product meal as a dietary protein source in fall chinook salmon diets. *Aquaculture*. **99**, 309-321.

Mutayoba S.K., Mutayoba B.M. and Okot P. (2003). The performance of growing pullets fed diets with varying energy and leucaena leaf meal levels. Livest. Res. Rural Dev. Available at:
http://www.lrrd.org.

Neijat M., house J.D., Guenter W. And Kebreab E. (2011). Calcium and phosphorus dynamics in commercial laying hens housed in conventional or enriched cage systems. *Poult. Sci.* **90**, 2383-2396.

Novak C., Yakout H.M. and Scheideler S.E. (2006). The effect of dietary protein level and total sulfur amino acid: lysine ratio on egg production parameters and egg yield in Hy-Line W-98 hen. *Poult. Sci.* **85**, 2195-2206.

Omer M.I. (2000). Utilization of some local animal and plant protein supplements for poultry in Sudan. Ph D. Thesis. University of Khartoum, Sudan.

Omer Dar-elgalal Ah Y. (2012). Biochemical and microbial analyses of wasted and disposed fish meal subjected to different heat treatments to be used as poultry feed. MS Thesis. Gezira Univ., Sudan.

Park Jae H. and Kyeong S.R. (2011). Relationship between dietary protein and Betaine supplementation in laying hens. *Japan Poult. Sci. Assoc.* **48**, 217-222.

Perez-Bonilla A., Novoa S., Garc J., Mohiti-Asli M., Frikha M. and Mateos G.G. (2012). Effects of energy concentration of the diet on productive performance and egg quality of brown egg-laying hens differing in initial body weight. *Poult. Sci.* **91**, 3156-3166.

Rao S.V., Rama V., Ravindran T., Srilatha A.K., Panda M.V. and Raju L.N. (2011). Effect of dietary concentrations of energy, crude protein, lysine and methionine on the performance of White Leghorn layers in the tropics. J. Appl. Poult. Res. **20**, 528-541.

Salih G.E., Ibrahim E.A., Mutaz S.B. and Awad M.A.R. (2012). The effect of direct and indirect boiling on chemical composition and microbial load of disposed waste fish of White Nile State, Sudan. *Gezira J. Engin. Appl. Sci.* **7(1)**, 1-17.

Shim M.Y., Song E., Billard L., Aggrey S.E., Pesti G.M. and Sodsee P. (2013). Effects of balanced dietary protein levels on egg production and egg quality parameters of individual commercial layers. *Poult. Sci. Assoc. Inc.* **92**, 2687-2696.

Sirirat N., Jin-Jenn L., Alex Tsubg-Yu H. and Tu-Fa L. (2013). Effect of different levels of nanoparticles chromium picolinate supplementation on performance, egg quality, mineral retention and tissues minerals accumulation in layer chickens. *J. Agric. Sci.* **5(2)**, 150-159.

Sittiya J. and Koh-en Y. (2014). Effects of replacing corn with whole grain paddy rice in laying hen diets on egg production performance. *J. Adv. Agric. Technol.* **1**, 1-14.

SPSS Inc. (2012). Statistical Package for Social Sciences Study. SPSS for Windows, Version 21. Chicago SPSS Inc. USA.

27

Genetic Characterization of Hamra Goat Population in Two Different Locations of Morocco using Microsatellite Markers

B. Hilal[1,2], I. Boujenane[1*], S. El Otmani[2], M. Chentouf[2] and M. Piro[3]

[1] Department of Animal Production and Biotechnology, Institut Agronomique et Vétérinaire Hassan II, Rabat, Morocco
[2] Institut National de la Recherche Agronomique, Regional Centre of Tangier, Tanger, Morocco
[3] Department of Medicine, Surgery and Reproduction, Laboratoire d'Analyses Génétiques Vétérinaires, Institut Agronomique et Vétérinaire Hassan II, Rabat, Morocco

*Correspondence E-mail: i.boujenane@iav.ac.ma

ABSTRACT

In this study, genetic diversity of two different populations of Hamra goat breed of Morocco was investigated in 60 different samples (including 30 from Beni Arouss and 30 from Rommani) using fifteen microsatellite markers. A total of 145 alleles were detected with average number per locus of 8.67 and 8.07 in Beni Arouss and Rommani goats, respectively. The Shannon's information index ranged from 1.58 in Rommani goats to 1.66 in Beni Arouss goats. The expected and the observed heterozygosity average over loci varied from 0.62 to 0.72 in Rommani and from 0.64 to 0.75 in Beni Arouss goats. Six markers in Beni Arouss goats and five in Rommani goats showed a significant deviation from Hardy-Weinberg equilibrium. The F_{IS} values were 0.110 and 0.108 for Beni Arouss and Rommani goats, respectively. A low genetic differentiation was indicated by F_{ST} values across the two goat groups. The genetic distance of Nei between the two groups was 0.046 indicating a low genetic differentiation. This was confirmed by the analysis of molecular variance (AMOVA) that showed that 99.15% of variation was distributed within genetic groups. The presence of two clusters (K=2) for microsatellite markers suggested a high level of population admixture. It was concluded that both groups (Beni Arouss and Romani) presented a high similarity and may be considered as belonging to the same population.

KEY WORDS
genetic diversity, Hamra goat, microsatellite marker, Morocco.

INTRODUCTION

Goats have been domesticated 10000 years ago in the Middle East (Luikart et al. 2001; Fernandez et al. 2006). The breeds have been developed based on traditional knowledge and improved through human interventions and natural selection (FAO, 2009). However, there is worldwide recognition of the need for the conservation of livestock diversity and for the characterization of breeds and populations including their genetic differentiation and relationships. The goat diversity analysis based on microsatellite markers have been successfully used because of their high mutation rate, abundance and distribution throughout the genome, neutrality, co-dominance nature and easy automation of analytical procedures. Several studies (Bruno-de-Sousa et al. 2011; Dixit et al. 2012; Bosman et al. 2015; Nafti et al. 2016) have been conducted to investigate the genetic diversity of goats in the world. Dixit et al. (2012) compared the genetic diversity of 20 Indian goat breeds and Bosman et al. (2015) investigated the genetic diversity of the south African commercial dairy goat population. In addition, Bruno-de-Sousa et al. (2011) and Nafti et al. (2016) used the microsatellites to identify the population structure of Portuguese goat breeds and southern Tunisia goat populations, respectively.

In Morocco, 95% of goat inventory, estimated to 6.2 million goats (FAOSTAT, 2015), is represented by indigenous populations. Nowadays, four local breeds are officially characterized and recognized by the Ministry of Agriculture: Draa goat in oases and Barcha, Laghzalia and Atlas goats in Atlas Mountains. Among the remaining indigenous populations, the Hamra goat is in the process of recognition.

This population is located in two regions that are far from each other (about 400 km); Beni Arouss (BA) located in North and Rommani (RO) found in the centre of Morocco. However, animals of these two regions have a remarkable resemblance with regard to their external appearance (color and morphology), but the Ministry of Agriculture does not know if these two groups are belonging to the same Hamra population or not.

This paper reports on the first molecular investigation of Hamra goat population and the comparison of its level of diversity in two groups using fifteen microsatellite DNA markers in order to determine the level of similarity between the Beni Arouss and Rommani goat population.

MATERIALS AND METHODS

Description of Hamra population
The Hamra goat population is characterized by its red coat color. Both males and females of this population are horned. Usually, the females are not bearded, while the males have a beard. The height at withers averaged 63 cm and 65 cm for females and males, respectively (Hilal et al. 2014).

Sample collection and DNA extraction
Fresh blood was collected from a total of 60 unrelated Hamra goats (30 in Beni Arouss and 30 in Rommani regions) based on the information provided by farmers. Sampling was carried out from 15 flocks (9 in Beni Arouss and 6 in Rommani regions). The blood was drawn from the jugular vein in vacationer tubes coated with EDTA as blood anti-coagulants. DNA was isolated from the whole blood using a commercial extraction and purification kit (DNeasy™ Blood Kit, Qiagen®), according to the manufacturer's protocol.

Microsatellites and PCR conditions
Fifteen microsatellite loci chosen from the recommended FAO/ISAG list (FAO, 2011) were tested. These microsatellite markers were arranged into six duplexes PCR panels (SRCRSP23/SRCRSP09, TGLA53/MAF209, SRCRSP08/ SRCRSP05, OARFCB48/MAF65, ILSTS87/ILSTS11, M CM527/CRSD247) and three simplexes (MAF70,

OARFCB20 and ETH10) based on their annealing temperature and their marking (Table 1).

The panel of microsatellites used in the present study has been also used earlier to study the genetic variability in other goat breeds by other researchers (Hoda et al. 2011; Agaoglu and Ertugrul, 2012; Bosman et al. 2015). PCR amplifications were performed in a final volume of 28 μL on an Applied Biosystem 2700 Thermocycler containing 2 μL of genomic DNA, 1 μL of each primer and 22 μL of SuperMix PCR kit containing 22 mM Tris-HCl, 55 mM KCl, 1.65 mM MgCl$_2$, 22 U/mL Taq polymerase and 220 μM for each dNTP. The PCR protocol used consisted of an initial denaturation step at 95 °C for 5 min, followed by 30 cycles of 15 seconds at 94 °C, 45 seconds at 55-65 °C depending on the primers used and 90 seconds at 72 °C and a final extension step at 72 °C for 30 minutes. The amplified PCR products were resolved on a 3500 DNA fragment analyzer (Applied Biosystem) and fluorescently-labelled fragments were detected and sized using GeneMapper 5.0 software (Applied Biosystems, Foster City, CA).

Statistical analyses
The genotypic data were statistically analyzed using different softwares. The POPGENE software (version 1.32) (Yeh et al. 1999) was used to calculate the number of alleles, the effective number of alleles, the Shannon information index, the observed and the expected heterozygosity at each locus and genetic group. Wright's F-statistics (F_{IS}, F_{IT} and F_{ST}) and the gene flow between the two groups were also calculated according to Weir and Cockerham (1984) using POPGENE software. Nei's standard genetic distance (Nei, 1978) among the two groups were estimated by using GE-NETIX software (version 4.03) (Belkhir et al. 2001). Deviations of genotype frequencies from Hardy-Weinberg expectations were assessed by exact tests based on Markov chains (20 batches, 5000 iterations and a dememorization number of 10 000) as implemented by the GENEPOP software (version 4.2) (Raymond and Rousset, 1995). Analysis of molecular variance (AMOVA) was done with the AR-LEQUIN (version 3.0) (Schneider et al. 1997) in order to explain the partitioning of the level of genetic variation of the genetic groups. Finally, the genetic structure was performed by full Bayesian approach using the software package STRUCTURE version 2.3.4 (k=7) (Hubisz et al. 2009). The probability (or likelihood) of different values was tested by assessing LnPr (X|K), i.e., the likelihood of the observed distribution of genotypes given the assumed number of ancestral populations. All runs used a burn-in period of 100000 iterations and a data collection period of 100000 iterations under an admixture model with allele frequencies correlated.

Table 1 Description of microsatellite DNA markers used in the current study

Microsatellite	Chromosome	Primer sequence (forward and reverse)	Annealing temperature (°C)	Dye	Size range
Duplex 1					
MAF209	CHI17	gatcacaaaaagttggatacaaccgtg tcatgcacttaagtatgtaggatgctg	55	NED	100-104
TGL53	BTA16	gctttcagaaatagtttgcattca atcttcacatgatattacagcaga	55	VIC	126-160
Duplex 2					
SRCRSP05	CHI21	ggactctaccaactgagctacaag tgaaatgaagctaaagcaatgc	55	NED	156-178
SRCRSP08	Not reported	tgcggtctggttctgatttcac gtttcttcctgcatgagaaagtcgatgcttag	55	6-FAM	215-255
Duplex 3					
SRCRSP9	CHI12	agaggatctggaaatggaatc gcactcttttcagccctaatg	58	6-FAM	99-135
SRCRSP23	Not reported	tgaacgggtaaagatgtg tgtttttaatggctgagtag	58	VIC	81-119
Duplex 4					
ILSTS087	BTA6	agcagacatgatgactcagc ctgcctcttttcttgagag	58	NED	135-155
ILSTS11	BTA14	gcttgctacatggaaagtgc ctaaaatgcagagccctacc	58	6-FAM	250-300
Duplex 5					
CSRD247	OAR14	ggacttgccagaactctgcaat cactgtggtttgtattagtcagg	58	VIC	220-247
MCM527	OAR5	gtccattgcctcaaatcaattc aaaccacttgactactccccaa	58	NED	165-187
Duplex 6					
MAF65	OAR15	aaaggccagagtatgcaattaggag ccactcctcctgagaatataacatg	58	VIC	116-158
OARFCB48	OAR17	gagttagtacaaggatgacaagaggcac gactctagaggatcgcaaagaaccag	58	PET	149-173
Simplex 1					
OARFCB20	OAR2	ggaaaacccccatatatacctatac aaatgtgtttaagattccatacatgtg	58	NED	93-112
Simplex 2					
ETH10	CH15	gttcaggactggccctgctaaca cctccagcccactttctcttctc	55	PET	200-210
Simplex 3					
MAF70	BTA4	cacggagtcacaaagagtcagacc gcaggactctacggggcctttgc	65	6-FAM	134-168

BTA: *Bos taurus*; OAR: *Ovis aries* and CHI: *Capra hircus*.

RESULTS AND DISCUSSION

Genetic variation among groups

The various measures of genetic variation evaluated by the number of observed (NO) and effective alleles (NE), Shannon information index (I), and the observed (HO) and expected heterozygosity (HE) for Beni Arouss and Rommani genetic groups across all loci are presented in Table 2. The microsatellites studied in Beni Arouss and Rommani genetic groups were successfully amplified and were found to be polymorphic. Total numbers of alleles observed per group were 127 and 118 alleles in Beni Arouss and Rommani groups, respectively. The observed number of alleles per locus ranged from 5 (MAF209, TGL53 and MCM527) to 13 (MAF65) for Beni Arouss group and from 5 (MAF20

9, MCM527 and ETH10) to 12 (SRCRSP23) for Rommani group. The mean number of alleles in Beni Arouss group seems to be higher than the Rommani group (8.67 *vs.* 8.07 alleles). Similar results had been observed in Noire-Rahalli (8.33 alleles) and Draa Moroccan goat populations (7.83 alleles) (Ouafi-Tadlaoui *et al.* 2002). The mean number of alleles of Beni Arouss and Rommani groups were higher than the three populations in South Africa: Saanen (6.80), Toggenburg (6.44) and British Alpine (6.84) (Bosman *et al.* 2015). Moreover, the allelic variation between the two groups was almost the same in different locus. The value of Shannon information index, which measures the polymorphism across loci, was highly informative with an overall mean of 1.58 in Rommani group and 1.66 in Beni Arouss group.

Table 2 Various measures of genetic variation at different loci of Beni Arouss and Rommani genetic groups of Hamra goat

Locus	Beni Arouss group					Rommani group				
	NO	NE	I	H_o	H_e	NO	NE	I	H_o	H_e
MAF209	5	1.87	0.82	0.30	0.47	5	1.99	0.83	0.47	0.51
TGL53	5	1.77	0.86	0.33	0.44	9	2.66	1.42	0.53	0.63
SRCRSP09	12	5.11	1.98	0.77	0.82	9	5.32	1.91	0.83	0.83
SRCRSP23	11	6.82	2.16	0.89	0.87	12	5.77	2.06	0.77	0.84
ILSTS087	8	2.99	1.48	0,47	0,68	7	1.80	0.99	0,43	0,45
ILSTS11	11	5.86	1.89	0.87	0,84	9	4.09	1.68	0.57	0,77
CSRD247	10	5.36	1.96	0.81	0,83	9	5.79	1.95	0.92	0,86
MCM527	5	3.09	1.23	0.29	0,69	5	1.42	0.65	0.07	0,20
ETH10	6	3.22	1.31	0.83	0.70	5	2.69	1.08	0.50	0.64
MAF65	13	6.36	2.12	0.73	0.86	11	7.83	2.23	0.80	0.89
OARFCB48	9	6.29	1.98	0.80	0.85	9	7.47	2.11	0.69	0.87
SRCRSP05	9	5.26	1.84	0.76	0.82	8	5.06	1.80	0.83	0.81
SRCRSP08	7	3.80	1.56	0.63	0.75	8	4.04	1.61	0.67	0.76
MAF70	11	5.03	1.93	0.50	0.81	9	5.01	1.81	0.43	0.81
OARFCB20	8	4.24	1.73	0.83	0.78	6	4.00	1.61	0.72	0.75
Mean±SD	8.67±2.66	4.47±1.62	1.66±0.44	0.64±0.22	0.75±0.13	8.07±2.12	4.33±1.97	1.58±0.49	0.61±0.22	0.72±0.19

NO: observed number of alleles; NE: effective number of alleles; I: Shannon's information index; H_o: observed heterozygosity and H_e: expected heterozygosity. SD: standard deviation.

The present Shannon's index values were comparable to those of the Nigerian West African dwarf goats (Awobajo et al. 2015), but slightly higher than those reported for Zalawadi (1.25), Gohilwadi (1.44) and Surti (1.33) Indian goat breeds (Shadma et al. 2008). Despite the small sample size of Beni Arouss and Rommani goat groups, they exhibited high levels of polymorphism at all studied loci. The genetic diversity in the two genetic groups was very similar and high varying from 72% in Rommani to 75% in Beni Arouss groups (Table 2). The high values of heterozygosity detected for Beni Arouss and Rommani genetic groups may be attributed to the large number of alleles detected in the two groups. These values are compared to those reported for six Portuguese native goat breeds (65%-71%) (Bruno-de-Sousa et al. 2011), lower than those of Nafti et al. (2016) on four indigenous goat populations of Tunisia, but higher than those reported for Saanen (65%), Toggenburg (62%) and British Alpine (64%) (Bosman et al. 2015). Based on the expected heterozygosity across the two groups, there was no appreciable difference in the level of genetic diversity. Moreover, six markers (ILSTS087, MCM527, CSRD247, MAF65, SRCRSP05 and MAF70) and five markers (TGL53, MCM527, MAF65, OARFCB48, and MAF70) showed a significant deviation from Hardy-Weinberg equilibrium in Beni Arous and Rommani groups, respectively. This indicates that differences between observed and expected number of genotypes were significant (P<0.05) for Beni Arouss and Rommani groups.

Fixation indices

Table 3 shows Wright's F-statistics for each locus in the two genetic groups. The within-group heterozygosity deficit (F_{IS}) was almost similar in both groups (11% and 10.8%).

Thus, large differences among loci were observed in estim-ated F_{IS} of the two groups, with values ranging from -0.221 (ETH10) to 0.551 (MCM527) and from -0.022 (OARFCB20) to 0.800 (MCM527) in Beni Arouss and Rommani groups, respectively. However, most of loci of the two groups showed a heterozygote deficit as depicted by the positive F_{IS} value. In the Beni Arouss and Rommani groups, 3 and 5 markers, respectively out of the 15 markers had negative F_{IS} values. This deficiency may be related to management conditions of Hamra goats in the two regions. The positive F_{IS} value indicated an increased homozygosity or heterozygosity deficit (Mishra et al. 2013). This excess of homozygotes in the two genetic groups was moderate in comparison to that found by Dixit et al. (2009) for Kutchi (26%), Mehsana (14%) and Sirohi (36%) Indian goat breeds, and by Gour et al. (2006) for Jamunapari goat breeds (19%), but higher than those observed by Mahrous et al. (2013) for three indigenous goat breeds (Barki and Zaraibi in Egypt and Ardi in Saudi Arabia) (5.3%). The heterozygote deficiency found in the Beni Arouss and Rommani goats groups could be due to the higher rate of inbreeding, to the population subdivision (Wahlund effect), and to the presence of "null alleles" (non-amplifying alleles). The overall genetic differentiation indicated by F_{ST} values was similar and very low in the two genetic groups. The F_{ST} value over all loci averaged 0.021 in Rommani group and 0.026 in Beni Arouss group, while the global deficit of heterozygotes (F_{IT}) varied from 0.128 in the Rommani group to 0.133 in the Beni Arouss group (Table 3). The F_{ST} value for the two groups was low compared to two Mexican cattle meta-populations (Iberian-like and Mexican Criollo) (Ulloa-Arviz et al. 2008) and lower than those found in three south African dairy goat populations (Saanen, British Alpine and Toggenburg) (Bosman et al. 2015).

Table 3 Wright's F-statistics for Beni Arouss and Rommani genetic groups for each of the 15 microsatellite markers

Locus	Beni Arouss group			Rommani group		
	F_{IS}	F_{IT}	F_{ST}	F_{IS}	F_{IT}	F_{ST}
MAF209	0.352	0.355	0.004	0.002	0.037	0.035
TGLA53	0.219	0.236	0.022	0.103	0.110	0.008
SRCRSP09	0.040	0.047	0.007	-0.002	0.003	0.005
SRCRSP23	0.001	0.023	0.022	0.069	0.084	0.016
ILSTS87	0.293	0.299	0.009	-0.018	-0.001	0.017
ILSTS11	-0.076	-0.045	0.029	0.282	0.291	0.013
CSRD247	0.004	0.057	0.053	-0.001	0.013	0.014
MCM527	0.551	0.606	0.121	0.800	0.805	0.026
ETH10	-0.221	-0.208	0.011	0.080	0.165	0.092
MAF65	0.157	0.169	0.015	0.017	0.033	0.016
OARFCB48	0.027	0.049	0.022	0.131	0.146	0.017
SRCRSP05	0.168	0.177	0.011	-0.017	0.025	0.041
SRCRSP08	0.131	0.140	0.011	0.079	0.092	0.014
MAF70	0.358	0.376	0.028	0.462	0.469	0.014
OARFCB20	-0.109	-0.090	0.017	-0.022	-0.002	0.020
Mean±SD	0.110±0.205	0.133±0.207	0.026± 0.029	0.108±0.227	0.128±0.222	0.022±0.021

SD: standard deviation.

F_{IS}: the within-group heterozygosity deficit; F_{IT}: the global deficit of heterozygotes and F_{ST}: the overall genetic differentiation.

The lack of differentiation observed between the two groups of Hamra goat population may be explained by the similarities in environment and breeding practices, as well as the maintained gene flow among the Hamra population although located in two different locations. The value of gene flow between the two genetic groups was 13.4 indicating a high rate of genetic flow between them. This may result from the higher mobility of goats from one region to the other and to the considerable exchange of genetic material among these goats. The low level of differentiation among the Hamra goat population of the two genetic groups was also confirmed by the standard genetic distance of Nei (0.046) showing a close relationship between them. Moreover, this reduction in genetic differentiation may result from the migration that had a greater effect than mutation or drift (Laval *et al.* 2000). Toro and Maki-Tanila (2007) suggested that the high genetic diversity observed within population groups could arise from overlapping generations and population mixtures from different geographical locations.

The analysis of molecular variation (AMOVA) within and among the two genetic goat groups of Hamra is shown in Table 4. The AMOVA showed that 99.15% and 0.85% of variation was distributed within and among the two genetic groups, respectively, suggesting that the two groups were submitted to the same geographical and environmental selections in recent history.

These results are lower than 4.8% among west African dwarf (WAD) goat populations (Awobajo *et al.* 2015) and than 7.5% among Spanish Guadarrama goat breed (Serrano *et al.* 2009). However, this analysis showed a high level of gene flow among the two groups, which is the most probable cause of this variation.

Table 4 Distribution of genetic variance in studied groups revealed by AMOVA

Source of variation	Degree of freedom	Sum of squares	Components of variation	Percentage
Among genetic groups	1	7.225	0.04089	0.85
Within genetic groups	118	563.017	4.77133	99.15
Total	119	570.242	4.81222	

To choose the appropriate number of inferred ancestral (K) populations, value of K varied from 2 to 7. However, the best value of the number of inferred ancestral populations [Ln Pr(X|K)] (-3312.7) was obtained for K= 2. A graphic representation of assumed ancestral populations is shown in Figure 1.

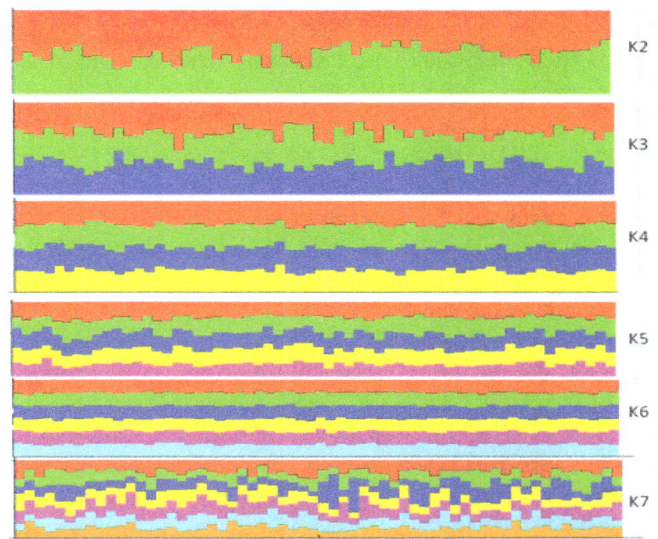

Figure 1 Clustering assignment of the two Hamra goat genetic groups provided by STRUCTURE analyses

The presence of two clusters suggested a high level of population admixture. In fact, genetic components seem to be shared by the two genetic groups which suppose a gene flow between them. The high level of admixture in the two genetic groups confirms their high similarity as indicated by genetic parameters.

CONCLUSION

All microsatellite markers used in the present study were shown to be highly polymorphic for the molecular characterization of Beni Arouss and Rommani goats. The results confirm a high similarity among Hamra goats in the two locations. The two goat groups were weakly structured, but they present an important pool of genetic diversity and high level of admixture. Therefore, the genetic information obtained in this study would help the Ministry of Agriculture in the process of breed recognition and would make a valuable contribution to the utilization of this goat population.

ACKNOWLEDGEMENT

The authors would like to thank farmers for providing animals and Mr. El Ouatiq and Mr. Daim for help in the field.

REFERENCES

Agaoglu O.K. and Ertugrul O. (2012). Assessment of genetic diversity, genetic relationship and bottleneck using microsatellites in some native Turkish goat breeds. *Small Rumin. Res.* **105**, 53-60.

Awobajo O.K., Salako A.E. and Osaiyuwu O.H. (2015). Analysis of genetic structure of Nigerian West African dwarf goats by microsatellite markers. *Small Rumin. Res.* **133**, 112-117.

Belkhir K., Borsa P., Chikhi L., Raufaste N. and Bonhomme F. (2001). GENETIX 4.03, logiciel sous windows TM pour la génétique des populations. Laboratoire Génome, Populations, Interactions, CNRS UMR 5000. Université de Montpellier II, Montpellier, France.

Bosman L., Marle-Köster E.V. and Visser C. (2015). Genetic diversity of South African dairy goats for genetic management and improvement. *Small Rumin. Res.* **123**, 224-231.

Bruno-de-Sousa C., Martinez A.M., Ginja C., Santos-Silva F., Carolino M.I., Delgado J.V. and Gama L.T. (2011). Genetic diversity and population structure in Portuguese goat breeds. *Livest. Sci.* **135**, 131-139.

Dixit S.P., Verma N.K., Aggarwal R.A.K., Kumar S., Chander R., Vyas M.K. and Singh K.P. (2009). Genetic structure and differentiation of three Indian goat breeds. *Asian-Australas J. Anim. Sci.* **22**, 1234-1240.

Dixit S.P., Verma N.K., Aggarwal R.A.K., Vyas M.K., Rana J. and Sharma A. (2012). Genetic diversity and relationship among Indian goat breeds based on microsatellite markers. *Small Rumin. Res.* **105**, 38-45.

FAO. (2009). Livestock keepers-guardians of biodiversity. Animal Production and Health Paper. Rome, Italy.

FAO. (2011). Draft guidelines on molecular genetic characterization of animal genetic resources. Commission of Genetic Resources for Food and Agriculture, Thirteenth Regular Session, Rome, Italy.

FAOSTAT. (2015). Food and Agriculture Organization of the United Nations the State of Food Insecurity in the World. Available at: http://faostat3.fao.org.

Fernandez H., Hughes S., Vigne J.D., Helmer D., Hodgins G., Miquel C., Hanni C., Luikart G. and Taberlet P. (2006). Divergent mtDNA lineages of goats in an Early Neolithic site, far from the initial domestication areas. *Proc. Natl. Acad. Sci.* **103**, 15375-15379.

Gour D.S., Malik G., Ahlawat S.P.S., Pandey A.K., Sharma R., Gupta N., Gupta S.C., Bisen P.S. and Kumar D. (2006). Analysis of genetic structure of Jamunapari goats by microsatellite markers. *Small Rumin. Res.* **66**, 140-149.

Hilal B., El Otmani S., Chentouf M. and Boujenane I. (2014). Morphological characterization of the local goat population "Beni Arrous". *Options Méditerranéennes, Séries A.* **108**, 433-437.

Hoda A., Hyka G., Dunner S., Obexer-Ruff G. and Econogene C. (2011). Genetic diversity of Albanian goat breeds based on microsatellite markers. *Arch. Zootec.* **60**, 607-615.

Hubisz M.J., Falush D., Stephens M. and Pritchard J.K. (2009). Inferring weak population structure with the assistance of sample group information. *Mol. Ecol. Res.* **9**, 1322-1332.

Laval G., Iannuccelli N., Legault C., Milan D., Groenen M.A.M., Giuffra E., Andersson L., Nissen P.H., Jorgensen C.B., Beeckmann P., Geldermann H., Foulley J.L., Chevalet C. and Ollivier L. (2000). Genetic diversity of eleven European pig breeds. *Genet. Sel. Evol.* **32**, 187-203.

Luikart G., Gielly L., Excoffier L., Vigne J.D., Bouuvet J. and Taberlet P. (2001). Multiple maternal origins and weak phylogeographic structure in domestic goats. *Proc. Natl. Acad. Sci.* **98**, 5927-5932.

Mahrous K.F., Alakilli S.Y.M., Salem L.M., Abd El-Aziem S.H. and El-Hanafy A.A. (2013). Genetic diversity in Egyptian and Saudi goat breeds using microsatellite markers. *J. Appl. Bio-Sci.* **72**, 5838-5845.

Mishra P., Ali A.S., Kuralkar S.V., Dixit S.P., Aggarwal R.A.K., Dangi P.S. and Verma N.K. (2013). Analysis of genetic diversity in Berari goat population of Maharashtra state. *Iranian J. Appl. Anim. Sci.* **3**, 553-559.

Nafti M., Khaldi Z. and Haddad B. (2016). Genetic relationships and structure among goat populations from southern Tunisia assessed using microsatellites. *J. New Sci. Agric. Biotechnol.* **27**, 1488-1497.

Nei M. (1978). Estimation of average heterozygosity and genetic distance from a small number of individuals. *Genetics.* **89**, 583-590.

Ouafi-Tadlaoui A., Babilliot J.M., Leroux C. and Martin P. (2002). Genetic diversity of the two main Moroccan goat breeds: phylogenetic relationships with four breeds reared in France. *Small Rumin. Res.* **45**, 225-233.

Raymond M. and Rousset F. (1995). GENEPOP: population genetics software for exact tests and ecumenicism. *J. Hered.* **86**, 248-249.

Schneider S., Roessli D. and Excoffier L. (1997). ARLEQUIN: A Software for Population Genetic Data Analysis. Genetics and Biometry Laboratory. University of Geneve, Geneve , Switzerland.

Serrano M., Calvo J.H., Martínez M., Marcos-Carcavilla A., Cuevas J., González C., Jurado J.J. and Tejada P.D. (2009). Microsatellite based genetic diversity and population structure of the endangered Spanish Guadarrama goat breed. *BMC Genet.* **10**, 61-68.

Shadma F., Bhonga D., Ranka N. and Joshi G. (2008). Genetic variability and bottleneck studies in Zalawadi, Gohilwadi and Surti goat breeds of Gujarat (India) using microsatellites. *Small Rumin. Res.* **77**, 58-64.

Toro M. and Maki-Tanila A. (2007). Genomics reveals domestication history and facilitates breed development. Pp. 75-102 in Utilization and Conservation of Farm Animal Genetic Resources. K. Oldenbroek, Ed. Wageningen Academic Publishers, Wageningen, Netherlands.

Ulloa-Arvizu R., Gayosso-Vázquez A., Ramos-Kuri M., Estrada F.J., Montaño M. and Alonso R.A. (2008). Genetic analysis of Mexican Criollo cattle populations. *J. Anim. Breed. Genet.* **125**, 351-359.

Weir B.S. and Cockerham C.C. (1984). Estimating F-statistics for the analysis of population structure. *Evolution.* **38**, 1358-1370.

Yeh F.C., Yang R.C., Boyle T.B.J., Ye Z.H. and Mao J.X. (1999). POPGENE: the user-friendly shareware for population genetic analysis. Molecular Biology and Biotechnology Centre, University of Alberta, Alberta, Canada.

PERMISSIONS

LIST OF CONTRIBUTORS

M. Raghebian
Department of Animal Science, Faculty of Agriculture and Natural Resources, Science and Research Branch, Islamic Azad University, Tehran, Iran

A. Babaei Yazdi, N. Dabiri, J. Shomeyzi and M.J. Bahrani
Department of Animal Science, Karaj Branch, Islamic Azad University, Karaj, Iran

A. Hajimohammadi
Department of Animal Science, Faculty of Agricultural Science, University of Guilan, Rasht, Iran

P. Hatami
Department of Animal Science, Faculty of Agriculture and Natural Resources, University of Tehran, Karaj, Iran

A. Raghebian
Department of Animal Science, Saveh Branch, Islamic Azad University, Saveh, Iran

V. Naseri and F. Kafilzadeh
Department of Animal Science, Faculty of Agriculture, Razi University, Kermanshah, Iran

H. Jahani-azizabadi
Department of Animal Science, College of Agriculture, University of Kurdistan, Sanandaj, Iran

A. Teimouri Yansari
Department of Animal Science, Faculty of Agricultural Science, Sari University of Agricultural Science and Natural Resources, Sari, Iran

A. Salemi and B.Asadi
Department of Animal Science, Behbahan Branch, Islamic Azad University, Behbahan, Iran

M. Vatankhah
Department of Animal Science Research, Chaharmahal and Bakhtiari Agricultural and Natural Resources Research and Education Center, AREEO, Shahrekord, Iran

K. Alemayehu and D. Kebede
Biotechnology Research Institute, Bahir Dar University, Bahir Dar, Ethiopia

S. Mirzaei Cheshmehgachi, M.M. Moeini, F. Hozhabri and M.E. Nooryan Soroor
Department of Animal Science, Faculty of Agriculture, Razi University, Kermanshah, Iran

A. Mokhtarpour
Research Center of Special Domestic Animals, Research Institute at University of Zabol, Zabol, Iran

F.Pourmollae
Agricultural Jihad Organization of Khorasan - Razavi, Mashhad, Iran

H. Hamed, A. El Feki and A. Gargouri
Departement des Science de la Vie, Faculte des Science de Sfax, Unite de Physiopathologie Environnementale, Valorisation des Molecules Bioactives et Modelisation Mathematique, Sfax, Tunisia

M.Mahdavi, A. Mohammadi and G.R.Dashab
Department of Animal Science, Faculty of Agriculture, University of Zabol, Zabol, Iran
Razi Vaccine and Serum Research Institute, Karaj, Iran

A. Teimouri Yansari
Department of Animal Science, Faculty of Agricultural Science, Sari University of Agricultural Science and Natural Resources, Sari, Iran

S. Arbabi, T. Ghoorchi and S. Ramzanpour
Department of Animal and Poultry Nutrition, Faculty of Animal Science, Gorgan University of Agricultural Science and Natural Resources, Gorgan, Iran

M. Simeonov
Agricultural Institute, Stara Zagora, Bulgaria

D. Pamukova
Trakia University, Faculty of Agriculture, Stara Zagora, Bulgaria

M.R. Mohammadabadi and F. Tohidinejad
Department of Animal Science, Faculty of Agriculture, Shahid Bahonar University of Kerman, Kerman, Iran

S. Azadbakht, A.A. Khadem and M.A. Norouzian
Department of Animal Science, College of Abouraihan, University of Tehran, Tehran, Iran

G.F.Baighi and A. Nobakht
Department of Animal Science, Maragheh Branch, Islamic Azad University, Maragheh, Iran

S. Menatian, H.R. Mirzaei Alamouti and R. Masoumi
Department of Animal Science, Faculty of Agriculture, University of Zanjan, Zanjan, Iran

F. Fatahnia
Department of Animal Science, Faculty of Agriculture, Ilam University, Ilam, Iran

M. Yari and M. Manafi
Department of Animal Science, College of Agriculture, Malayer University, Malayer, Iran
Iranian Grape and Raisin Institute, Malayer University, Malayer, Iran

M. Hedayati and R. Karimi
Iranian Grape and Raisin Institute, Malayer University, Malayer, Iran
Department of Landscape Engineering, College of Agriculture, Malayer University, Malayer, Iran

A. Jonker
Grasslands Research Centre, AgResearch Limited, Palmerston North, New Zealand

M.Tahmoorespur, N. Nazifi, Z. Pirkhezranian, A.A. Naserian M. Sadjadian, M. Danesh Mesgaran, A.R. Vakili, R. Valizadeh and M.H.Sekhavati
Department of Animal Science, Faculty of Agriculture, Ferdowsi University of Mashhad, Mashhad, Iran

A. Kiani
Department of Animal Science, Faculty of Agricultural Science, Lorestan University, Khoramabad, Iran

M.H. Gharoni
Department of Veterinary Science, Faculty of Veterinary Medicine, Lorestan University, Khoramabad, Iran

R. Shariati
Research and Development Section, Garrin Dam Simorg, Khorramabd, Iran

H. Daghigh Kia, Z. Blooki, S. Safa, H. Vaseghi Dodran and M. Mahdipour
Department of Animal Science, Faculty of Agriculture, University of Tabriz, Tabriz, Iran

I. Afrooznia, M. Zandi and M.R. Sanjabi
Department of Agriculture, Iranian Research Organization for Science and Technology (IROST), Tehran, Iran

M.H.Hadi Tavatori
Agricultural and Natural Resources Research Center of Qazvin Province, Qazvin, Iran

V. Bahrampour
Department of Animal Science, Technical and Vocational University, Kerman, Iran

R. Khandaker, M.K.I.Khan and M.M.Momin
Department of Genetics and Animal Breeding, Chittagong Veterinary and Animal Science University Khulshi, Chittagong - 4225, Bangladesh

T. Abbassi-Daloii, M. Tahmoorespur, G.E. Salih, A.E. Abd-elgabar and M.A. Daffallah
Department of Poultry Production, Faculty of Animal Production, University of Gezira, Sudan

M.A.E. Elkhairey
Department of Poultry Production, Faculty of Animal Production, University of Gezira, Sudan
College of Animal Science and Technology, Northwest A and F University, Yangling 712100, Shaanxi, China

M.S. Elnour
Department of Poultry Production, College of Animal Production, University of Bahri, Khartoum, Sudan

B. Hilal, I. Boujenane, S. El Otmani and M. Chentouf
Department of Animal Production and Biotechnology, Institut Agronomique et Vétérinaire Hassan II, Rabat, Morocco
Institut National de la Recherche Agronomique, Regional Centre of Tangier, Tanger, Morocco

M. Piro
Department of Medecine, Surgery and Reproduction, Laboratoire d'Analyses Génétiques Vétérinaires, Institut
Agronomique et Vétérinaire Hassan II, Rabat, Morocco

Index